International Association for Wind Engineering

Recent Advances in

WIND ENGINEERING

VOLUME VI

International Association for Wind Engineering

Recent Advances in

WIND ENGINEERING

VOLUME VI

PUBLISHING FOR ONE WORLD

NEW AGE INTERNATIONAL (P) LIMITED, PUBLISHERS

New Delhi • Bangalore • Chennai • Cochin • Guwahati

Hyderabad • Kolkata • Lucknow • Mumbai

Visit us at **www.newagepublishers.com**

Published by New Age International (P) Ltd., Publishers
First Edition: 1994
Reprint: 2012

Branches:

- 37/10, 8th Cross (Near Hanuman Temple), Azad Nagar, Chamarajpet **Bangalore**-560 018. Tel.: (080) 2675 6823, Telefax: 2675 6820 E-mail: bangalore@newagepublishers.com
- 26, Damodaran Street, T. Nagar, **Chennai**-600 017. Tel.: (044) 24353401 Telefax: 24351463, E-mail: chennai@newagepublishers.com
- CC-39/1016, Carrier Station Road, Ernakulam South, **Cochin**-682 016 Tel.: (0484) 2377004, Telefax: 4051303. E-mail:cochin@newagepublishers.com
- Hemsen Complex, Mohd. Shah Road, Paltan Bazar, Near Starline Hotel **Guwahati**-781 008. Tel.: (0361) 2513881. Telefax: 2543669 E-mail: guwahati@newagepublishers.com
- 105, 1st Floor, Madhiray Kaveri Tower, 3-2-19, Azam Jahi Road, Nimboliadda **Hyderabad**-500 027. Tel.: (040) 24652456, Telefax: 24652457 E-mail:hyderabad@newagepublishers.com
- RDB Chambers (Formerly Lotus Cinema)106A, 1st Floor, S.N. Banerjee Road **Kolkata**-700 014. Tel.: (033) 22273773, Telefax: 22275247 E-mail:kolkata@newagepublishers.com
- 16-A, Jopling Road, **Lucknow**-226 001. Tel.: (0522) 2209578, 4045297 Telefax: 2204098 E-mail: lucknow@newagepublishers.com
- 142C, Victor House, Ground Floor, N.M. Joshi Marg, Lower Parel **Mumbai**-400 013. Tel.: (022) 24927869. Telefax: 24915415 E-mail: mumbai@newagepublishers.com
- 22, Golden House, Daryaganj, **New Delhi**-110 002. Tel.: (011) 23262370, 23262368 Telefax: 43551305. E-mail: sales@newagepublishers.com

ISBN : 978-81-224-0718-1

₹ 695.00

C-11-11-5976

Printed in India at Glorious Printers, Delhi.

PUBLISHING FOR ONE WORLD

NEW AGE INTERNATIONAL (P) LIMITED, PUBLISHERS
4835/24, Ansari Road, Daryaganj, New Delhi-110002
Visit us at **www.newagepublishers.com**

FOREWORD

The proposal to prepare this State-of-the art volume on Wind Engineering was initiated on 19 September, 1992, the day Dr. Alan G. Davenport turned sixty. The purpose was to commemorate his sixtieth birth anniversary, by bringing together articles from leading experts of wind engineering covering the latest information in a field whose development during the last three decades is closely related to the name of this great personality.

The readiness with which my invitation for contributing articles to this volume was accepted, is a mark of high esteem and affection in which Prof. Davenport is held by the various experts in Wind Engineering. The chapter break-up for the volume has been developed through several iterations by a process of informal consultations with various experts within India and outside and hopefully covers the state-of-the-art on the established aspects of Wind Engineering as well as those which have come into focus recently. I am grateful to all the authors who have contributed chapters and appreciate their co-operation. Personally, it has been a pleasure in working with them to bring out this volume. Coming from specialists in related aspects of Wind Engineering, the technical quality of the chapters is indeed excellent and I have not attempted any editorial changes except in relation to the format in some cases. It is my earnest hope that the volume represents a landmark in the literature on Wind Engineering which will help in identifying the future areas of research.

I have received valuable help and advice from several of my colleagues. This is gratefully acknowledged along with the support received from the staff of the Wind Engineering Center at the University of Roorkee.

PREM KRISHNA
Chairman,
International Association of Wind Engineering

Alan Garnett Davenport

A Doyen of Wind Engineering

The name of Prof. A.G. Davenport is synonymous with the developments in Wind Engineering over the last three decades and to a readership in this area, he needs little introduction. Yet I consider it my privilege to write about his person and career. This is a daunting task, not because there is any difficulty in finding superlatives, but because these may not do justice to his stature and achievements. I had requested my friend Professor Barry Vickery, another well known expert in wind engineering and a long time associate of Professor Davenport, to help me write this citation. Through his good offices it has been possible to obtain the copy of a brilliantly written citation, prepared by Dr. G. Caldwell and Dr. R.K. Rowe of the University of Western Ontario, presenting Dr. Davenport for the Canada Gold Medal in Science and Engineering, awarded to him this year. With the kind permission of the authors of this citation I have quoted (all material within quotes) profusely from it and I gratefully acknowledge the same.

My first exposure to A.G. Davenport's work was through his early papers in the beginning of the 60s on account of my requirement to pursue doctoral work on cable structures, which are inherently flexible and responsive to wind loading. I met him first, as a matter of chance, at the ASCE speciality conference in Structural Engineering in April 1969 at Louisville, U.S.A. The manner of his introduction, before he made a presentation, made it clear that Dr. Davenport was already a famous person. This was justifiably so, since his work during those early years set a direction for research in wind engineering, and, a momentum which has become perennial. During the quarter of a century that has elapsed since, it has been a pleasure to know this fine gentleman and to follow his enormous contributions to wind engineering in particular and engineering in general, through his teaching, research and its extension. Much that is happening today in the world of Wind Engineering bears his hallmark.

''It would be absurd to claim that Davenport's interest in air-flow and wind behaviour was innate. Yet it almost seems so, given his preoccupation with building and flying model airplanes as a boy in India, where he was born, and in South Africa, where he grew up. The preoccupation was transferred from model to reality when, during his years at Cambridge (BA, MA, 1954-58), he flew with the University Air Squadron, and at Toronto (MASc, 1957) on a volunteer basis with the Canadian Navy. If it were Davenport's wish to come to Canada in 1957, it was serendipity that took him to the University of Toronto where he came under the influence of Professor Carson Morrison. Despite his own lack of knowledge of the field, Morrison had the wit to capitalize on Davenport's interests and set him the problem of explaining why the new Rimouski television antenna was literally being blown to bits. Davenport secured the antenna with cables and proceeded to make a wide range of measurements, which when analysed revealed clearly that the structure was much too flexible and weak to withstand the more or less constant winds that swept its exposed site. Study of the Rimouski antenna influenced Davenport's subsequent career in two important ways: analysis of the voluminous measurements introduced him to the computer — the University of Toronto's Ferranti and the first in Canada; and the results led to one of the first papers ever written on the modelling of wind dynamics. That paper won Davenport the Engineering Institute of Canada's Duggan Medal for 1960.

Then followed a stint with the National Research Council's Building Research Group in Ottawa — only a short eighteen months, but a critical and providential time in directing Davenport's career. He was set the task of assessing all aspects of wind loading for the national building code, which as he has said, "led me discover all of the problems and none of the answers." Be that as it may, it exposed for him in clear and unequivocal terms the vast amount of original investigation that was needed on the implications of turbulent air-flow for the stability of engineering structures — structures that ranged in scale from the innocent chimney pot to the mighty skyscraper. The NRC experience was highly instructive, not because he was influenced by any individual with particularly germane expertise but because he was given the freedom to enquire openly and widely and, indeed, was encouraged to do so by the director and his likeminded colleagues. At this time, he learned Russian, tutored by a member of the NRC staff, so that he could read the leading world literature on turbulence theory which, until the rise of NASA, was Russian dominated.

Rewarding as the NRC experience was, it persuaded him that he would have to move if he were to solve some of the fundamental problems he had identified. He rejected offers from MIT, the University of Illinois, and other prestigious American graduate schools to return to England and read for the Ph.D. degree (1961) at the University of Bristol, under the direction of Sir Alfred Pugsley.

Davenport's thesis addressed the statistical treatment of wind loads on tall masts and long bridges. It included, as an illustrative example, the much-touted new road bridge across the Firth of Forth in eastern Scotland, which then had been designed and planned as a key link in Britain's new highway system to cope with the post-war boom in private automobile traffic. The thesis and four accompanying papers all broke new ground. Most noteworthy among the papers was, "The application of statistical concepts to the wind-loading of structures" (Proc. Inst. Civil Engineers, 1961), which became the basis for all subsequent studies of the effects of wind on civil engineering structures and for the application and development of wind tunnels in the design of tall buildings and long bridges in particular. Judging by the demand for copies, it was the most influential paper Davenport has written to date."

In 1961 Dr. Davenport joined the faculty of the University of Western Ontario. There he began a "series of field-based experiments on how air turbulence was translated into forces that acted on man-made structures of various shapes and sizes — from the simple billboard to the favoured television antenna. These early experiments were a useful stepping stone to subsequent more sophisticated experiments, but more to the point they demonstrated that the approach was an unsatisfactory and wasteful way of undertaking systematic studies. What was needed was a wind tunnel.''

''A further major stimulus in this respect came from Davenport's involvement, in 1963, in his first major building project — the design of no less a structure than the twin towers of the World Trade Centre (WTC) in New York. Numerous factors combined to make the design problems complex and multiple : the unique twinning of the two buildings and their magnitude and shape; the buildings' unprecedented size (together about ten times the floor space of the Empire State Building); their lightness (proportionately about one-third the weight of some older New York skyscrapers) thus making them more vulnerable to the harbour-front winds; the recognition that the buildings would have to be damped to minimize sway; and the crudeness and inadequacy of the existing guidelines for coping with wind dynamics. After modelling the twin towers in wind tunnels in England and the US (these tunnels far from ideally suited for the purposes) and conducting meteorological experiments to understand the wind effects nearly 500 m above ground, Davenport played the key role in devising the damping system for the WTC. The WTC shaped what was to come; it was the precursor of more refined damping techniques, developed on the same principles and now used to counteract not only the effects of regularly recurrent, variable winds but also of potentially disastrous, episodic earthquakes.

The Boundary Layer Wind Tunnel at the University of Western Ontario was designed by Alan Davenport and was constructed and commissioned in November 1965. It may be unbelievable that it has cost less than $ 10,000 and has served well for almost 30 years till date. ''With the completion of the BLWT and Davenport's appointment as its Director, the demand for his services and for the testing of models of a wide variety of structures, rose exponentially. An early example was the Sears Building in Chicago, a somewhat later one the CN Tower in Toronto. By Davenport's own admission. ''Finding new problems to solve was easy; 'the phone just kept ringing!' Less easy was coping with the demand, and so it came that he assembled, and Western hired, the permanent core of his laboratory staff. Although he has always been in charge and the key figure in deciding what research would be conducted and how each project would be undertaken, he would be the first to stress that his co-workers have been indispensible in executing the pioneering studies that have brought him and his BLWT global acclaim. That that permanent core remains intact to this day is a tribute to Davenport as leader and manager, to the loyalty of his team, and to the exciting and rewarding careers all of them have enjoyed in this laboratory. Following rapidly upon its opening, BLWT became widely regarded as the leading contributor internationally to the methodology and technology now used for defining wind-action on wind-sensitive structures, such as low and high buildings, long-span bridges and roofs, communication towers, transmission lines, and off-shore drilling platforms. It retains that reputation to this day.''

''The number and variety of investigations Davenport and his team have conducted at BLWT through the last 25 years are legion. In the 1970s alone, more tall buildings were erected throughout the world than in any other decade, and a staggering two-thirds of them were analysed and tested at the University of Western Ontario BLWT! It is impracticable, therefore, to list all of Davenport's projects, far less describe in detail what all his contributions have been. Collectively and in essence, Davenport's contributions may be described in terms of (a) basic scientific knowledge concerned broadly with climate, more specifically wind-force, in the Earth's boundary layer (the first 1000 m or so above ground level); (b) the aerodynamic and aeroelastic behaviour of bluff shaped structures in turbulent boundary-layer flow, and the dynamic response of such structures to that flow; (c) the development of laboratory and statistical techniques for realistic assessment of the likely performance of civil engineering structures subjected to wind loading, based on small scale modelling in wind tunnels; (d) the codification of acceptable design and construction criteria for these structures; and (e) remarkable innovation in the design of a large number of these landmark structures world-wide. The emphasis on each component has varied dependent on the demands of the project in which he has been engaged. As a corollary, so proportionately has the growth of his knowledge, and thus the magnitude of his contribution, in each category.

In the early years of BLWT, before design testing for wind had come into vogue, Davenport was called upon to investigate structures that had partly or wholly failed (for example, the Tacoma Narrows Bridge, WA; the Cooling Towers at the Ferrybridge Power Station, England; and the John Hancock Building, Boston, MA). With wind testing now an integral component of design, which in itself is a substantial tribute to Davenport's unique contribution to national and international building codes and design standards, his work on failed structures has almost wholly given way to that on relevant aspects of feasibility studies and early preconstruction design and planning. Among these studies, the Japanese Millennium Tower for Tokyo harbour may be worthy of special note because it exemplifies a high degree of architectural and engineering futurism. It is part of a conceptual design to accommodate a small city in a building nearly 1 km-high and built off-shore-a means of reconciling competing demands to house and service an increasingly dense population and to preserve scarce arable land. It is uncommon to invite outsiders to work on a Japanese project within Japan, especially one sponsored by three major national companies. Hence it is a tribute to Davenport, who has trained a number of Japanese engineers.''

Many tall buildings and long bridges the world over, bear the hallmark of Alan Davenport's innovative work. Not only that, he has lent the benefit of his rich experience and knowledge in encouraging wind engineering development wherever sought. In 1985 it was mooted that a four-yearly Symposium on Wind Engineering be held in the Asia-Pacific Region, and, we were delighted that Professor Davenport found the time to travel to Roorkee and delivered the keynote address at the first Asia-Pacific Symposium on Wind Engineering in December 1985. This was an example of his abiding interest in wind engineering development. Through his sustained effort in research, standardisation and the solution of intricate problems of the design of complex and wind sensitive structures, he has contributed in raising the quality of engineering of such structures. It may not, however, be thought that his attention has not been drawn to the needs of uplifting the standards of engineering related to structures whose safety is of paramount importance for disaster reduction for the safety of common man in the developing world. Ever since the United Nations conceived the idea of the current decade as an "International Decade of Natural Disaster Reduction," Prof. Davenport has been a champion of its activities, particularly as related to wind disaster reduction.

In bringing out this State-of-the-Art Volume, I am sure, the International Association of Wind Engineering is honouring one of its most worthy members.

Roorkee
19 September 1994

PREM KRISHNA

CONTENTS

Development of Wind Tunnels for Physical Modelling of the Atmospheric Boundary Layer 1
Cermak, J.E.

Methods of Fluctuating Pressure Measurement in Wind Engineering 26
Holmes, J.D.

Bluff Body Aerodynamics for Wind Engineering 47
Melbourne, W.H.

Applications of CFD to Bluff Body Aerodynamics 65
Murakami S.

Field Measurements of High Winds and their Interpretation 90
Jeary, Alan P.

Estimation of Extreme Wind Speeds 109
Simiu, E.

Instrumentation of Prototype Structures 124
Rusheweyh, H.

Measuring the Dynamic Characteristics of Prototype Structures 133
Littler, J.D. and Ellis, B.R.

Evaluation of Wind Loads on Low Buildings A Brief Historical Review 155
Stathopoulos, T.

Aerodynamics of Tall Buildings 180
Kwok, K.C.S

Wind Induced Response of Chimneys 205
Vickery, B.J.

A Review of Some Key Developments in the Aeroelasticity of Long-Span Bridges 234
Scanlan, R.H.

Full Model Testing of Large Cable-Supported Bridges 249
Miyata. T.

Suppression of Wind-Induced Vibrations of Structures 281
Ito, M.

Buffeting Response Analysis and Control of Long-Span Bridges 295
Xiang, H.F.

Non-linear Effects in the Buffeting Problem 320
Niemann, H.J. and Hoffer, R.

Wind Damage Studies 341
McDonald J.R.

Wind Flow and City Planning 371
Gerhardt, H.J and Kramer C.

Small Scale Modelling of Combined Wind and Buoyancy Effects 393
Poreh, Michael

DEVELOPMENT OF WIND TUNNELS FOR PHYSICAL MODELLING OF THE ATMOSPHERIC BOUNDARY LAYER (ABL)

J.E. CERMAK

Colorado State University
Fort Collins, Colorado 80523, USA

SUMMARY

Developments that trace the evolution of physical-modelling techniques to simulate wind and wind effects are presented. Characteristics of ABLs are reviewed and features that require modelling for wind-engineering applications are defined. Similarity criteria are presented for "exact" similarity and the necessary approximations are discussed. Special features of wind tunnels needed to achieve the modelling requirements are identified. Comparisons are made of wind and temperature data measured in the field and in a boundary-layer wind tunnel. Wind tunnels are described that have been designed and built during the last 25 years with special capabilities for modelling the ABL.

INTRODUCTION

Efforts to determine wind effects on objects and dispersion of air pollutants using small-scale models in air flows generated by various means are reported in publications dating back to the decade of 1750-1760. Two centuries elapsed before serious efforts were made to produce air flows that modelled characteristics of natural wind in the ABL. However, measurement of wind pressures on buildings by Irminger (1893) and Irminger et al (1930 and 1936) in Denmark, at the National Physical Laboratory by Stanton (1924) and Bailey et al (1943), development of a warm-cold air tunnel (16 m long, 1 m wide, 0.25 m high) at the Kaiser Wilhelm Institute for Flow Research by Prandtl et al (1934), studies of dispersion of stack effluent at the University of Michigan by Sherlock et al (1941), and studies of evaporation from plane boundaries at the University of Iowa and Colorado State University by Albertson (1948) and Cermak (1952), respectively, clearly demonstrated that the ABL should be modelled for wind-effect studies. During the two centuries, 1750-1950, laboratory flow facilities for such studies evolved from the rotating arm of Robins (1742) and Smeaton (1759) shown in Fig. 1 to the type shown in Fig. 2 which was constructed by the writer in 1949.

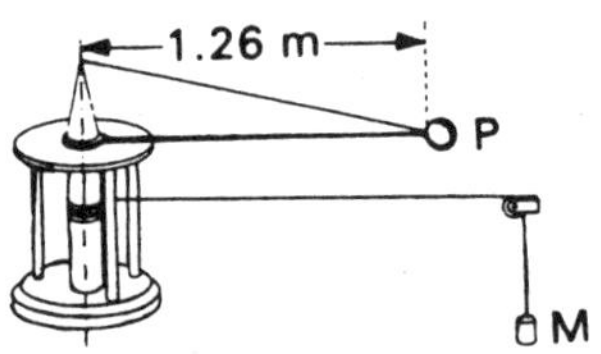

Fig. 1. Air resistance test apparatus -- Robins (1742).

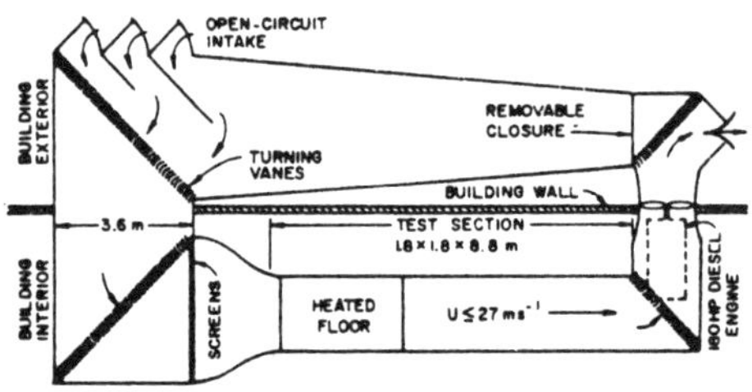

Fig. 2. Colorado State University Wind Tunnel constructed in 1949 -- Cermak (1952).

Although the wind tunnel shown in Fig. 2 was designed for studies of heat and mass transfer by forced convection in which simulation of the ABL was of no concern, significant information for modelling of the ABL was obtained, Cermak et al (1956). A historical summary of the foregoing developments is given by Cermak (1975) and examples of wind tunnel use for industrial aerodynamics studies, the precursor of wind engineering, are presented by Scruton (1960).

Following a summary description of primary features of the ABL, similarity criteria for modelling the ABL are presented. Wind-tunnel characteristics needed to satisfy the dominant similarity criteria are discussed and a boundary-layer wind tunnel (BLWT) having these characteristics, the Colorado State University meteorological wind tunnel (MWT), is described. Confirmation of similarity is shown by comparisons of flow measurements in the MWT with micrometeorological data.

During the last decade BLWTs with multiple and/or special capabilities have been designed and constructed. Several wind tunnels that mark significant advances in the evolution of natural wind modelling for wind-engineering purposes are described.

THE ATMOSPHERIC BOUNDARY LAYER (ABL)

Wind in the planetary boundary layer (PBL) is driven primarily by horizontal pressure differences caused by nonuniform solar heating of Earth's surface. Vertical distributions of mean velocities and turbulence in the PBL are determined by the surface shear stress, vertical gradients of temperature and rotation of Earth.

A model developed by Lettau (1962), based on Reynolds equations for turbulent boundary layers, predicts vertical distributions of mean wind velocity. Numerical integration for an atmosphere that is steady, horizontally homogeneous, dry, adiabatic, and without vertical motion gives a vertical distribution shown in Fig. 3 when the surface roughness length z_o is 1 m, the geostrophic wind speed U_{geo} is 34.5 ms^{-1}, and the latitude is 30° north.

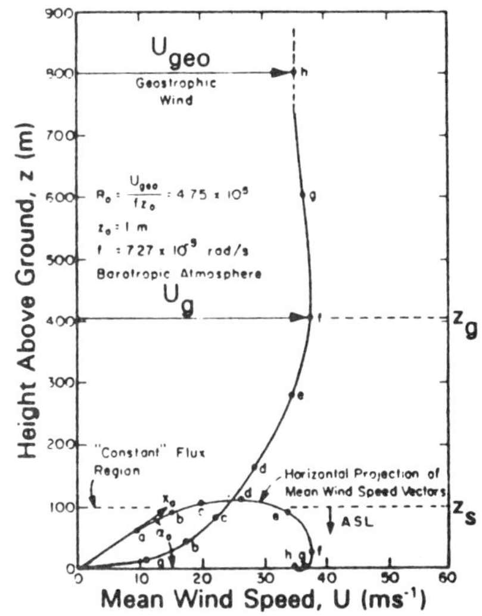

Fig. 3. A PBL according to model of Lettau (1962).

Mean Wind Speeds Distributions

The ABL is defined for wind-engineering purposes to be a region between the surface and the gradient wind height z_g where the mean wind speed U becomes a maximum U_g. During strong winds (neutral thermal stratification) the vertical distribution of U is closely approximated by a "power-law"

$$U/U_g = (z/z_g)^{1/n}. \quad (1)$$

Both the power-law exponent 1/n and the gradient wind height z_g depend upon the surface-roughness length z_o as indicated by Fig. 4. A more detailed gradation of z_o with respect to surface type developed by Wieringa (1992) is given in Table 1.

Equation 1 and the relationship given by Fig. 4 and Table 1 are restricted to flat surfaces of uniform roughness.

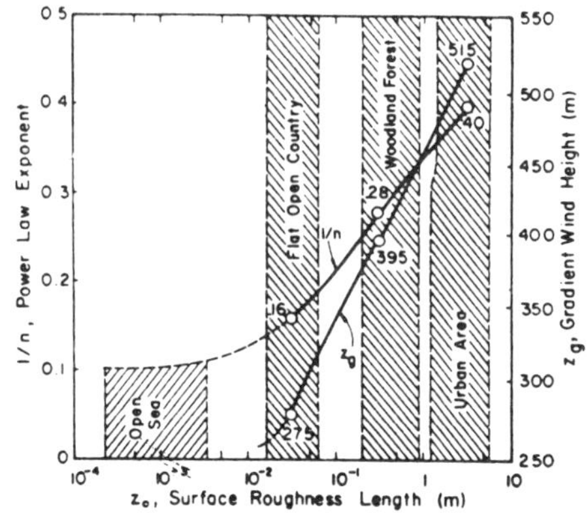

Fig. 4. Wind-profile parameters for strong winds over various surfaces -- Davenport (1960).

Table 1. Range of homogeneous surface roughness z_o -- Wieringa (1992).

Surface type	Roughness length (m)
Sea, loose sand and snow	≈0.0002 (V-dependent)
Concrete, flat desert, tidal flat	0.0002-0.0005
Flat snow field	0.0001-0.0007
Rough ice field	0.001-0.012
Fallow ground	0.001-0.004
Short grass and moss	0.008-0.03
Long grass and heather	0.02-0.06
Low mature agricultural crops	0.04-0.09
High mature crops ("grain")	0.12-0.18
Continuous bushland	0.35-0.45
Mature pine forest	0.8-1.6
Dense low buildings ("suburb")	0.4-0.7
Regularly-built large town	0.7-1.5
Tropical forest	1.7-2.3

Low wind speeds must be considered in the lower part of the ABL, the atmospheric surface layer (ASL) shown in Fig. 1, when dispersion of air pollutants is considered. Thermal stratification then perturbs the logarithmic distribution for neutral flows to give mean distributions of wind speed U and temperature T that can be expressed by log-linear relationships according to Monin and Obukhov (1954) similarity as follows:

$$U(z) - U(z_{ref}) = (u_*/\kappa)\,[\ell n(z/z_{ref}) + B_U(z - z_{ref})/L]; \text{ and} \tag{2}$$

$$T(z) - T(z_{ref}) = T_*\,[\ell n(z/z_{ref}) + B_T(z - z_{ref})/L]. \tag{3}$$

In Eqs. 2 and 3 the shear velocity u_*, friction temperature T_* and Monin-Obukhov length L are defined as follows: $u_* = (\tau_o/\rho)^{1/2}$, $T_* = -H_o/(\rho_o C_{po} \kappa u_*)$, and $L = -u_*^3/[(\kappa g H_o/(T_o \rho_o C_{po})]$.

Equations 2 and 3 provide a basis for comparison of mean wind speed and temperature distribution in the ASL for wind-tunnel and field measurements.

Turbulence Characteristics in the ABL

For wind-engineering applications turbulence measures of primary concern are the turbulence intensities, spectra, length scale and turbulent fluxes. The following expressions for these quantities over flat uniformly rough surfaces when the flow is near neutral give a framework for evaluation of the ABL and ASL models.

A simple form for intensity of the longitudinal component of turbulence I_u in the ASL is given by Tieleman (1991) as

$$I_u = \sigma_u/U = \kappa A_u/\ell n(z/z_o) \tag{4}$$

where $\kappa = 0.4$ (von Kármán constant) and $A_u = 2.5$. Accordingly, a determination of z_o from Table 1 provides a direct estimate of I_u. Upon matching scaling laws for the "inner" and "outer" layers of the ABL Harris et al (1980) have developed equations for I_u that are applicable throughout the ABL.

Olesen et al (1984) have presented two models for the turbulence spectra--a "pointed" model for flat uniformly rough surface and a "blunt" model for upwind approach with small perturbations of flatness and/or roughness. These models have been developed by Tieleman (1992) to provide spectral forms for all three turbulence components within the ASL which are listed in Table 2. Widely used spectral models of von Kármán (1948) and Davenport (1961) are discussed in a report by Tieleman (1991). The strong dependence of velocity spectra on thermal stratification of the ASL is evident in data reported by Kaimal et al (1972) as shown in Fig. 5.

A distinguishing feature of the neutral ASL is near constancy of the turbulent shear stress -- the "constant" flux layer indicated in Fig. 3. Data in Fig. 6 extracted from a report by Izumi (1971) on the 1968 Kansas field program illustrate this property.

Table 2. Universal velocity spectra for the surface layer -- Tieleman (1992).

"Pointed" Model	"Blunt" Model
(a) $\frac{nS_{uu}(n)}{\sigma_u^2} = \frac{20.53\ F}{1 + 475.1\ F^{5/3}}$	(d) $\frac{nS_{uu}(n)}{\sigma_u^2} = \frac{40.42\ F}{(1 + 60.62\ F)^{5/3}}$
(b) $\frac{nS_{vv}(n)}{\sigma_v^2} = \frac{6.83\ F}{1 + 75.84\ F^{5/3}}$	(e) $\frac{nS_{vv}(n)}{\sigma_v^2} = \frac{13.44\ F}{(1 + 20.16\ F)^{5/3}}$
(c) $\frac{nS_{ww}(n)}{\sigma_w^2} = \frac{1.67\ F}{1 + 7.23\ F^{5/3}}$	(f) $\frac{nS_{ww}(n)}{\sigma_w^2} = \frac{3.28\ F}{(1 + 4.92\ F)^{5/3}}$

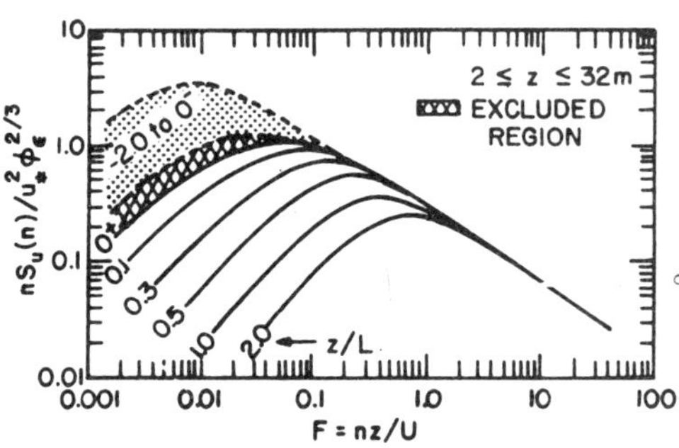

Fig. 5. Dependence of u spectra on z/L -- Kaimal et al (1972).

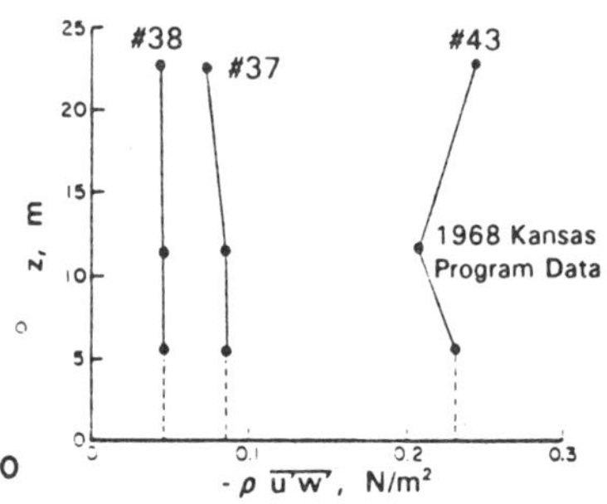

Fig. 6. Shear in ASL (neutral stability) -- Izumi (1971).

SIMILARITY CRITERIA FOR MODELLING THE ABL

General similarity criteria may be obtained directly by inspectional analysis, Ruark (1935), of the equations for conservation of mass, momentum and energy in turbulent flow. By use of inspectional analysis the following nondimensional equations are given by Cermak (1975) in which similarity parameters are contained in brackets:

$$\frac{\partial \rho^*}{\partial t^*} + \frac{\partial\left(\rho^* u_i^*\right)}{\partial x_i^*} = 0 \tag{5}$$

$$\frac{\partial U_i^*}{\partial t^*} + U_j^* \frac{\partial U_i^*}{\partial x_j^*} + \left[\frac{L_o \Omega_o}{U_o}\right] 2\epsilon_{ijk}\Omega_j^* U_k^* = -\frac{\partial P^*}{\partial x_i^*} + \frac{\partial\left(\overline{-u_i' u_j'}\right)^*}{\partial x_j^*} - \left[\frac{\Delta T_o}{T_o}\frac{L_o g_o}{U_o^2}\right]\Delta T^* g^* \delta_{i3} + \left[\frac{\nu_o}{U_o L_o}\right]\frac{\partial^2 U_i^*}{\partial x_k^* \partial x_k^*} \tag{6}$$

$$\frac{\partial T^*}{\partial t^*} + U_i^* \frac{\partial T^*}{\partial x_i^*} = \left[\frac{k_o}{\rho_o C_{po}\nu_o}\right]\left[\frac{\nu_o}{L_o U_o}\right]\frac{\partial^2 T^*}{\partial x_k^* \partial x_k^*} + \frac{\partial\left(\overline{-\theta' u_i'}\right)}{\partial x_i^*} + \left[\frac{\nu_o}{U_o L_o}\right]\left[\frac{U_o^2}{C_{po} T_o}\right]\phi_\epsilon^* \tag{7}$$

The variables are scaled as follows to form the nondimensional quantities:

$$U_i^* = \frac{U_i}{U_o},\ u_i^* = \frac{u_i}{U_o},\ (u_i')^* = \frac{u_i'}{U_o},\ x_i^* = \frac{x_i}{L_o},\ t^* = \frac{tU_o}{L_o},$$

$$\Omega_j^* = \frac{\Omega_j}{\Omega_o},\ P^* = \frac{P}{\left(\rho_o U_o^2\right)},\ T^* = \frac{T}{T_o},\ \Delta T^* = \frac{\Delta T}{\Delta T_o},\ \text{and } g^* = \frac{g}{g_o}$$

Although Eqs. 5, 6 and 7 are very general, there are some restrictions. Because the temperature dependence of density has been linearized in Eq. 6

(Boussinesq approximation) flows are limited to cases with $\Delta T \ll T_o$. This restriction does not impose serious limitations on the use of Eq. 6 for ABL flow. Equation 7 does not allow for an exchange of energy by radiation or phase changes of water. Therefore, a strict interpretation of Eq. 7 limits application to ABLs without cloud content or precipitation.

Equations 5, 6 and 7 give the requirements for kinematic, dynamic, and thermic similarity when proper boundary conditions are satisfied. Equation 5 is invariant for this transformation in which all lengths are scaled equally; i.e., $x_i^* = x_i/L_o$. Therefore, geometric similarity results in kinematic similarity. Equation 6 states that dynamic similarity can be achieved if each of the following parameters is equal for two systems: $U_o/(L_o\Omega_o)$, Rossby number (Ro); $(\Delta T_o/T_o)(L_o g_o/U_o^2)$, Richardson number (Ri); and $U_o L_o/\nu_o$, Reynolds number (Re). Thermic similarity will be obtained if, according to Eq. 7, the following dimensionless groups are equal for two systems: $\nu_o\rho_o C_{po}/k_o$, Prandtl number (Pr); $U_o^2/(C_{po}T_o)$, Eckert number (Ec); and Re.

The foregoing requirements for geometric, dynamic, and thermic similarities must be augmented by boundary specifications to complete requirements for similarity. These external conditions are: (1) similarity of surface roughness and temperature distributions; (2) similarity of flow structure above the ABL; (3) similarity of the horizontal pressure gradient; and (4) sufficient flow development to establish equilibrium of ABL for conditions (1) and (2). Boundary-layer studies by Cermak et al (1956) and Peterka et al (1974) in the wind tunnel of Fig. 2 reveal the need for condition (4). These conditions motivate consideration of a long wind-tunnel test section with features shown in Fig. 7.

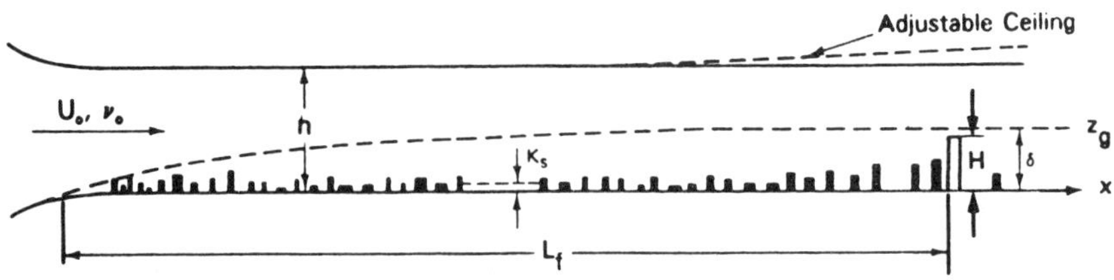

Fig. 7. Concept of BLWT test section.

The foregoing requirements for "exact" similarity of the ABL, result in the necessity that physical modeling of the ABL in a wind tunnel be considered on the basis of "approximate" similarity. The primary compromises that must be made are relaxation of the requirements that $(Re)_m = (Re)_p$ and $(Ro)_m = (Ro)_p$ because the nominal length scale is 1:500. Since Ec is proportional to the Mach number squared, equality of Ec for model and prototype need not be satisfied for wind speeds experienced in the ABL.

For "aerodynamically rough" surfaces inequality of Re for model and prototype is not a deterrent to achievement of flow similarity provided Re exceeds a value that depends upon the relative surface roughness k_s/L_f. Values of Re required for Reynolds-number independence (RNI) for neutral thermal stability are given by Fig. 8. The "model law" established by Jensen (1958) gives a RNI criterion for mean pressures on low-rise buildings. This criterion requires equality of the ratio of building height to the aerodynamic roughness length H/z_o for model and prototype. Further research is needed to establish RNI criteria for thermally stratified flow, particularly in stable conditions.

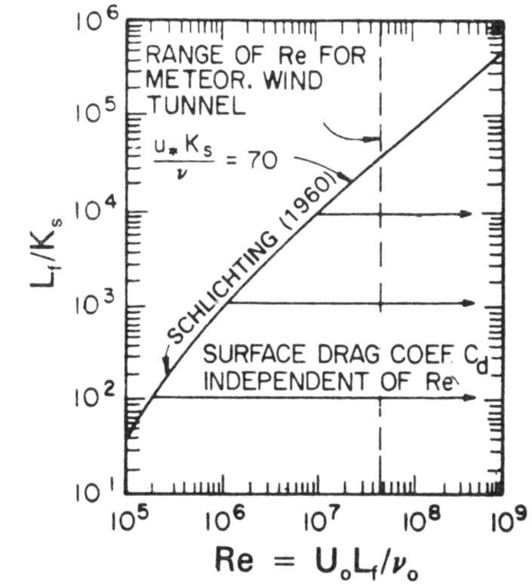

Fig. 8. Reynolds-number independence for C_d in neutral flow -- Cermak (1975).

In wind tunnels or channels which do not rotate, $(Ro)_m$ will be approximately 500 times larger than $(Ro)_p$; therefore turning of the mean flow velocity with height shown in Fig. 3 will not be replicated. Development of a technique to model Rossby-number effects and maintain RNI remain a challenge. The approach used for studies of flow over obstacles by Alessio et al (1983), a water channel (8 m long x 2 m wide x 0.6 m depth) placed on the 14-m diameter rotating platform of the Institut de Mécanique, Université de Grenoble, offers an interesting concept.

THE METEOROLOGICAL WIND TUNNEL (MWT)

Based on research in the wind tunnel shown in Fig. 2 by Cermak et al (1956) and the similarity considerations presented in the previous section, the design of a wind tunnel following the concept illustrated by Fig. 7 was reported by Cermak (1958). This BLWT, commonly identified as the MWT, is shown in Fig. 9 and described by Plate et al (1963) and Cermak (1981).

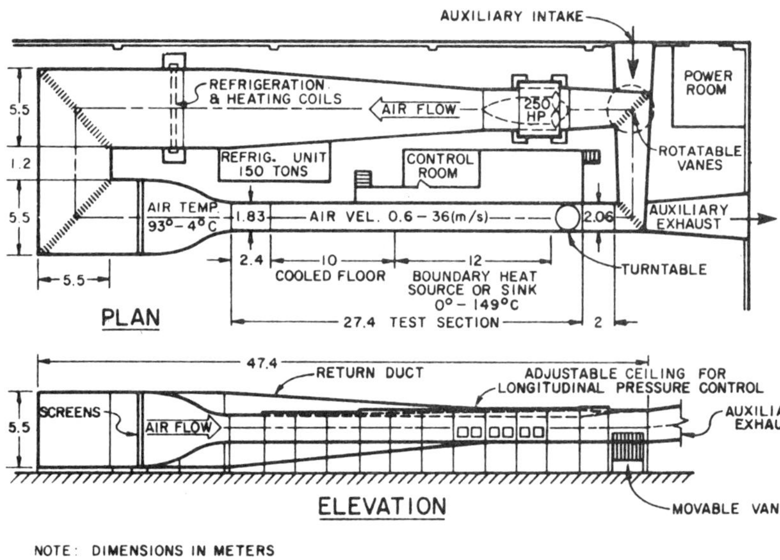

Fig. 9. Meteorological Wind Tunnel (MWT), Colorado State University -- Cermak (1958) and Plate and Cermak (1963).

Special Features of MWT for ABL Simulation

A major difference between the wind tunnel shown in Fig. 2 and the MWT is the test-section length. The long test section serves to increase the ranges of surface roughness and wind speed for which Reynolds-number independence can be achieved as indicated by Fig. 8, to produce significant distances over which the boundary layer is in statistical equilibrium (shown by constant shape factor H in Fig. 10), and to provide large boundary-layer thicknesses indicated in Fig. 11.

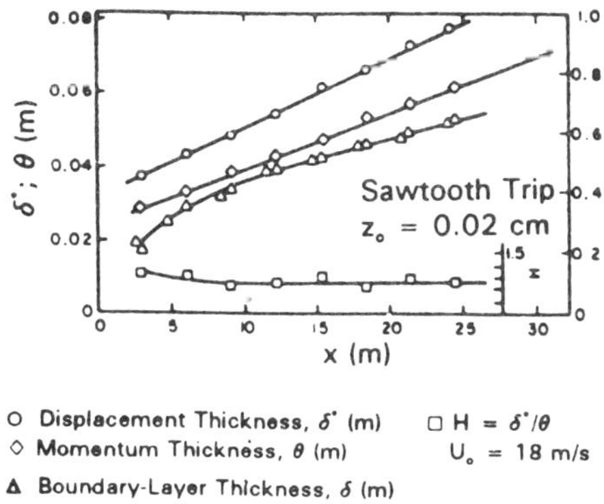

Fig. 10. Mean-velocity profile parameters for MWT -- Zoric and Sandborn (1972).

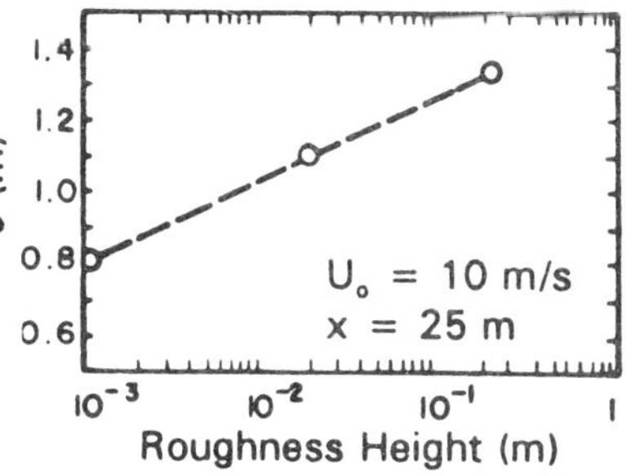

Fig. 11. Dependence of δ on roughness height in MWT -- Cermak et al (1966).

The requirement that the pressure gradient $\partial P/\partial x$ be zero for ABL similarity is met by vertical adjustment of the test-section ceiling. Continuous adjustment along the test-section length, as shown in Fig. 9, allows a maximum height increase of 0.5 m at the downwind end.

Matching of Richardson numbers for the ASL is made possible in the MWT by the heating and cooling systems indicated in Fig. 9. Large magnitudes of Ri are achieved for low wind speeds with large temperature differences. Values of Ri in the range -1 to 1 can be developed. This range includes most of the values found in field measurements such as those given in Fig. 5.

For studies of wind effects on low-rise buildings and/or low-level dispersion, depth of the simulated ASL should be increased to permit length scaling in the range of 1:50 to 1:100. This motivates partial modelling of the ABL in which the characteristic feature of "constant" shear stress (see Fig. 6) is generated over a depth of approximately 0.5 m. The introduction of inclined horizontal vanes at the test-section shown in Fig. 12 promotes downward transport of the Reynolds stress $-\rho\overline{u'w'}$ to produce the desired flow. Figure 13 gives a vertical distribution of shear stress for the horizontal vane configuration.

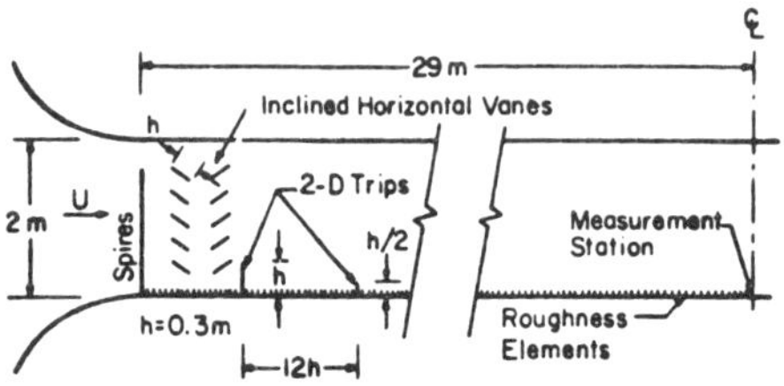

Fig. 12. Test section of MWT with flow-conditioning system for ASL simulation -- Cermak et al. (1993).

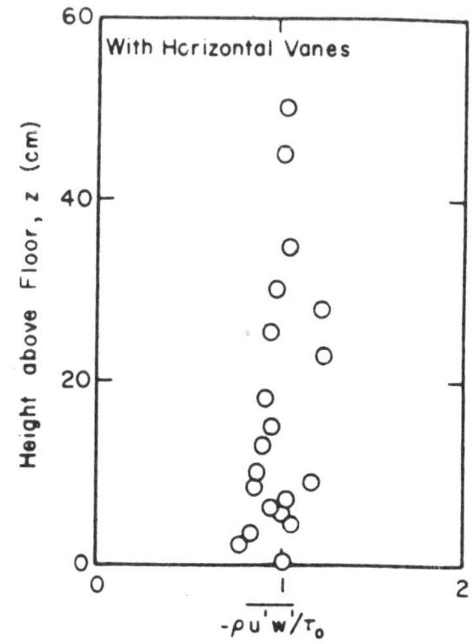

Fig. 13. Vertical distribution of shear stress with horizontal vanes -- Cermak et al. (1993).

Comparison of Measurements in MWT and Field

Comparisons of wind-tunnel and field measurements are essential for checking the validity of modelling criteria and procedures. Micrometeorological data used for the subsequent comparisons are primarily from measurements in the ASL.

Stably stratified flow over a smooth surface with only a sawtooth trip at the test-section entrance was investigated in the MWT by Arya et al (1969). A graph showing the measured values of Ri as a function of z/L ($0 < z < 0.26$) accompanied by a relationship for the atmosphere reported by Monin et al (1969) is given in Fig. 14. The agreement indicates good similarity for the two scales of flow. This is further confirmed by measurements of mean velocity and temperature for both stable and unstable flows

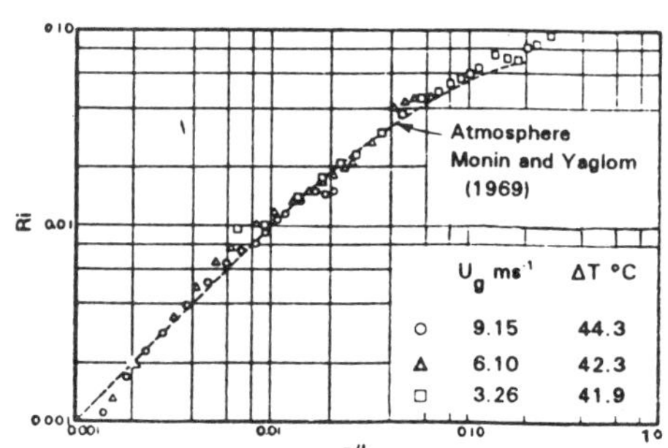

Fig. 14. Ri as a function of z/L -- Arya and Plate (1969).

in the MWT by Chuang et al (1966). On Fig. 15 vertical profiles of these data are compared to micrometeorological data from Project Prairie Grass reported by Barad (1952) using Eqs. 2 and 3 to form the coordinates. It is evident that data from both sources follow the log-linear relationships in a similar way.

Various devices are frequently placed at the test-section entrance to increase δ. Effects of the following three configurations in the MWT are reported by Cermak (1982) for neutral flow: (1) a 4-cm sawtooth trip for "natural" boundary-layer development, (2) 61-cm high vortex generators designed by Counihan (1969), and (3) 1.8-m high spires designed by Standen (1972). Profiles of mean and turbulent velocities were measured over three surface roughness (z_o = 0.02 cm, 0.15 cm and 0.35 cm) for each entrance configuration at x = 3.3 m, 6.1 m, 12.2 m, 18.3 m, and 25.9 m downwind of the entrance. Because of space limitations, only profiles of longitudinal turbulence spectra with z_o = 0.35 cm at x = 18.29 m are presented. These are shown in Fig. 16. At x = 18.29 m all profiles approach a common form indicating that the initial flow signature introduced by the various devices is erased if the test-section length is sufficient. The spires and vortex generators increased the boundary-layer thickness δ by 33% of δ for the sawtooth trip. Of greater significance is the agreement with spectral model (a) in Table 2 which is based on atmospheric data.

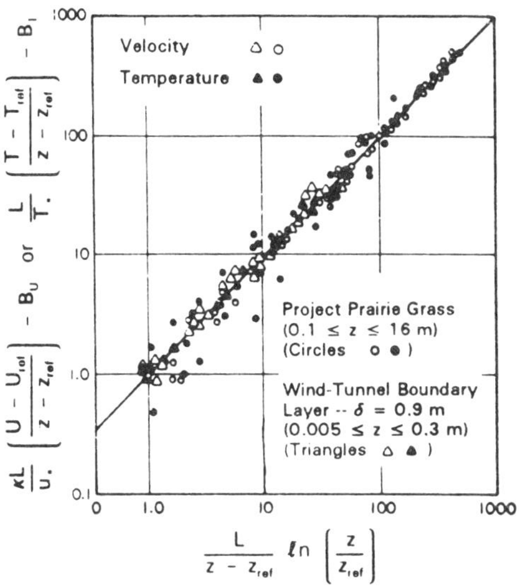

Fig. 15. Mean-velocity and temperature profiles from MWT and atmosphere compared to log-linear profile -- Chuang and Cermak (1966).

Vertical profiles of mean velocity and turbulence intensities measured by Cochran (1992) in the thick "constant" shear layer are presented in Fig. 13. These

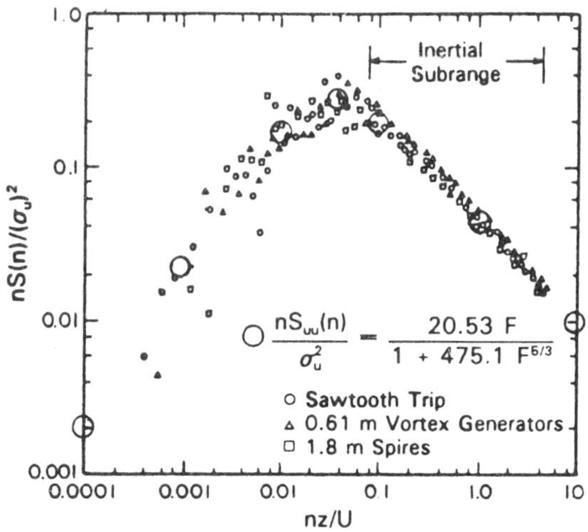

Fig. 16. Spectra of longitudinal turbulence: x = 18.29 m, z_o = 0.35 cm, U_g = 18 m/s, and z/z_g = 0.05 -- Cermak (1982). Spectrum given by model (a) of Table 2 is indicated by circles.

data are compared with field data measured at the Texas Tech University field site as reported by Chok (1988) in which model heights are related to full-scale heights by a length scale of 1:50. The mean velocity profiles presented in Fig. 17 and the longitudinal turbulence intensity given in Fig. 18 for the modelled and field ASL confirm that the simulation is satisfactory. The longitudinal turbulence spectra at a height of 4 m are presented in Fig. 19 for the TTU field site and the MWT. Model and full-scale data are in fair agreement excepting at high frequencies for which the field instrumentation has inadequate response. However, the wind-tunnel data agree very well with model (d) of Table 2.

The foregoing comparisons give good evidence that satisfactory modelling of the ASL can be achieved. Comparisons such as those shown by Figs. 14 and 15 have provided justification for development of many BLWTs for modelling of the ABL and ASL.

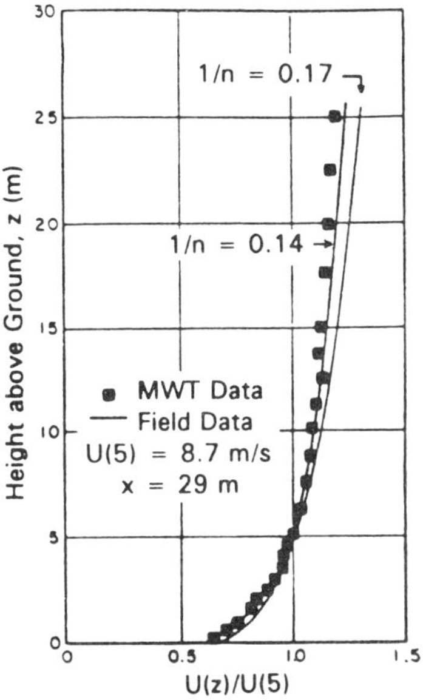

Fig. 17. Mean velocities in MWT (see Fig. 12) and at TTU field site -- Cermak et al (1993), Cochran (1992) and Chok (1988).

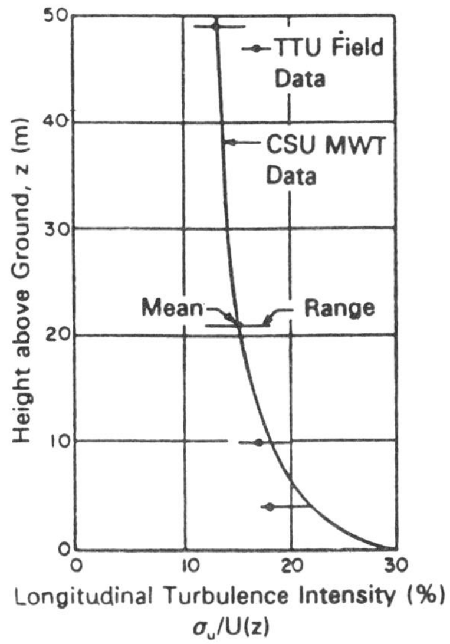

Fig. 18. Longitudinal turbulence intensities in MWT (see Fig. 12) and at TTU field site -- Cermak et al (1993), Cochran (1992) and Chok (1988)

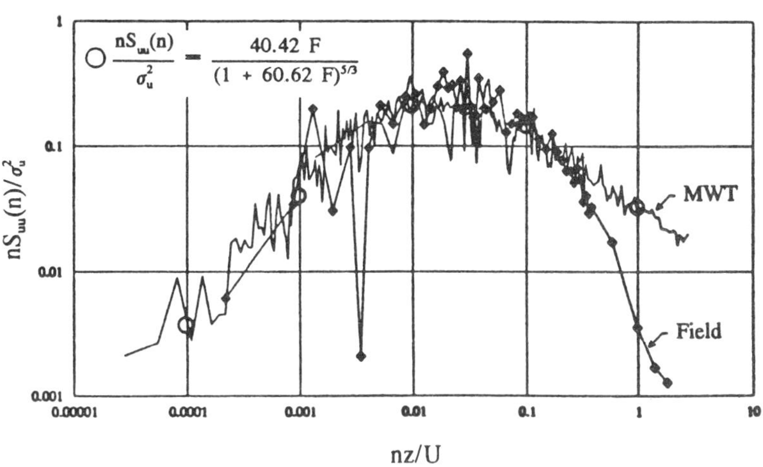

Fig. 19. Longitudinal turbulence spectra in MWT and at TTU field site (z = 4 m) -- Cermak et al (1993), Cochran (1992) and Thomas et al (1993). Circles denote model (d) spectrum of Table 2.

EVOLUTION OF BLWT CONCEPTS AND FUNCTIONS

During the last three decades many BLWTs have been constructed -- some of the closed-circuit type shown in Fig. 9 with and without thermal stratification capabilities and others of an open-circuit configuration. A comprehensive world-wide list of low-speed wind tunnels which includes many in the BLWT class has been published by Rae et al (1984). Some BLWTs have been constructed in the last 25 years with innovative features for special or multiple functions. Several of these are described that represent distinct progress in physical modelling capabilities.

The National Bureau of Standards (NBS) -- Arizona State University Facility

The unsteady flow wind tunnel illustrated in Fig. 20 was designed and constructed by Dr. Philip Klebanoff at the NBS (now the National Institute for Standards and Technology) in Gaithersburg, Maryland during the late 1960s. In 1984 it was dismantled and moved to Arizona State University, Tempe, Arizona. Basic flow characteristics of the facility are described by Saric et al (1988).

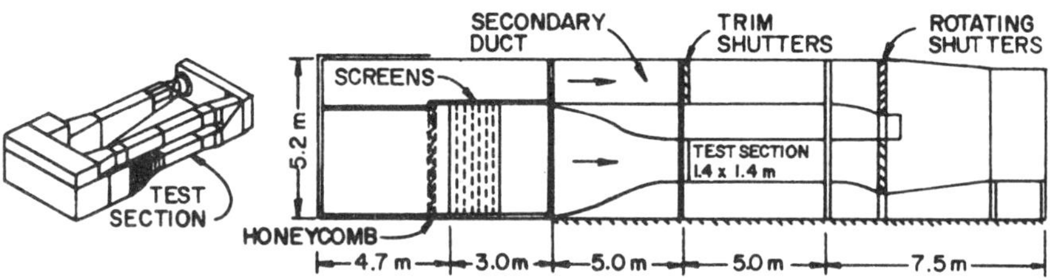

Fig 20. NBS-Arizona State University Unsteady Flow Wind Tunnel (test-section side elevation) -- Saric et al (1988).

Although the test-section is too short for classification as a BLWT, the concept used to generate unsteady flow rates in the test section without changing flow rate through the fan is noteworthy. With no flow through the secondary duct, steady flow speeds range from 1 to 36 ms^{-1}. Flow speed fluctuations in the test section are generated by rotating the two sets of shutters 180 degrees out of phase. Oscillation frequencies range from 0.1 to 25 Hz. Amplitudes from near zero to 100 percent of the mean speed can be achieved at a frequency of 0.1 Hz and mean speed of 14 ms^{-1} At higher frequencies the rate of amplitude modulation is controlled by the 0.2 s time constant for the facility. The shutters can also be programmed to produce "gusts" through the test section. This feature in a BLWT would be very useful for wind-engineering studies.

The University of Western Ontario BLWT II

Traditional BLWTs such as shown in Fig. 9 employ a diffuser as the flow-return section to minimize energy losses. In the University of Western Ontario BLWT II (Fig. 21), the flow-return section is reconfigured to form a working section of constant cross section. This section serves dual functions for studies of flow over a water surface or over a solid floor when the tank is covered. This development tends to maximize the utility of a facility with respect to cost of the facility. Another innovative feature of BLWT II is addition of peg-like roughness elements in the high-speed test section that can be computer programmed to change height. This feature greatly facilitates adjustment of roughness as a model is turned to change wind direction.

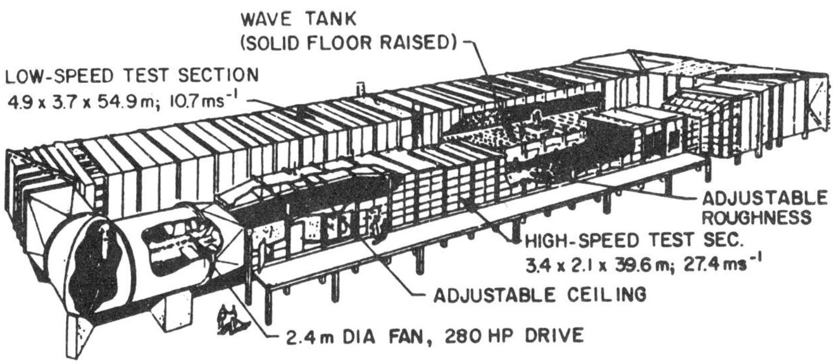

Fig. 21. The University of Western Ontario BLWT II -- Davenport et al (1985).

The Monash University 1-MW Environmental Wind Tunnel

The multi-purpose wind tunnel constructed at Monash University (Fig. 22) within the last few years represents an advancement in size and wind speed, as well as concept, for BLWTs. Three working sections are included in the closed-circuit facility. The first working section downwind of the fans enables model studies at Reynolds numbers higher than in most BLWTs. The next working section is being developed to serve as an open-jet, motor-vehicle anechoic test section. Configuration of the upper working section creates an unique approach for development of thermally stratified boundary layers. Heating of the floor results in unstable stratification while heating of the ceiling gives an upside-down, stably-stratified boundary layer. The former case can develop an instability of Pasquill (1974) category A and the later a stability of category E when the ambient wind speed is 0.5 ms^{-1}.

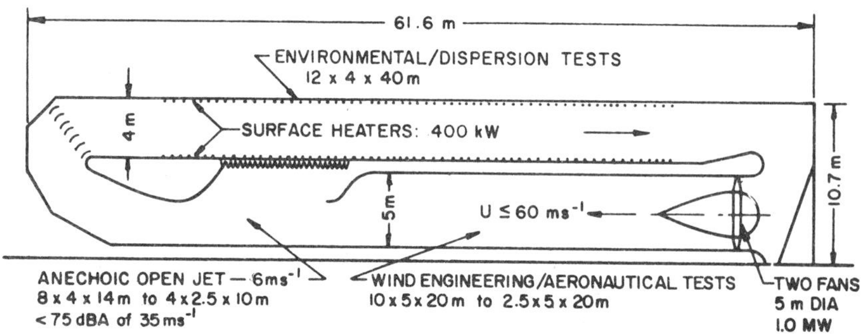

Fig. 22. The Monash University 1-MW Environmental Wind Tunnel (vertical section) -- Courtesy Professor W. H. Melbourne (1994).

Special acoustical treatment is provided to control noise level at the higher wind speeds. Acoustic lining is installed in the corner immediately upwind of the fans and downstream of the fans. With addition of acoustic wedges in the open jet, noise levels less than 75 dB are expected for a maximum wind speed of 60 ms^{-1}.

The University of Karlsruhe Convective BLWT

A new wind tunnel specifically designed to simulate convective boundary layers has been constructed at the University of Karlsruhe (Fig. 23) by Rau et al (1993). The design shown in Fig. 23 is unique in that the wind tunnel has ten channels with independent fans and heaters. The 10-m long working section is 1.5 m high and 1.5 m wide. Extreme thermal conditions are as follows: inversion height, 0.5 m;

temperature jump, 40°C; temperature gradient, 100°C m^{-1}; and surface heat flux, 1.5 kW m^{-2}. The minimum wind speed is 0.5 ms^{-1}.

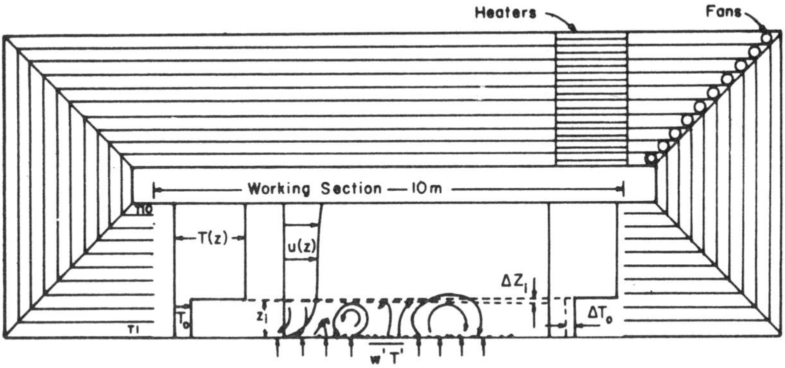

Fig. 23. The University of Karlsruhe Convective BLWT (vertical section) -- Rau and Plate (1993).

Public Works Research Institute of Japan BLWT

Motivated by the need for aeroelastic tests on full models of long-span bridges, a very wide test-section BLWT was designed and built by the Public Works Research Institute, Ministry of Construction, Japan in cooperation with the Honshu-Shikoku Bridge Authority. The wind tunnel as described by Miyata et al (1992) is shown in Fig. 24. Flow through the 4 x 41 m test section is developed by 36 1.8-m diameter fans. Mean wind speeds range from 0.5 to 12 ms^{-1} with a turbulence intensity of about 1 percent. Benefits of the 41-m width are clearly demonstrated through testing of a full-bridge model of the 3910 m long Akashi Kaikyo Bridge at a scale of 1:100 as reported by Sumiyoshi et al (1993).

CONCLUDING REMARKS

Creating wind in the laboratory has progressed in two centuries from moving an object through still air to physical modelling of the ABL in BLWTs. Modelling of the entire ABL for steady flow with neutral stability is sufficiently validated by comparisons with full-scale data to serve as a standard for wind-effect studies on

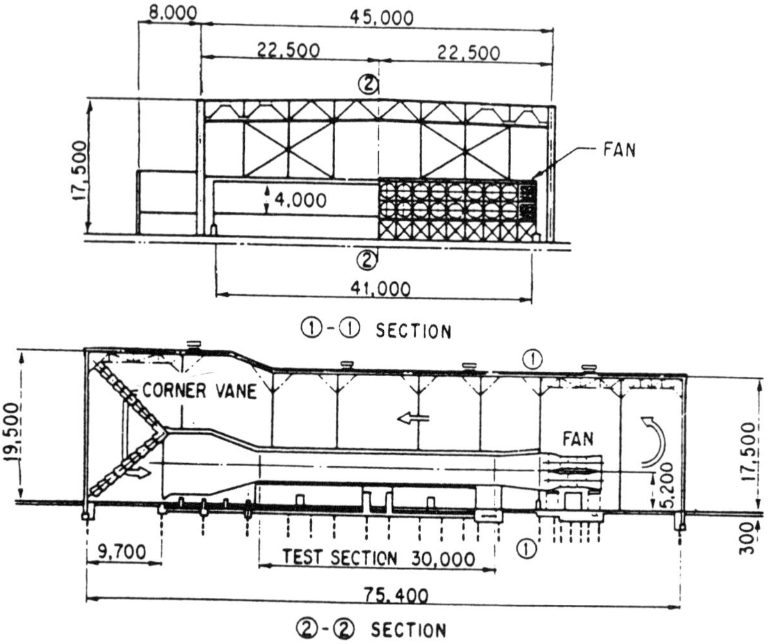

(Dimensions in millimeters)

Fig. 24. The Public Works Research Institute BLWT for long-span bridge tests, Tsukuba, Japan -- Miyata et al (1992).

high-rise structures. Physical modelling of the ASL for steady flow with and without thermal stratification has been thoroughly validated and provides reliable flows for wind-effect studies on low-rise structures and low-level dispersion. Within the past 25 years the basic BLWT has been adapted to perform multiple testing functions or to extend capabilities for special types of investigations such as wind and wave forces on offshore structures, convective flow and dispersion over urban areas and wind-effects on long-span structures. Further BLWT development is needed for modelling of unsteady flows (including large-scale gusts) and for generating upper boundary conditions (above the ABL) with various turbulence scales and intensities. Although the primary use of BLWTs is acquisition of wind-effect data for wind-engineering applications, their use for basic research on perturbed boundary layers and validation of codes for computational fluid mechanics has been significant.

SYMBOLS AND NOTATION

A_u = dimensionless coefficient
B_T = dimensionless coefficient
B_U = dimensionless coefficient
C_d = surface drag coefficient
C_p = specific heat at constant pressure
Ec = Eckert number, $U_o^2/(C_{po}T_o)$
F = reduced frequency, nz/U
f = Coriolis parameter, $2\ \Omega_e \sin\phi$
g = gravitational acceleration
H = height of building or boundary-layer shape factor
H_o = surface heat flux
h = height of test section ceiling or trip fence
k = thermal conductivity
K_s = equivalent sand roughness length
L = Monin-Obukhov length, $-u_*^3/[\kappa g H_o/T_o\rho_o C_{po})]$
L_f = length of upwind fetch
n = frequency
n^{-1} = power law exponent
P = mean pressure
Pr = Prandtl number, $\nu_o\rho_o C_{po}/k_o$
Re = Reynolds number, $U_o L_o/\nu_o$
Ri = bulk Richardson number, $(\Delta T_o/T_o)/(L_o g/U_o^2)$
Ro = Rossby number, $U_o/(L_o\Omega_o)$
Ro_s = surface Rossby number, $U_{g\infty}/(f\ z_o)$
S_{ii} = spectrum of ith turbulence component
T = local mean temperature
T_* = $-H_o/(\rho_o C_{po}\kappa u_*)$
ΔT = mean temperature difference
t = time
u_i' = ith component of velocity fluctuation from the mean
u_* = friction velocity, $(\tau_o/\rho)^{1/2}$
U_i = ith component of local mean velocity
U_g = mean velocity at z = δ or z_g
$U_{g\infty}$ = geostrophic wind speed
x = distance downwind from test section entrance
x_i = ith space coordinate
z = vertical space coordinate
z_o = roughnes length
z_g = height of gradient wind
δ = boundary-layer thickness
δ_{ij} = Kronecker delta
ϵ_{ijk} = permutation tensor
θ = average potential temperature
θ = local temperature fluctuation
κ = von Kármán constant
ν = kinematic viscosity
ρ = mass density
σ_i = rms of ith turbulence component
τ_o = surface shear stress
ϕ = latitude
ϕ_ϵ = dissipation function
Ω_e = angular velocity of Earth
Ω_i = ith component of angular velocity
$\overline{(\ )}$ = time average
$(\ )_o$ = reference quantity
$(\ )_g$ = value of quantity at the gradient wind height
$(\ )^*$ = nondimensional quantity

REFERENCES

Albertson, M. L. (1948), "Analysis of Evaporation as a Boundary Layer Phenomenon", PhD Dissertation, State University of Iowa, Iowa City, IA, 50 pp.

Alessio, S., Briatore, L., Loughetto, A., D'Hières, G. C. and Didelle, H. (1983), "Laboratory Simulation of Rotating Flows Over Obstacles", Il Nuovo Cimento, Vol. 6C, N. 4, pp. 401-428.

Arya, S. P. S. and Plate, E. J. (1969), "Modeling of the Stably Stratified Atmospheric Boundary Layer", Jl. of Atmos. Sciences, Vol. 26, No. 4, July, pp. 656-665.

Bailey, A. and Vincent, N. D. G. (1943), "Wind-Pressures on Buildings Including Effects of Adjacent Buildings", Jl. Institution of Civil Engineers, Vol. 20, No. 8, October, pp. 243-275.

Barad, M. L. (ed.) (1952), "Project Prairie Grass -- A Field Program in Diffusion", Geophysical Research Paper 59, Geophysical Research Directorate, Bedford, MA.

Cermak, J. E. (1952), "Application of Model Techniques to Mass Transfer and Evaporation Studies", Paper Presentation, Centennial of Engineering, Chicago, IL, September, CER53JEC2.

Cermak, J. E. and Spengos, A. C. (1956), "Heat Transfer by Forced Convection From a Horizontal Flat Plate Into a Turbulent Boundary Layer", Proc. Heat Transfer and Fluid Mechanics Institute, Stanford University, Stanford, CA, June, CER56JEC21.

Cermak, J. E. (1958), "Wind Tunnel for the Study of Turbulence in the Atmospheric Surface Layer", Technical Report Air Force Cambridge Research Center, Air Research and Development Command, Bedford, MA, (Contract AF19(604)-1706), November, CER58JEC42.

Cermak, J. E., Sandborn, V. A., Plate, E. J., Binder, G. H., Chuang, H., Meroney, R. N. and Ito, S. (1966), "Simulation of Atmospheric Motion by Wind-Tunnel Flows", Fluid Dynamics and Diffusion Lab., Technical Report No. CER66JEC-VAS-EJP-GJB-HC-RNM-SI17, May, 101 pp.

Cermak, J. E. (1975), "Applications of Fluid Mechanics to Wind Engineering", 1974 Freeman Scholar Lecture, ASME, Jl. of Fluids Engineering, Vol. 97, Series 1, No. 1, March, CEP74-75JEC7.

Cermak, J. E. (1981), "Wind-tunnel Design for Physical Modeling of Atmospheric Boundary Layers", Special Volume on Advances in Fluid Mechanics, ASCE Jl. of Engineering Mechanics, Vol. 107, No. EM3, June, pp. 623-642, CEP80-81JEC12.

Cermak, J. E. (1982), "Physical Modeling of the Atmospheric Boundary Layer (ABL) in Long Boundary-layer Wind Tunnels (BLWT)", Proc. of International Workshop on Wind Tunnel Modeling Criteria and Techniques in Civil Engineering Application, National Bureau of Standards, Gaithersburg, MD, Cambridge University Press, Cambridge, U.K., April, pp. 97-137.

Cermak, J. E., Cochran, L. S. and Leffler, R. D. (1993), "Wind-tunnel Modelling of the Atmospheric Surface Layer", Proc. of the Third Asia-Pacific Symposium on Wind Engineering, Hong Kong, December.

Chuang, H. and Cermak, J. E. (1966-1967), "Similarity of Thermally Stratified Shear Flows in the Laboratory and Atmosphere", The Physics of Fluids Supplement, Vol. 10, Part II, No. 9, Proc. of an International Symposium on Boundary Layers and Turbulence With Geophysical Applications, Kyoto, Japan, September, pp. S255-S258

Chok, C. V. (1988), "Wind Parameters of Texas Tech University Field Site", Master of Science Thesis, Civil Engineering Department, Texas Tech University, Lubbock, TX.

Cochran, L. S. (1992), "Wind-tunnel Modelling of Low-rise Structures, 'PhD Dissertation, Civil Engrg. Dept., Colorado State Univ., Ft. Collins, CO.

Counihan, J. (1969), "An Improved Method of Simulating an Atmospheric Boundary Layer in a Wind Tunnel", Atmos. Environ., Vol. 3, pp. 197-214.

Davenport, A. G. (1960), "A Rationale for Determination of Design Wind Velocities", Proc. ASCE, Jl. of the Structural Div., Vol. 86, pp. 39-66.

Davenport, A. G. (1961), "The Spectrum of Horizontal Gustiness Near the Gravel in High Winds", Quarterly Jl. Royal Meteorological Soc., Vol. 81, pp. 194-211.

Davenport, A. G., Isyumov, N., King, J. P. C., Novak, M., Surry, D. and Vickery, B. J. (1985), "BLWT II: The Design and Performance of a New Boundary Layer Wind Tunnel", Proc. of the 5th U.S. National Conference on Wind Engrg., Texas Tech University, Lubbock, TX, pp. 2A35-39.

Harris, R. I. and Deaves, D. M. (1980), "The Structure of Strong Winds", Paper 4, CIRA Conference on Wind Engineering in the Eighties, Construction Industry Research and Information Assoc., London.

Irminger, J. O. V. (1893), "Experiments on Wind Pressure," Proc. Inst. of Civil Engrs., Vol. 118, pp. 468-472.

Irminger, J. O. V. and Nϕkkentved, Chr. (1930), "Wind-pressures on Buildings," Series 1, Ingeniϕrvidenskabelige Skrifter, A Nr. 23, Kϕbenhaven.

Irminger, J. O. V. and Nϕkkentred, Chr. (1936), "Wind-pressure on Buildings," Series 2, Ingeniϕrvidenskabelige Skrifter, A Nr. 42, Kϕbenhaven.

Izumi, Y. (ed.) (1971), "Kansas 1968 Field Progress Report", Environmental Research Paper No. 79, AFCRL-72-0041, Air Force Cambridge Research Laboratories, L. G. Hanscom Field, Bedford, MA, 75 pp.

Jensen, M. (1958), "The Model-Law for Phenomena in Natural Wind," Ingeniϕren, Vol. 2, No. 4, pp. 121-128.

Kaimal, J. C., Wyngaard, J. C., Izumi, Y. and Coté, O. R. (1972), "Spectral Characteristics of Surface-layer Turbulence", Quarterly Jl. Royal Meteorological Soc., Vol. 98, pp. 563-589.

Kármán, T. von (1948), "Progress in the Statistical Theory of Turbulence", Proc. National Academy of Sciences U.S., Washington, D.C., Vol. 34.

Lettau, H. H. (1962), "Theoretical Wind Spirals in the Boundary Layer of a Barotropic Atmosphere", Beitr. Phys. Atmos., Vol. 35, pp. 195-212.

Miyata, T., Yamada, H., Yokoyama, K., Kanazaki, T., Iijima, T. and Tatsumi, M. (1992), "Construction of Boundary Layer Wind Tunnel for Long-span Bridges", Jl. of Wind Engrg. and Ind. Aerodyn., Vol. 42, pp. 885-896.

Monin, A. S. and Obukhov, A. M. (1954), "Basic Laws of Turbulent Mixing in the Ground Layer of the Atmosphere", Trudy Geofizs. Inst. An SSSR, 2, No. 24 (15 p.), pp. 163-187, (see Monin, A. S. and Yaglom, A. M., "Statistical Fluid

Mechanics: Mechanics of Turbulence: Vol. 1, MIT Press, Cambridge, MA, 1971, pp. 425-442).

Monin, A. S. and Yaglom, A. M. (1971), "Statistical Hydrodynamics, Part I: The Mechanics of Turbulence", Nauka Press, Moscow, 1969, Chapter 4, pp. 327-427 (English transl. by U. S. Dept. of Commerce) also Statistical Fluid Mechanics: Mechanics of Turbulence, The MIT Press, (John L. Lumley, ed.).

Olesen, H. R., Larsen, S. E. and Højstrup, J. (1984) "Modelling Velocity Spectra in the Lower Part of the Planetary Boundary Layer", Boundary Layer Meteorology, Vol. 29, pp. 285-312.

Pasquill, F. (1974), "Atmospheric Diffusion", John Wiley & Sons, New York, NY.

Peterka, J. A. and Cermak, J. E. (1974), "Simulation of Atmospheric Flows in Short Wind Tunnel Test Sections", Technical Report CER73-74JAP-JEC32, Fluid Dynamics and Diffusion Laboratory, Colorado State University, Fort Collins, CO.

Plate, E. J. and Cermak, J. E. (1963), "Micrometeorological Wind Tunnel Facility, Description and Characteristics", Final Report for U.S. Army Electronic Research and Development Activity, Fort Huachuca, AZ, (Contract No. DA-36-0309SC-80371), February, CER63EJP-JEC9.

Prandtl, L. and Reichardt, H. (1934), "Einfluss von Wärmeschichtung auf de Eigenschaften einer Turbulenten Strömung", Deutsche Forschung, No. 21, Berlin, pp. 110-121.

Rae, W. H., Jr. and Pope, A. (1984), "Low-Speed Wind Tunnel Testing", John Wiley & Sons, New York, NY.

Rau, M. and Plate, E. J. (1993), "Wind Tunnel Modeling of Convective Boundary Layers", Proc. of the NATO Advanced Study Institute on Wind Climate and Wind Engineering in Cities, Karlsruhe, Germany (to be published by Kluwer Academic Publishers in 1994).

Robins, B. (1742), "New Principles of Gunnery", J. Nourse, London.

Ruark, A. E. (1935), "Inspectional Analysis: A Method Which Supplements Dimensional Analysis", Jl. Elisha Mitchell Sci. Soc., Vol. 51, pp. 127-133.

Saric, W. S., Takagi, S. and Mousseux, M. (1988), "The ASU Unsteady Wind Tunnel and Fundamental Requirements for Freestream Turbulence Measurements", Paper No. AIAA-88-0053, AIAA 26th Aerospace Sciences Meeting, Reno, NV.

Schlichting, H. (1960), "Boundary Layer Theory", McGraw-Hill Book Company, New York, NY.

Scruton, C. (1960), "The Use of Wind Tunnels in Industrial Aerodynamics Research", National Physical Lab., Great Britain, NPL/Aero/411, 53 pp.

Sherlock, R. H. and Stalker, E. A. (1941), "A Study of Flow Phenomena in the Wake of Smokestacks", Engr. Res. Bull. No. 29, Univ. of Mich., 49 pp.

Smeaton, J. (1759-1760), "An Experimental Investigation Concerning the Natural Powers of Water and Wind", Phil. Trans. Roy. Soc. of London, Vol. 51, pp. 100-174.

Standen, N. M. (1972), "A Spire Array for Generating Thick Turbulent Shear Layers are Natural Wind Simulation in Wind Tunnels", Technical Report LTR-LA-94, National Aeronautical Establishement, Ottawa, Canada.

Stanton, T. E. (1924-1925), "Report on Measurement of the Pressure of the Wind on Structures", Proc. Inst. of Civil Engr., Vol. 219, pp. 125-158.

Sumiyoshi, Y., Endo, T., Miyata, T., Sato, H. and Kitagawa, M. (1993), "Experiments for the Akashi Kaikyo Bridge in a Large Boundary Layer Wind Tunnel", Proc. of the International Seminar on Utilization of Large Boundary Layer Wind Tunnel, Tsukuba, Japan, pp. 101-120.

Tieleman, H. W. (1991), "The Atmospheric Surface Layer and Its Simulation for Wind-load Studies on Low-rise Structures", Technical Report VPI-E-91-04, Department of Engineering Science and Mechanics, Virginia Polytechnic Institute and State University, Blacksburg, VA.

Tieleman, H. W. (1992), "Universality of Velocity Spectra", Proc. of the 11th Australasian Fluid Mechanics Conference, Hobart, Tasmania.

Thomas, G., Cochran, L. S., Cermak, J. E. and Mehta, K. C. (1993), "Comparison of Full-scale and Wind-tunnel Measured Spectra", Proc. ot the 7th U.S. National Wind Engineering Conference, University of California, Los Angeles, CA.

Wieringa, J. (1992), "Updating the Davenport Roughness Classification", Jl. of Wind Engrg. and Ind. Aerodyn., Elsevier, Vol. 41, pp. 357-368.

Zoric, D. and Sandborn, V. A. (1972), "Similarity of Large Reynolds Number Boundary Layers", Boundary-Layer Met., Vol. 2, No. 3, Mar., pp. 326-333.

METHODS OF FLUCTUATING PRESSURE MEASUREMENT IN WIND ENGINEERING

J.D. HOLMES

Project Area Manager and Chief Research Scientist, CSIRO Division of Building, Construction and Engineering, Melbourne, Australia

SUMMARY

The principles behind electrical pressure transducers and sensors, and pressure scanning systems are first described. Three types of connecting tubing systems for single-point measurements are discussed. Discussion of each type is based on calculations of amplitude and phase response, from a theoretical model that has been well validated by experiment. A discussion of parallel-tube/manifold systems for the measurement of average fluctuating pressures at several measurement points is also given.

INTRODUCTION

The accurate measurement of transient, fluctuating and peak fluid pressures on a surface is a practical problem which has been faced by engineers and scientists in a number of disciplines, such as aeronautical engineering and acoustics, as well as wind engineering. Nowadays, this is normally accomplished by means of electronic pressure sensors or transducers, whose outputs are sampled by digital computers. The characteristics of modern pressure transducers and electronic-mechanical and electronic switching systems will be covered briefly in this review paper.

A common situation is that in which, for reasons of cost or geometric constraint, it is required to mount a pressure sensor or transducer remotely from the point where the pressure measurement is required. Then the fluctuating pressure must be transmitted by tubing between the measurement and sensing points. The transmission of pressure fluctuations is affected by the mass inertia, compressibility, and energy dissipation in the air. Standing waves can produce unwanted résonant peaks in the amplitude frequency response characteristics of the system, and a nonlinear variation of phase lag with frequency.

The paper reviews studies, carried out within the last few years, that have greatly improved the frequency response, in both amplitude and phase, of such measurement systems. An ideal system would have an amplitude response which is constant over the frequencies of interest, and a linear phase variation with frequency. The latter characteristic guarantees that there is no distortion of transient pressure "signatures" by the system.

As well as pressure measurement at a single point, systems in which pressures from a number of points are connected to a common manifold or "pneumatic averager" have become widely used recently. In wind engineering, this arrangement has been used to obtain fluctuating and peak pressures appropriate to a finite area or panel on a building model in a turbulent wind-tunnel flow (e.g.Surry and Stathopoulos 1977, Holmes and Rains 1981, Gumley 1984, Holmes 1987, Kareem *et al.* 1989). A similar arrangement has been used to derive total fluctuating forces from the local pressures, on a section of a two-dimensional bluff body (Bearman and Luo 1988; Luo and Bearman 1990). Optimum techniques for this type of measurement are discussed in the paper.

PRESSURE TRANSDUCERS

The basic requirement of a pressure transducer is a device which accurately transforms a pressure difference to an electrical voltage. Two basic phenomena are used: (i) the change in resistance of conducting or semiconducting element as it is strained; and (ii) the change of capacitance as one plate of a capacitor is distorted. A microphone is a very sensitive form of pressure transducer usually designed to measure only fluctuations at high frequencies, and hence is not suitable for wind engineering applications.

Pressure transducers are available in three forms : absolute (the rear of the diaphragm is exposed to a fixed reference pressure, usually a hard vacuum), gage (referenced to atmospheric pressure), and differential, for which the reference pressure is selected by the user. In wind engineering, the latter type is invariably used. The performance of pressure transducers has improved greatly since they were first introduced in the nineteen-fifties. Modern instruments can have a response from zero to several thousand Hertz with very stable sensitivities. The desirable pressure difference range is ± 1500 Pascals (N/m^2), for both wind-tunnel and full-scale measurements. This is at the low pressure end (high sensitivity) of most designs of transducer, so the choice for wind engineering purposes is fairly limited.

Figure 1 shows an older type based on the capacitance principle which was designed to fit inside a 'Scanivalve' pressure scanning device (see later discussion). These transducers had high sensitivity, but were expensive and had variable frequency response characteristics. Figure 2 shows a modern type of solid state piezo-resistive pressure 'sensor' which has now become the norm for wind engineering. These sensors are fabricated in high volume using silicon processing techniques from the semiconductor industry, and are hence cheap and have very repeatable properties. Strain-sensitive resistors are ion implanted into silicon wafers on one side, and a diaphragm is created by chemically etching the silicon on the other side (Sensym, 1991). Typical diaphragm thicknesses are 5 to 200 microns, with values at the lower end for wind engineering applications. Many of these devices, including that shown in Figure 2, contain built-in amplifiers.

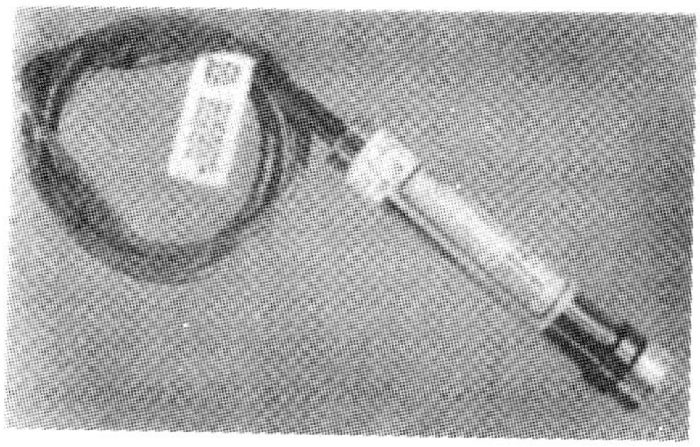

Figure 1. An older type of capacitative pressure transducer.

SCANNING DEVICES

Scanning devices for measurement of pressures at a large number of points, were developed for, and are still mainly used, in the aerospace industry. The requirements in this industry are different to those in wind engineering. In aeronautical applications, *steady* pressures from many measurement positions are required to be rapidly sampled. A reasonably short response or settling time is required to maximise the scanning speed. However, in wind engineering the requirement is to measure rapidly fluctuating pressures sequentially for a defined sample time. The physical characteristics of the scanning device form part of the measurement system and should be included in any calibration of the frequency response of the system. Two types of scanning devices are available : the electro-mechanical type, and the electronic type.

Figure 2. Example of a modern piezo-resistive pressure sensor.

In the older electro-mechanical type, a single pressure transducer is sequentially connected to many pressure tappings. The pressure transducer may be inserted in the device, or mounted externally. The internal tubing connections are complex in geometry and vary between different types. The tapered tubing and right angled bends cannot be included in theoretical models, and experimental calibration of the frequency response of measurement systems including these devices, is required.

A recent development is the electronic pressure scanner. This consists of a module containing a number (typically 16 or 32) of silicon piezo-resistive pressure sensors, one for each pressure port, which are electronically multiplexed at up to 20,000 Hertz, through an amplifier. The amplifier and multiplexing circuitry are contained inside the module, which is quite small in volume and can often be mounted within wind-tunnel models. These devices have largely replaced the electro-mechanical scanners in the aeronautical industry, and are starting to be adopted in wind engineering. The multiplexing rate is sufficiently greater than the frequencies of interest in wind engineering, thus enabling effectively simultaneous sampling of pressures from many pressure taps on wind-tunnel models.

PREDICTION OF RESPONSE TO FLUCTUATING PRESSURES

Determination of the response to fluctuating pressures of a proposed measurement system may be required for either of two reasons. The first reason is to determine the amplitude and phase response of a system with the intent of correcting the measurements after sampling. Although correction circuits have been used (Irwin, 1988), it is very difficult to simultaneously correct both amplitude and phase, and digital methods are more effective (e.g. Irwin, 1979). The second reason is to develop a system with characteristics such that measurements can be carried out over the required bandwidth without correction. Systems of this type are the main ones of interest in this paper.

Theoretical and experimental methods of determination of frequency response are available, and are discussed following.

Theoretical Model

The response of a measurement system to fluctuating pressures depends on the characteristics of all the elements of the system, including pressure tapping, manifold if used, connecting tubing, pressure sensor and/or scanner. Although experimental calibration should always be carried out for confirmation, a theoretical model of the response of systems is extremely useful for generic studies, and for the initial design stage of proposed measurement systems.

For many years the definitive model has been that of Bergh and Tijdeman (1965). This followed a number of earlier studies, notably that of Iberall (1950). Bergh and Tijdeman derived a recursion formula for the sinusoidal pressure fluctuation in a volume j connected by lengths of tubing to a preceding volume j – 1, and to the following volume, j + 1. It is based on a laminar viscous flow model of the fluctuating flow in the tubes, and has been shown to model well the behaviour of real measurement systems, when the tube lengths and diameters, and transducer volume and diaphragm flexibility, are accurately known and incorporated into the calculations (Holmes and Lewis 1986,1987a, Letchford 1987). The diaphragm flexibility, which can be significant for many transducers, can often be conveniently represented by an equivalent additional volume which is a constant for a given transducer. Bergh and Tijdeman (1965) and Holmes and Lewis (1987a) describe methods of measuring this flexibility experimentally.

An example of the agreement between the theory and experiment is given in Figure 3. For one of the transducers, it was necessary to model the *dynamic* diaphragm flexibility (i.e. inertial effects on the response of the diaphragm) to obtain agreement with the measurements.

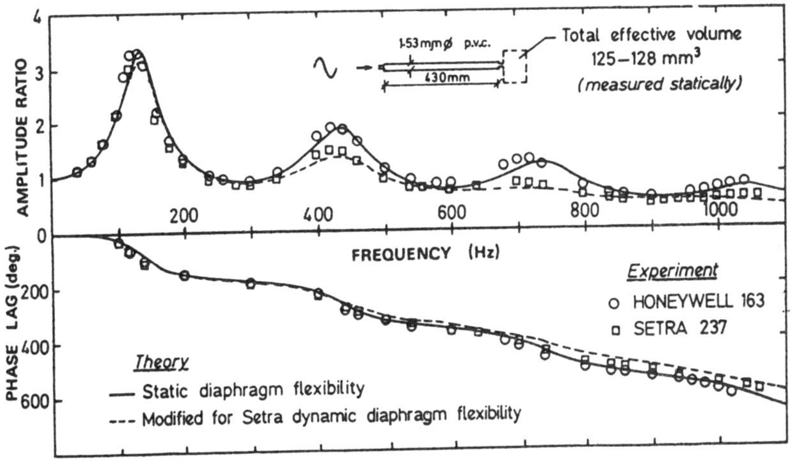

Figure 3. Comparison of theoretical and experimental response predictions.

The original work of Bergh and Tijdeman was applied to series connected tubes and volumes. The theory has since been extended to parallel-tube/ manifold systems by Gumley (1983), to ring-connected reference systems by Waldeck (1986), and to the leaked-tube system by Holmes and Lewis (1989).

Experimental Calibration

Experimental calibration of the frequency response of measurement systems to be used in practice should always be carried out. Small changes in dimensions, especially tubing diameters can have significant effects on the response characteristics, and cannot always be measured with sufficient accuracy.

The fundamental requirements for the calibration equipment are (Holmes and Lewis 1987a):

(a) a source of fluctuating pressure

(b) a reference or coupling cavity

(c) a reference flush-mounted transducer which is used to measure the pressure in the cavity

(d) amplitude and phase measuring equipment for the output of the reference transducer, and that of the system under test. This could be a digital computer.

Although random pressure fluctuations, in some cases derived from a wind tunnel, have been used, sinusoidal fluctuations are recommended for more accurate calibrations. These may be obtained by use of a fluidic function generator as used by Irwin *et al.* (1979), and Holmes and Lewis (1987a), or a loudspeaker (Holmes and Lewis 1987b).

Upper Frequency Limit

Since wind pressure fluctuations on buildings are non-Gaussian, there is no simple relationship between the peak pressures and the standard deviation, and peak pressures or peak factors must be determined experimentally. Unfortunately the values of recorded peak pressures are highly dependent on the characteristics of the measurement system at the high frequency end. Several investigations have been carried out of the frequency response requirements for both point pressures and area-averaged pressures, e.g. Durgin (1982), Marshall and Reinhold (1982), Holmes (1984), and Letchford *et al.* (1992).

Irwin (1988), based on earlier work, suggested that to avoid attenuation of peaks in negative pressure regions for tall building tests in wind tunnels the measurement system should have a flat amplitude response to a frequency equal to approximately $3U_g/B$, where U_g is the mean *gradient* wind speed and B is the building width. This limit was supported by the work of Letchford *et al.* (1992), using full-scale data from the Texas Tech low-rise building. However, they converted the desirable upper frequency limit to $10U_h/B$, where U_h is the mean wind speed at the top of the building.

Although the upper frequency limit will depend on building shape, wind direction and the nature of the flow in the vicinity of the measurement postion (i.e.

stagnation, separation, wake), the above criteria can be used in the absence of better values.

TUBING SYSTEMS FOR SINGLE-POINT MEASUREMENTS

Examples of the theoretical amplitude and phase response characteristics of two simple measurement systems, each consisting of a length of tubing of constant diameter connecting the point of measurement to a pressure transducer, are shown in Figure 4. The pressure transducer is represented by a volume which will comprise the actual physical volume to which the transducer diaphragm is exposed, plus the equivalent additional volume associated with the flexibility of the diaphragm (Holmes and Lewis 1987a). The systems shown in Figure 4 would only be suitable for measurements of static pressures, and dynamic pressures at low frequencies (less than 20 Hertz), unless analogue or digital corrections were applied to the measured pressures.

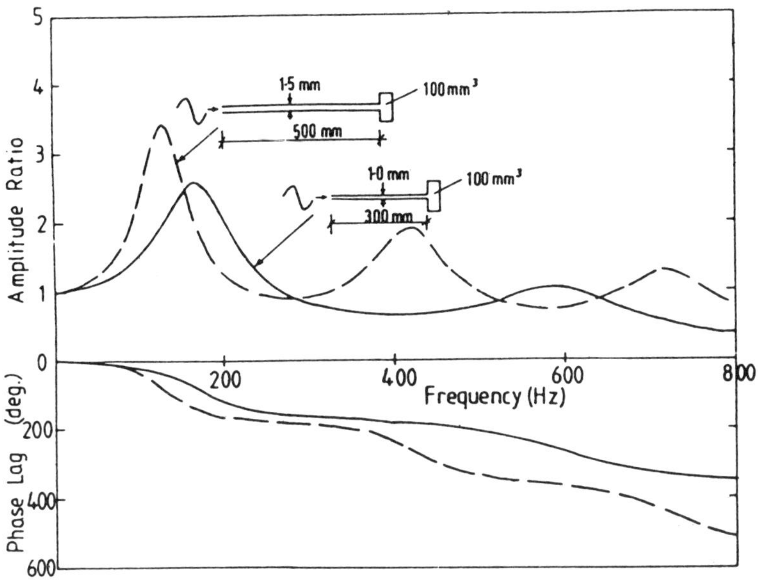

Figure 4. Frequency response of long-tube systems.

Three common practical systems for single-point measurements, that require no correction will be discussed:

(a) "Short" tube systems;

(b) "restricted" tube systems; and

(c) "leaked" tube systems.

Discussion of each of (a),(b) and (c), will be supported by computations of the response using a computer program based on the Bergh and Tijdeman (1965) theoretical model.

A comparison of the relative merits of each system for typical measurement situations will be given.

Short-tube Systems

This system uses a relatively short length of tubing to connect the measurement point to the sensor. Typically, for wind-tunnel application, this may consist of tubing 20–100 millimetres long, and 1–2 mm internal diameter. The short tube lengths will result in resonant frequencies that are high, hopefully well above the range of interest for the measurements. However, the short tube also results in low dissipation of energy, and the amplitude response rises to a high value at the peak. The amplitude ratio may exceed 1.1 at frequencies as low as 20% of the resonant frequency. The transducer volume may also be of the same order of magnitude as the tubing volume; in this case, the frequency response will be quite dependent on this variable. Normally, an electronic low-pass filter will be required for this system.

Figure 5 shows the computed responses for two systems of this type with typical tube lengths and diameters. A transducer volume of 100 mm^3 has been assumed. The usable frequency range for both the systems shown is 0 Hertz to about 120 Hertz. This may be adequate for many applications, but in many cases a much higher upper frequency limit is required. Then restricted-tube systems or leaked-tube systems are more appropriate.

Restricted-tube Systems

Restricted-tube systems may be defined as those involving one or more changes in internal diameter along the tube length. Such systems often allow location of pressure sensors at distances of 150–500 mm from the measurement point, with good amplitude and phase characteristics up to 200 Hertz, or more.

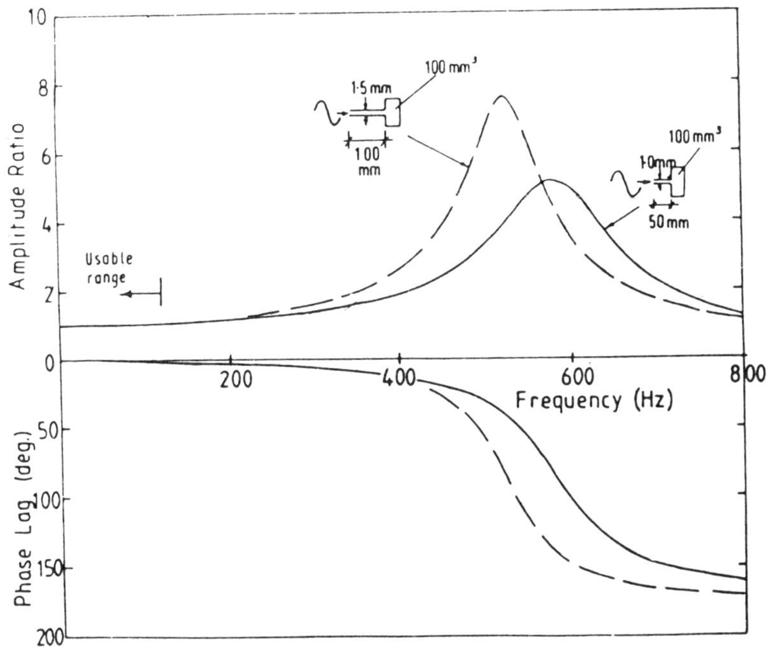

Figure 5. Examples of short-tube systems.

The simplest system of this type is the two-stage type, in which a section of narrower tube is inserted between the main tube section and the transducer. Figure 6 (from Holmes and Lewis 1987a) shows the computed response characteristics for a two-stage system with a second stage tube diameter equal to one-third of that of the first stage. Keeping the overall tube length constant, the length of the second stage tube has been chosen to give the least departure from unity of the amplitude response. The figure also shows the contributions of each stage to the amplitude and phase response. The resonance in the first stage tube is damped by the second stage, consisting of the narrow tube with transducer volume, to give a nearly flat combined amplitude response, and near-linear phase response.

Figure 7 shows the response for three systems, including that shown in Figure 6, giving similar response characteristics. They each have the same overall tubing length (500 mm), but have the small diameter tubing arranged differently with respect to the main tube. The three systems shown have two-, three- and four-stage tubing arrangements, and have acceptable frequency response characteristics to frequencies approaching 300 Hertz.

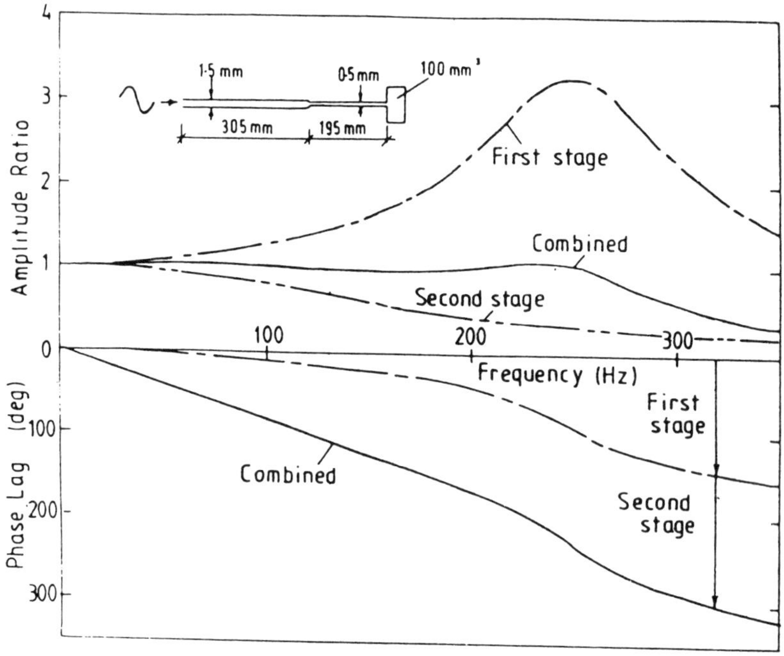

Figure 6. A two-stage, restricted-tube system with contributions from each stage.

Systems which are most popular in wind engineering studies, i.e. model studies of wind loads on structures, consist of one or more short lengths of very small diameter capillary tubing or "restrictors", usually made from metal, inserted in the flexible main tube line, thus forming a three-stage system. Restrictors were originally used by Surry and Isyumov (1975), and their effects first investigated in detail by Irwin *et al*. (1979).

Holmes and Lewis (1987a) describe an optimisation process for systems of this type. Figure 8, derived from results in that paper, shows the response characteristics for three-stage systems with restrictors of three different internal diameters; in each case, the amplitude response has been optimised by moving the restrictor to the position to produce a flat (within ±5%) amplitude response to the highest frequency, whilst keeping the overall tube length constant. The three-stage system from Figure 7 is included in Figure 8.

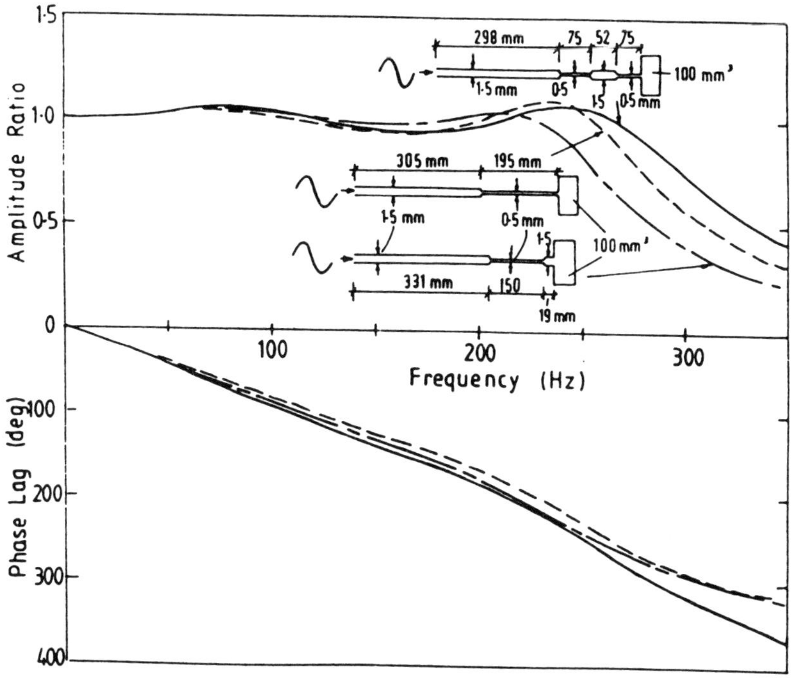

Figure 7. Optimum two-, three- and four-stage systems.

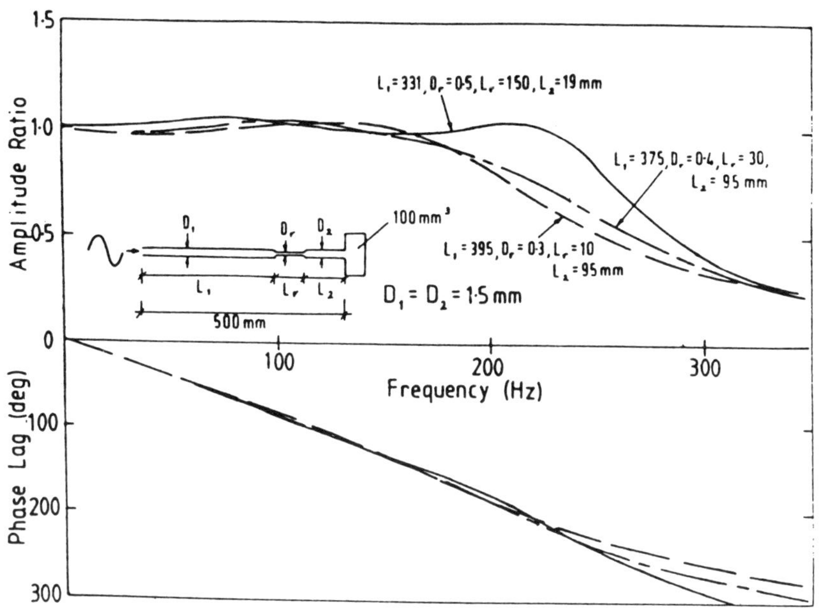

Figure 8. Optimum three-stage systems.

There is an optimum length of restrictor, of a given internal diameter, to give the best response. This tends to be one whose optimum position is close to the transducer, rather than close to the tube inlet, even though the latter position produces a larger damping effect. Figure 9 shows an example of the relationship between tne optimum position of a restrictor and its length for a fixed internal diameter (0.5 mm). As the length of the restrictor increases, its optimum position becomes independent of its length. For a given main tube length and diameter, there is an optimum restrictor internal diameter (Holmes and Lewis 1987a).

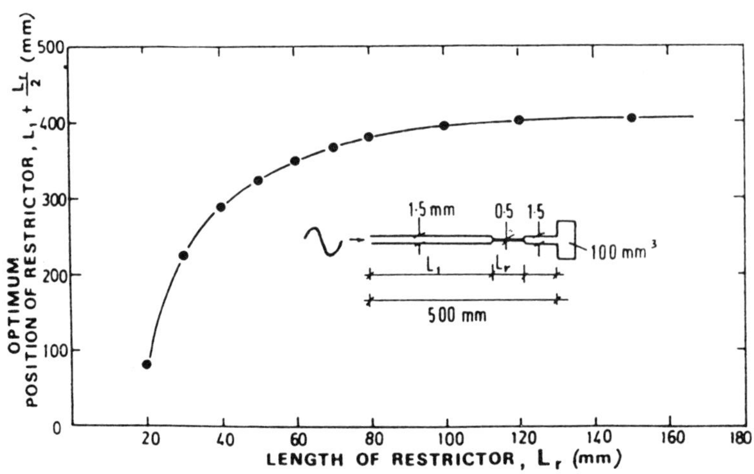

Figure 9. Optimum position of restrictor versus restrictor length.

Comparing Figures 7 and 8 with Figure 4, which includes the response of a system with the same overall tube length, main tube diameter and transducer volume, it can be seen that restricted-tube systems are very effective in removing resonant peaks and giving linear phase response characteristics. Comparison with Figure 5 shows that an effective frequency range can be obtained which is better than that for a constant diameter tubing with a fraction of the length.

An even better frequency range can be obtained with a leaked-tube system, as described following.

Leaked-tube Systems

The leaked-tube system was proposed by Gerstoft and Hansen (1987). A relatively flat amplitude frequency response to frequencies of 500 Hertz, with 1 metre of connecting tubing, is possible with a system of this type. This is achieved by inserting a controlled side leak part-way along the main connecting tube, usually close to the transducer. It has the effect of attenuating the amplitude response to low frequency fluctuations, and to steady pressures, to the level of a conventional closed system at higher frequencies. Thus, the leak effectively introduces a high-pass filter into the system. The amplitude ratio at frequencies approaching zero, is simply a function of the resistance to steady laminar flow of the main tube and leak tube.

Figure 10 shows the amplitude response of a leaked tube system computed using a modified version of the Bergh and Tijdeman theory (Holmes and Lewis 1989). The response of the closed system with the same tube length and diameter is also shown in this Figure. The low-frequency attenuation effect of the leak, as described above, is clearly shown.

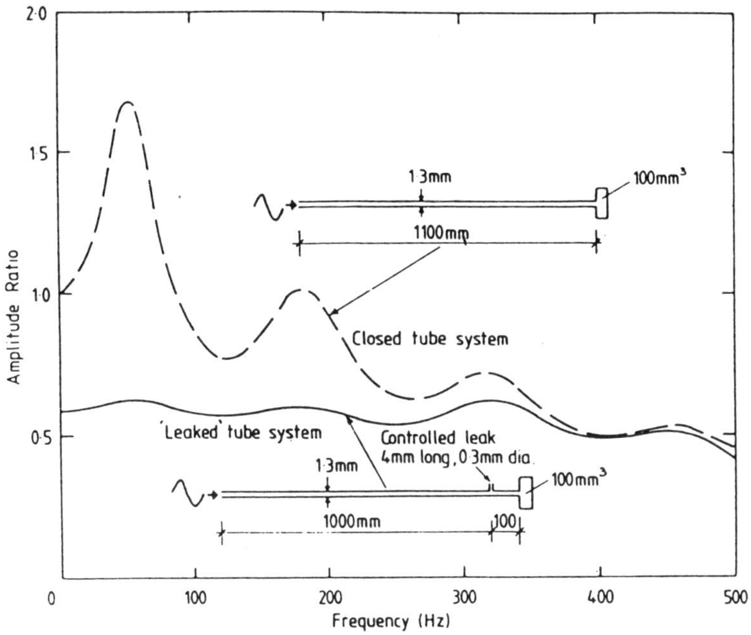

Figure 10. Principle of the leaked-tube system.

The response characteristics of several practical realisations of this system, with and without a 'Scanivalve' scanning device, have been described by Holmes and Lewis (1989). A requirement when using these systems in practice, is to ensure that the outlet of the leak tube is exposed to an environment with a negligible level of pressure fluctuations, otherwise the system will try to average any pressure fluctuations there with those being measured at the inlet to the main tube.

PARALLEL-TUBE/MANIFOLD SYSTEMS

As discussed earlier, systems which attempt to average the pressure fluctuations from a number of measurement points have become popular recently, after being first used in wind engineering by Surry and Stathopoulos (1977).

Figure 11 shows the type of parallel tube and manifold arrangement that has been commonly used in wind engineering work. Provided that the inlet tubes are identical in length and diameter, such a system should provide a true average in the manifold, of the fluctuating pressures at the entry to the m input tubes, assuming that laminar flow exists in the tubes. Gumley (1983) extended the Bergh and Tijdeman theory to cover the type of element shown in Figure11. Holmes and Lewis (1987b) also used the theory to optimize the design of systems of this type.

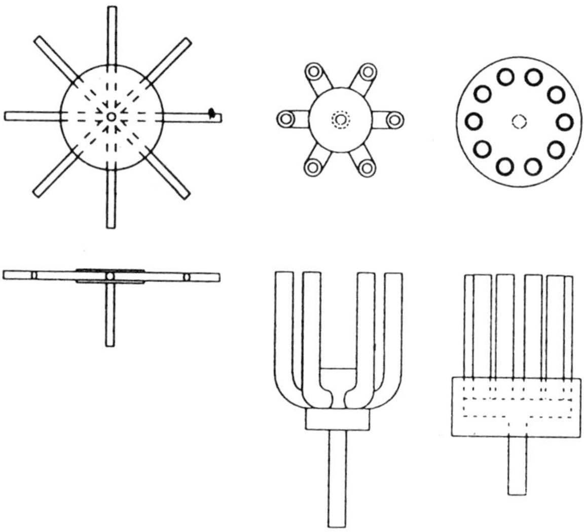

(a) (b) (c)

Figure 11. Manifolds for pressure averaging.

Figure 12 shows the amplitude and phase response for optimized systems of this type with ten input tubes supplying a manifold of 100 mm^3 volume. The response of a single input system of the same overall length, and using the same restrictor tube, is shown for comparison purposes.

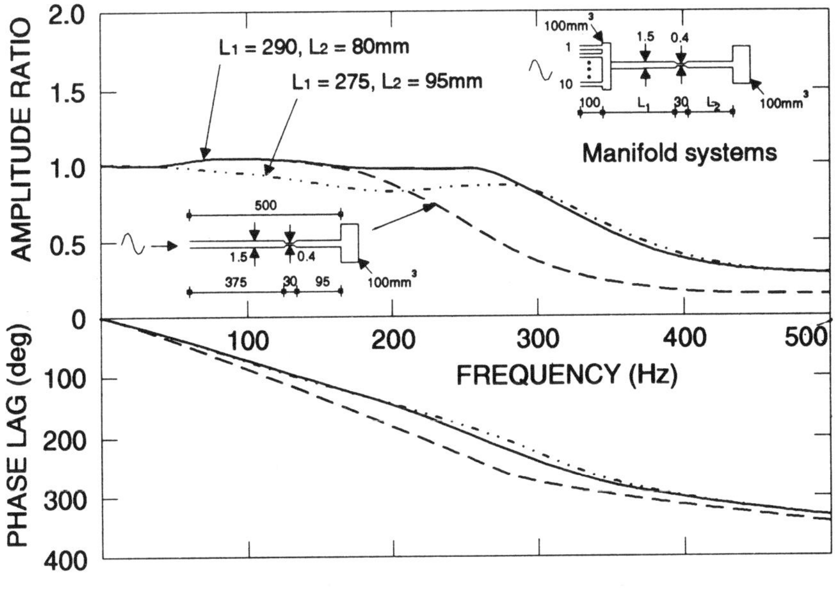

Figure 12. Optimum parallel-tube/manifold systems, and comparison with single-tube system of the same overall length.

Usually, flatter amplitude response curves out to higher frequencies, can be obtained with the multi-tube-manifold systems, due to the reinforcement of the higher frequencies in the input tubes. However, once the number of input tubes exceeds about five, there is little change to the response characteristics (Holmes and Lewis 1987b). The response is also not greatly sensitive to the volume of the averaging manifold.

Gumley (1981) and Letchford (1987) investigated systems with more than one manifold, both theoretically and experimentally. However, it is difficult to see any advantage of such systems over the single manifold system, except for the convenience of making tube connections.

The assumption that the average of discrete fluctuating point pressures, sampled within a finite area of a surface, adequately approximates the continuous average aerodynamic load on the surface requires consideration. This has been carried out by Surry and Stathopoulos (1977) and Holmes and Lewis (1987b).

Figure 13, reproduced from Holmes and Lewis (1987b), shows the ratio of the variance of the averaged panel force to the variance of the point pressure, using firstly, the correct continuous averaging over the panel denoted by R_c, and secondly, the discrete averaging approximation performed using the pneumatic averaging system with the ten pressure tappings within a panel, denoted by R_d. Calculations of these ratios were made, assuming a correlation coefficient for the fluctuating pressures of the form, exp(–Cr), where r is a separation distance, and C is a constant. The variance of the local pressure fluctuations across the panel of dimensions B by B/2, were assumed constant.

It can be seen that R_d exceeds R_c for all values of CB. This is due to the implied assumption, in the discrete averaging, that the pressure fluctuations are fully correlated in the tributary area around each pressure tap. Clearly, the error increases with increasing C due to the lower correlation of the pressure fluctuations, and with increasing panel size, B. However, the errors can be decreased by increasing the number of pressure tappings within a panel of a certain size. However, it should be noted that the errors are larger at higher frequencies than at lower frequencies; a more detailed analysis of the errors requires knowledge of the coherence of the pressure fluctuations.

CONCLUSIONS

This paper has reviewed recent developments in the measurement of fluctuating surface pressures, with particular reference to wind engineering. Modern pressure sensor technology enables such measurements to be carried out for a large number of points at a low cost.

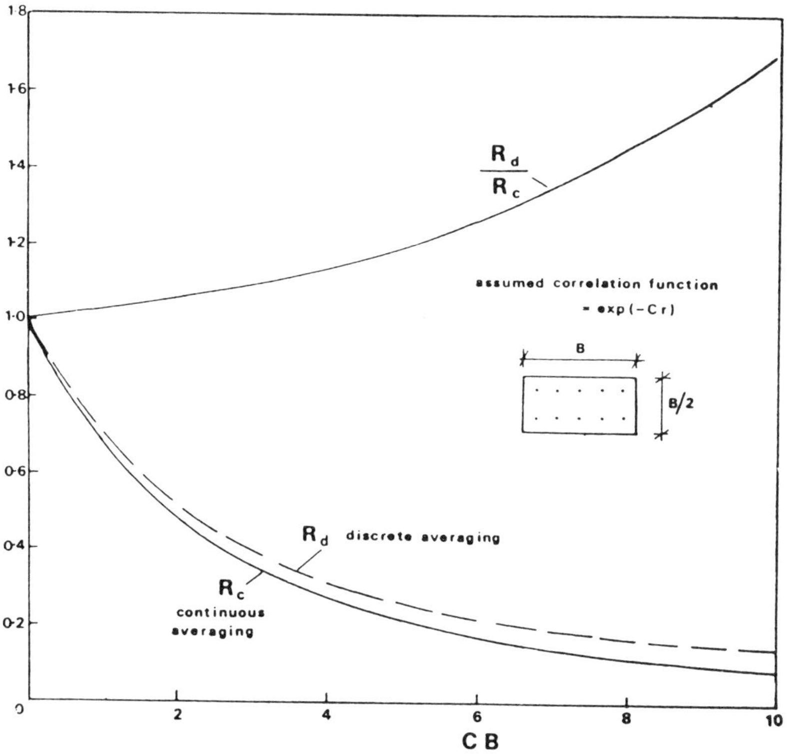

Figure 13. Comparison of discrete and continuous averaging of fluctuating pressures.

Three possiblities are available for the measurement of pressures from single points on a surface: short-tube, restricted-tube and leaked-tube systems. Taking these in order they give increasing upper frequency limits, at the expense of more complicated tubing arrangements.

A brief discussion of parallel-tube manifold systems used for the measurement of average pressures of measurement points has been given. It has been found that higher frequency response characteristics can be obtained for these systems with closed restrictor-tube systems, compared with single-tube systems of the same overall length.

REFERENCES

Bearman, P.W. and Luo, S.C. (1988), 'Investigation of the Aerodynamic Instability of a Square-section Cylinder by Forced Oscillation', Journal of Fluids and Structures, Vol. 2, pp.161–176.

Bergh, H. and Tijdeman, H. (1965), 'Theoretical and Experimental Results for the Dynamic Response of Pressure Measurement Systems', National Aero- and Astronautical Research Institute (Netherlands), Report NLR-TR-F.238, January.

Durgin, F.H. (1992), 'Instrumentation Requirements for Measuring Aerodynamic Pressures and Forces on Buildings and Structures', in T. Reinhold (ed.) Wind Tunnel Modeling for Civil Engineering Applications, Cambridge University Press, Cambridge, pp.329-348

Gerstoft, P. and Hansen, S.O. (1987), 'A New Tubing System for the Measurement of Fluctuating Pressures', Journal of Wind Engineering and Industrial Aerodynamics, Vol. 25, pp.335-354, 1987.

Gumley, S.J. (1981), 'Tubing Systems for the Measurement of Fluctuating Pressures in Wind Engineering', D.Phil. Thesis, University of Oxford.

Gumley, S.J. (1983), 'Tubing Systems for Pneumatic Averaging of Fluctuating Pressures', Journal of Wind Engineering and Industrial Aerodynamics, Vol. 12, 189–228.

Gumley, S.J. (1984), 'A Parametric Study of Extreme Pressures for the Static Design of Canopy Structures', Journal of Wind Engineering and Industrial Aerodynamics, Vol 16, 43–56.

Holmes, J.D. (1984), 'Effect of Frequency Response on Peak Pressure Measurements', Journal of Wind Engineering and Industrial Aerodynamics, Vol. 17, pp.1–9.

Holmes, J.D. (1987), 'Distribution of Peak Wind Loads on a Low-rise Building, Journal of Wind Engineering and Industrial Aerodynamics, Vol. 29, 59–67.

Holmes, J.D. and Lewis, R.E. (1986), 'The Dynamic Response of Pressure-measurement Systems', 9th. Australasian Fluid Mechanics Conference, Auckland, 8–12 December.

Holmes, J.D. and Lewis, R.E. (1987a), 'Optimization of Dynamic-pressure-measurement Systems. I. Single point measurements', Journal of Wind Engineering and Industrial Aerodynamics, Vol. 25, 249–273.

Holmes, J.D. and Lewis, R.E. (1987b), 'Optimization of Dynamic-pressure-measurement systems. II. Parallel tube-manifold systems', Journal of Wind Engineering and Industrial Aerodynamics, Vol. 25, 275–290.

Holmes, J.D. and R.E.Lewis, R.E. (1989), 'A Re-examination of the Leaked-tube Dynamic Pressure Measurement System'. 10th. Australasian Fluid Mechanics Conference, Univ. Melbourne, December, pp.5.39–5.42.

Holmes, J.D. and Rains, G.J. (1981), 'Wind Loads on Flat and Curved Roof Low Rise Buildings', Colloque "Construire avec le Vent", Nantes, France, July.

Iberall, A.S. (1950), 'Attenuation of Oscillatory Pressures in Instrument Lines', Journal of Research National Bureau of Standards, Vol. 45, pp.85–108.

Irwin, P.A. (1988), 'Pressure Model Techniques for Cladding Loads', Journal of Wind Engineering and Industrial Aerodynamics, Vol. 29, pp.69–78.

Irwin, H.P.A.H., Cooper,K.R. and Girard, R. (1979), 'Correction of Distortion Effects Caused by Tubing Systems in Measurements of Fluctuating Pressures', Journal of Industrial Aerodynamics, Vol. 5, 93–107.

Kareem, A. Cheng, C-M. and Lu, P.C. (1989), 'Pressure and Force Fluctuations on Isolated Circular Cylinders of Finite Height in Boundary Layer Flows', Journal of Fluids and Structures, Vol. 3, 481–508.

Letchford, C.W. (1987), 'Pneumatic Averaging in Wind Engineering', D.Phil. Thesis, U.Oxford.

Letchford, C.W., Sandri, P., Levitan, M.L., and Mehta, K.C. (1992), 'Frequency Response Requirements for Fluctuating Wind Pressure Measurements', Journal of Wind Engineering and Industrial Aerodynamics, Vol. 40, pp.263–276.

Luo, S.C. and Bearman, P.W. (1990), 'Predictions of Fluctuating Lift on a Transversely Oscillating Square-section Cylinder', Journal of Fluids and Structures, Vol. 4, 219–228.

Marshall, R.D. and Reinhold, T. (1982), Discussion of 'Effective Wind Loads on Flat Roofs', by T.Stathopoulos, D.Surry, and A.G.Davenport, J. Struct. Div., ASCE, Vol. 108, pp.495-498.

Surry, D. and Isyumov, N. (1975), 'Model Studies of Wind Effects – A Perspective on the Problems of Experimental Technique and Instrumentation', 6th. International Congress on Aerospace Instrumentation, Ottawa.

Sensym (1991), 'Solid-state Sensor Handbook', Sensym Inc., Sunnyvale, California.

Surry, D and Stathopoulos,T. (1977), 'An Experimental Approach to the Economical Measurement of Spatially-averaged Wind Loads', Journal of Industrial Aerodynamics, Vol. 25, 275–290.

Waldeck, J.L. (1986), 'Effects of the Reference-pressure System on the Dynamic Response of Pressure Transducers in Full-scale Experiments', Journal of Wind Engineering and Industrial Aerodynamics, Vol. 23, 37–50, Elsevier.

STATE OF THE ART VOLUME

BLUFF BODY AERODYNAMICS FOR WIND ENGINEERING

W.H. MELBOURNE

Department of Mechanical Engineering
Monash University, Clayton, Victoria 3168, Australia

SUMMARY

The flow field around a bluff body is differentiated from that around a streamlined body by the presence of separated flow and the unsteady oscillation of the shear layers. For wind engineering applications the aerodynamics of bluff bodies is complicated by the presence of freestream turbulence which controls the pressure fluctuations on the upstream face and the shear layer behaviour which in turn controls the mean and fluctuating response of both along-wind and cross-wind motions of a structure under wind action.

This paper describes the primary features of bluff body aerodynamics, and, in particular, establishes the critical role of freestream turbulence and its influence on the shear layer behaviour and on the generation of surface pressures.

The paper concludes with some wind engineering examples of the development of cladding pressures and the response of structures which are particularly dependent on the unique characteristics of bluff body aerodynamics.

INTRODUCTION

The term "bluff body aerodynamics" is intended to cover the study of flow fields around, and force on, a body from which boundary layers are separating to form unsteady vortex flows in a wake region. The great majority of body shapes in the field of wind engineering involve bluff body aerodynamics as distinct from the attached flow aerodynamics associated with streamlined aeronautical shapes.

The study of wind engineering phenomena has as its basis the study of bluff body aerodynamics. In a paper "Turbulence and the Leading Edge Phenomenon", Melbourne, 1992, discussed the behaviour of shear layers separating from the leading edge of a bluff body and the influencing role played by freestream turbulence, in particular the effect on the pressures developed under the reattaching shear layer. This review will lean heavily on this earlier work to describe the origins of bluff body aerodynamics to provide support for the wind engineering applications presented in this volume.

The significant characteristics of the flow field around a bluff body can be described with reference to Fig. 1 showing the time averaged pattern of flow past a square cylinder. The upstream flow divides around a stagnation streamline and then separates either at some sharp edge or where the pressure gradient becomes sufficiently adverse. Upstream and outside the separated shear layers the flow is essentially irrotational whilst under the shear layers an unsteady, low velocity wake region is formed. Reynolds number, and surface shape, and roughness, along with freestream turbulence, also play a significant part in determining how the shear layer behaves.

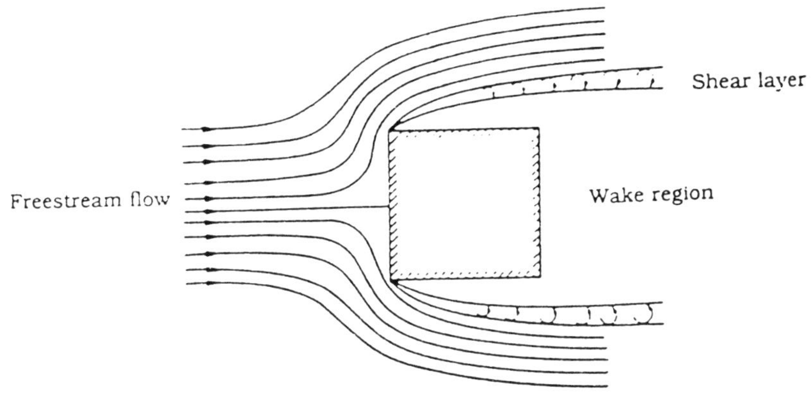

Fig. 1 Time average pattern of flow past a square cylinder

The aerodynamic forces on a bluff body in a turbulent flow can be conveniently divided into those which cause along-wind response and those that cause cross-wind response.

From the work primarily of Davenport (1967) and Vickery (1971) it can be concluded that the along-wind response of most structures originates almost entirely from the action of the incident turbulence of the longitudinal component of the wind flow (superimposed on a mean displacement due to the mean drag). Their analytical methods, using spectral and spatial correlation considerations to predict the along-wind response of many structures has become highly developed, to the point where the gust factor approach is included in a number of wind loading codes. By comparison, the mechanisms causing the cross-wind response have proved to be so complex that as yet there is no simple generalised analytical method available to calculate the cross-wind response of structures. However the various mechanisms causing cross-wind response can be categorised as being associated with

(i) a periodicity in the separated wake flow.
(ii) the cross-component of the incident turbulence, and
(iii)aerodynamic instabilities related to the motion of the body.

The details of the various modes of the response of structures to bluff body aerodynamic input and their prediction will be covered in other chapters in this volume. This chapter will essentially deal with the flow and pressure fields associated with bluff aerodynamics and the primary influence of freestream turbulence.

THE EARLY HISTORICAL DEVELOPMENT OF BLUFF BODY AERODYNAMICS

The first known documented appreciations of the effects of turbulence on the pressure on a streamwise face under a separated shear layer had their origins in wind engineering studies, as they came from a comparison of full scale and model scale pressure measurements on small buildings. Bailey (1933), working at the National Physical Laboratory in England measured wind pressures on a railway shed in full scale and compared them with pressures measured on a wind tunnel model of the shed in relatively low turbulence uniform flow. The measured pressure coefficients were not in good agreement, particularly near the windward eaves. The difference can be attributed to the difference in the approach roughness and consequent turbulence of the incident flow. Bailey et al (1943) undertook some more wind tunnel measurements on a model immersed in the boundary layer on the floor of a large wind tunnel and there was much better agreement between the model and full scale in the boundary layer flow than in the low turbulence uniform flow. In particular they noted the effect of the ratio of the height of the building, h, and the boundary layer roughness length, Zo, which Jensen (1958) later proposed as a significant modelling scaling ratio. Jensen demonstrated the dependence of the pressure distributions around a bluff body on the boundary layer, or more specifically as we know now on the turbulence parameters, in a very similar experiment to that of Bailey in that he measured pressure distributions on a shed in full scale and in four scaled boundary layers.

The effect of increasing the turbulence level at eaves height, as the model boundary layer roughness length increased, was shown to move the peak low pressure forward towards the leading edge. For low turbulence the flow separates and does not reattach such that the low pressure over the roof (and leeward wall) is nearly constant in the recirculating wake region. For high turbulence the flow separates at the leading edge and reattaches on the roof and in this region very low pressures were seen to occur (as they did in full scale) with a significant rise in pressure beyond reattachment.

In the 1960's the work of Jensen and Davenport led to the development of boundary layer wind tunnels in which models of buildings and structures were tested. Using Jensen's Law meant that the model measurements of pressure and force on bluff bodies were being conducted in properly scaled turbulent flow, and hence the effects of turbulence on these parameters were being reasonably well covered by those using these modelling techniques. There were several studies in this period which did lay the foundations for much of the more recent research. The first of these was a study by Vickery (1966) of the effects of freestream turbulence on the flow around a square cylinder. Whilst this study was aimed primarily at the fluctuating forces (and it was found that the addition of turbulence greatly reduced the fluctuating lift) it was shown that the incident turbulence had a significant effect on the mean and fluctuating pressure fields and that the spanwise correlation of pressures was also significantly lower in turbulent flow. Bearman (1972) showed that the base pressure behind square and circular flat plates decreased with increasing turbulence (and integral scale of turbulence) to present the paradox with Vickery's measurements of base pressure behind a square cylinder which increased with increasing turbulence.

It was Gartshore (1973) who provided the first incisive description of the effects of turbulence of flow around bluff bodies and who provided an explanation of the above paradox. The work of Gartshore is worth dwelling upon, as it was not widely published at the time, and to this day the critical part of the effect he demonstrated to be due to the fine scale turbulence is not well understood.

Gartshore (1973) suggested that the effect of increasing freestream turbulence is to increase the turbulent mixing in the shear layers and hence to increase the rate of entrainment from the wake and decrease the radius of curvature of the shear layer. Gartshore further showed that fine scale turbulence, produced by a thin rod on the stagnation streamline, is sufficient to cause these effects. Gartshore's graphical descriptions of the paradoxical effects of freestream turbulence are given in Fig. 2 and his conclusions are as follows:

> "Small scale free stream turbulence can influence the drag, fluctuating lift, base pressure coefficient and galloping behaviour of bluff bodies by increasing the rate of entrainment of the shear layers shed from them. The freestream turbulence, after severe distortion by the mean flow field, produces the increased mixing and entrainment by its presence in the shed shear layers, probably in the form of local concentrations of streamwise vorticity.
>
> Only turbulence approaching a bluff body along its front stagnation line is required to produce the major effects of freestream turbulence on the flow around the body, so that a narrow wake effectively replaces grid-produced turbulence for experimental purposes. This observation also emphasises the importance of nearby upstream structures when considering the flow around tall buildings in the natural wind.
>
> Turbulence can produce increased drag and unstable galloping oscillations in shallow rectangular bodies (b/d < 0.5) which are stable in the absence of turbulence.
>
> It can also decrease the drag and stabilise bodies of deeper section b/d > 0.6 which are unstable without turbulence. These effects are explicable qualitatively in terms of the increased entrainment resulting from turbulence.

The effect of turbulence scale is small provided this scale is significantly smaller than the body dimension."

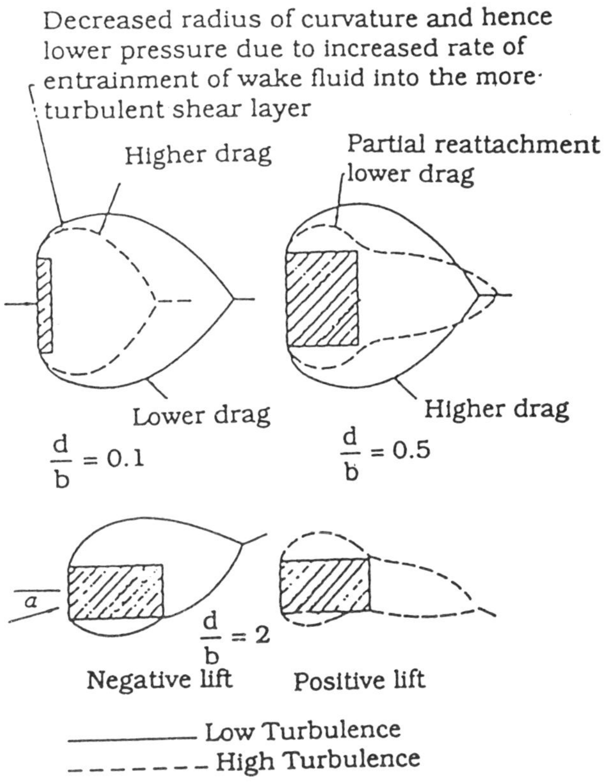

Fig. 2 Sketches of Expected Effect of Turbulence on Dividing Streamlines as it Affects Drag and Lift of Rectangular Sections (after Gartshore, 1973).

The key observation of Gartshore was the recognition that increasing turbulence caused the radius of curvature of the separated shear layer to reduce. This has the most important effect of causing earlier re-attachment of the flow on a streamwise surface, if it is present, which leads to lower pressures on the streamwise surface near the leading edge and higher base pressures (and lower drag) and can change lift curve slopes from negative to positive, thus impacting on the galloping instability phenomenon. Virtually all the streamwise surface and cross-wind effects of turbulence on bluff body aerodynamics can be attributed directly or indirectly to this process.

Lee (1975) extended Vickery's pressure measurements on a square cylinder in increasing freestream turbulence and in particular noted that increasing turbulence caused the shear layers to thicken and reattach intermittently onto the streamwise surfaces in support of Gartshore's conclusion. He also noted that the axial correlation of the streamwise surface pressures was relatively unaffected by increasing turbulence intensity up to $\sigma_u/u = 0.08$, but that there was then a significant reduction in correlation as the turbulence intensity increased to 0.12, (presumably because the highly three dimensional intermittent reattachment process was becoming established). Laneville et al (1975) in flow visualisation experiments also showed that an increase in turbulence intensity caused earlier reattachment of the separated shear layers on rectangular prisms.

Melbourne (1975) after several years of observing the occurrence of very low intermittent, peak pressures on streamwise surfaces near the leading edge of tall building models, and influenced by the conclusions of Gartshore (1973), advanced an hypothesis suggesting that the very low pressures which occurred under the reattaching shear layer under conditions of high freestream turbulence were due to an instability process under certain conditions, with reference to Fig. 3, as follows

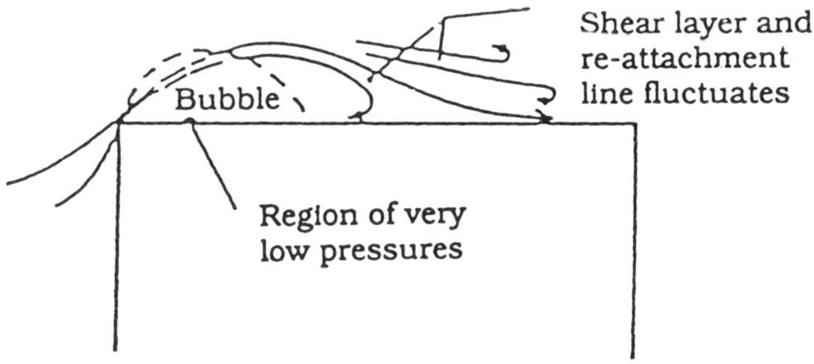

Fig. 3 Diagrammatic representation of Melbourne's (1975) hypothesis on the cause of the very low intermittent pressures near the leading edge under a separating and reattaching shear layer.

> "The shear layers and pressure underneath fluctuate. For certain conditions of freestream turbulence, angle of attack, (and radius of leading edge curvature) the shear layer commences to occasionally reattach onto the wall. As this occurs the cavity region under the shear layer is no longer vented and any decrease in pressure causes the initial radius of curvature of the shear layer to decrease further which in turn increases the local freestream velocity outside the shear layer and further decreases the pressure. This is an unstable process which proceeds until the shear layer breaks up into a complete separation again and the cavity becomes vented. During this unstable process, very low, intermittent pressures can occur on the surface under the shear layer near the leading edge."

This hypothesis was largely based on what could be seen with crude flow visualisation at the time and whilst the description of the dynamic behaviour of the shear layer with the occurrence of the very low pressures was correct the instability idea was shown by Saathoff to be incorrect and that it was the presence of strong vortices convecting downstream under the shear layer which caused the very low pressures and the appearance of early reattachment at the same time; this will be discussed in much more detail in the next section. Melbourne (1979) further investigated

Gartshore's conclusions with respect to the importance of the fine scale turbulence, and in so doing developed a Small Scale Spectral Density Parameter to quantify the spectral density of the incident freestream flow at scales the size of the shear layer width. In particular he showed that increase of the small scale turbulence within flows with the same large scale turbulence, significantly increased the magnitude of the very low pressures under reattaching shear layers. He also showed that venting the bubble under the reattaching shear layer with a leading edge slot significantly reduced the magnitude of the very low pressures which had applications in reducing cladding wind loads on roofs and the response of cantilevered roofs.

THE DETAILED EFFECTS OF TURBULENCE AND THE DEVELOPMENT OF SURFACE PRESSURES

Through the 1980s there was a much more concerted attack on the fundamental fluid mechanics associated with the leading edge phenomenon as distinct from measurements of the global effects of turbulence on bluff body aerodynamics, which by then were becoming quite well known, particularly in the wind engineering community.

Hillier et al (1981) and Cherry et al (1984) started to have a much closer look at the effects of the freestream turbulence on the separation bubble on a long flat plate with a blunt leading edge. They showed that mean and fluctuating pressures under the separation bubble were very dependent on turbulence intensity, that the mean flow-field responded strongly to turbulence intensity but with little effect of integral scale, and that the mean reattachment length (X_R , the distance from the leading edge) decreased with increasing freestream turbulence intensity. Kiya et al (1983, a, b, 1985) extended the measurements on the flat plate and also concentrated on the turbulence effects on the mean and standard deviation pressures, the reattachment length and the flow structure. Apart from the valuable parametric information obtained from the flat plate the key contribution of the work of Cherry, Hillier, Kiya and Sasaki was the recognition that the primary source of the pressure fluctuations in a separation bubble is the convection of vortices through the region, which is characterised by a flapping motion of the shear layer. Melbourne and Gartshore had previously seen this flapping as the reattachment region oscillating backwards and forwards, alternately tightening up to create early reattachment and then lifting off the surface. The frequency of the shedding of the vortices from the bubble was identified as $0.11\bar{u}$ or $0.6\bar{u}/X_R$ respectively by the two groups. Kiya et al (1983) looked at the three dimensional flow in the bubble and suggested that pressure fluctuations in the reattachment zone were mainly caused by the convection of large "hairpin" vortices.

Nakamura et al (1984, 1987,1988) investigated the effects of freestream turbulence on the flow around short rectangular cylinders (from square to 5:1). In particular they investigated the effects of scale of turbulence by changing the model size (which also changed the blockage). They concluded, like Gartshore, that small-scale turbulence increased the shear layer growth and caused earlier reattachment, and that large scale turbulence weakens the vortex shedding by reducing spanwise correlation and changing the mean flow considerably.

In the studies on long and short after body rectangular sections, just described, the maximum values of turbulence intensity used were about 0.15, and only mean and standard deviation pressures were measured, with particular interest in the maximum values of the standard deviation pressure which occur around the mean reattachment region. However the main interest for wind engineering applications involves very much higher turbulence intensities and peak pressures. This need gave rise to the work of Saathoff et al (1987, 1989 and Saathoff's PhD Thesis 1988) who concentrated their studies on the peak pressures which occurred under the reattaching shear layer and the effects of turbulence intensity up to 0.25. Much of their work revisited that of earlier workers in a parametric sense and then went on to establish the cause of the very low peak pressures that had been measured in many wind engineering studies. The first of these studies involved pressure measurements along the streamwise side of an axisymmetric circular cylinder with the blunt face normal to the flow; this work established the blockage corrections for this type of measurement and then went on to determine the effect of turbulence intensity, $\sigma_u\sqrt{u}$, and turbulence scale L_x/D. on these pressure measurements

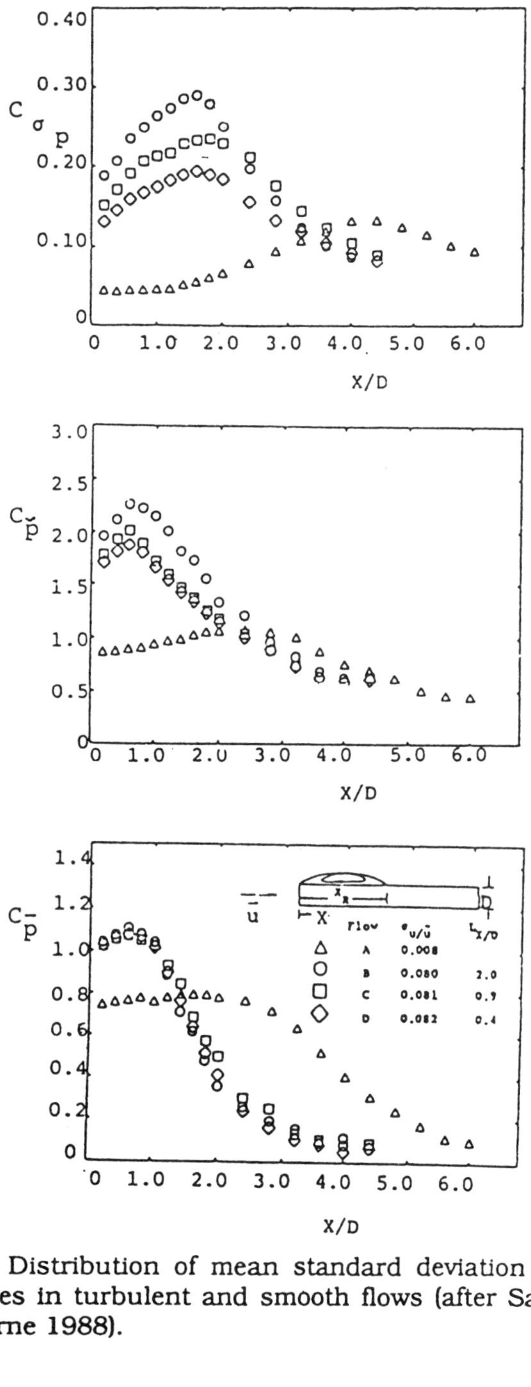

Fig. 4 Distribution of mean standard deviation and peak pressures in turbulent and smooth flows (after Saathof and Melbourne 1988).

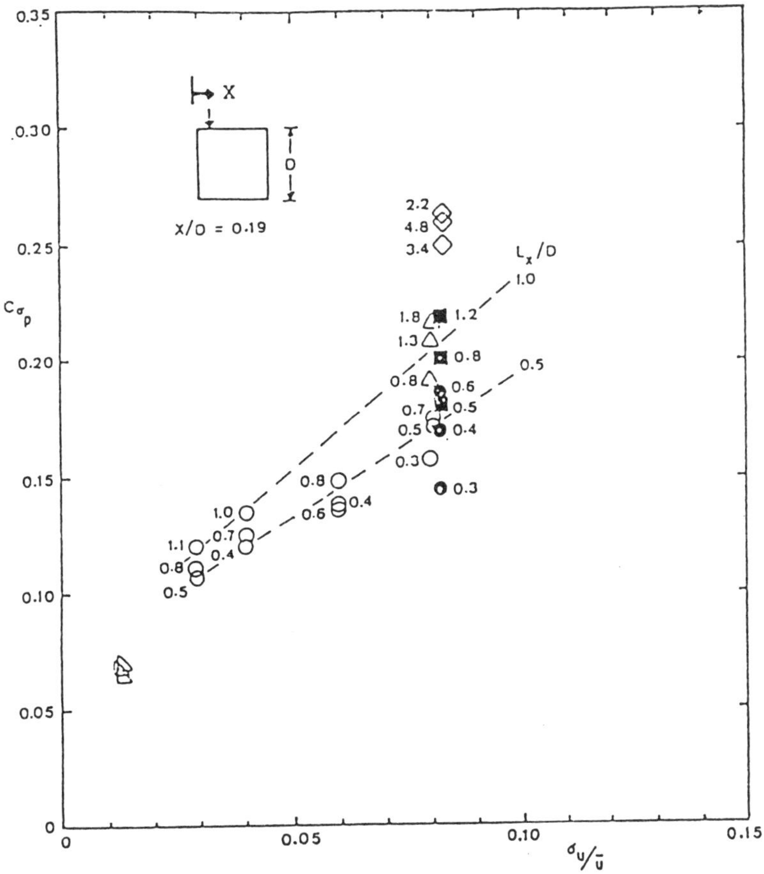

Fig. 5 Fluctuating pressure coefficient at X/D = 0.19 as a function of turbulence intensity and scale (data are corrected for blockage) (after Saathoff and Melbourne 1989).

as shown, for example in Fig. 3. It was found that a reasonable collapse of these standard deviation pressures could be achieved if plotted as a function of $\frac{\sigma_u}{\bar{u}} (L_x/\bar{D})^{0.15}$ The earlier measurements of mean, standard deviation and peak pressures on a flat plate (two dimensional rectangular with a long afterbody) are given in Fig. 4. However the most significant finding of this work was the finding that the passage of the vortex under the shear layer as it convects downstream, causes a very large pressure drop beneath it, and that this is the cause of the very low peak pressures measured under separated shear layers. Saathoff et al also pursued the lateral correlation of the peak pressures under the separated shear layer using conditional sampling techniques; it was shown that increasing turbulence scale significantly increases the correlation of these peak pressures which is very significant in considering the response of large bridge and canopy structures. This correlation observation is the opposite to that observed by some earlier researchers, but the correlation depends on whether the pressures are measured under the bubble or in the reattachment region and whether only the peak pressure occurrences are being considered.

WIND ENGINEERING APPLICATIONS

Whilst many of the chapters in this volume will deal with detailed aspects of wind loading a brief summary of wind engineering applications which have particular dependence on the unique characteristics of bluff body aerodynamics will be given here.

LOCAL CLADDING PRESSURE

The occurrence of very low pressures on the cladding of buildings and structures is the cause of the most frequent failures in strong wind conditions. Some of the world's wind codes have tried to come to terms with this with the introduction of peak factors for local design pressures for cladding. However to avoid being over-conservative for conventional buildings most leave the determination of design pressures to particular wind tunnel studies, particularly where significant edge discontinuities occur. Although we may still not fully understand the mechanism causing these very low pressures, the boundary conditions necessary for their occurrence is very clear. Simply, if very high freestream turbulence ($\sigma_u/\bar{U}$ >> 0.2) combines with an edge discontinuity the scene can be set for the occurrence of very low pressures under the reattaching shear layer. Whilst the role of the turbulence is now reasonably well understood, the extreme pressues which can be generated when the vortex is anchored at one end by an edge discontinuity is less clear. Without the edge discontinuity maximum peak pressure coefficients ($C_p = \frac{p}{1/2\ \sigma \bar{u}^2}$) -3 to -4 are common. With an edge discontinuity there have been a number of instances of peak pressure coefficients occurring around -8 on the faces of tall buildings, podium roof areas below the face of tall buildings and on the corner areas of roofs on low rise buildings. Examples of these can be found in Cheung et al (1991).

CROSS-WIND RESPONSE OF TALL BUILDING

The vortex shedding process from bluff bodies with sharp edges and streamwise faces is significantly effected by the reattachment or non-reattachment of these shear layers. As described before, turbulence plays a major role in this process and as described by Gartshore (1973) and Kwok et al (1975) can have a significant effect on the galloping instability process of slender rectangular structures. However it is not so well appreciated just how significant this process is in driving the cross-wind response of buildings where square sectioned buildings can be shown to have a relatively much greater cross-wind response than circular buildings or square buildings on which the corners are chamfered. Examples of mode generalised cross-wind force spectra for a variety of shapes are given in Melbourne (1989).

BRIDGE DECKS

The response of slender box girder bridge sections is dominated by the mechanism of incident turbulence producing a fluctuating angle of attack and hence fluctuating lift on the deck which is transformed into response through the resonance process. Underlying all this is the secondary (or may be primary) effect of the way in which the turbulence modifies the flow over the deck via the reattachment process and hence modifies both the lift characteristics of the section and the spanwise correlation of these forces which has been discussed by Melbourne (1982).

LARGE CANOPIES

The response of large cantilevered canopies over grandstands has been described by Melbourne et al (1988) including the amelioration of response by using a slot in the leading edge to vent the separation bubble. Less well known is the effect of the pressure distributions under and downstream of a separation bubble on lightweight roof systems on which the major response is due to dynamic overshoot of a driven mode as distinct from the resonance process. Recent aeroelastic model studies of two arch roof systems have revealed asymmetric driven responses, and in one case to magnitudes in excess of that which would be expected through the resonance process as described by Melbourne (1993).

ACKNOWLEDGEMENT

Acknowledgement is made to Elsevier Science for permission to reproduce material from 'Turbulence and the Leading Edge Phenomenon' (W.H. Melbourne), Journal of Wind Engineering and Industrial Aerodynamics, 49 (1993) 45-64.

REFERENCES

Bailey, A. (1933), "Wind pressures on buildings", Inst. Civ. Eng. Selected Engineering Papers, No. 139.

Bailey, A. and Vincent, N.D.C. (1943), "Wind pressure experiments at the Severn Bridge", Journal Inst. Civil Engineers, Vol. 11, 1938-39, 363p, March 1939. Discussion: G.S. Gough, Vol. 12, 1938-39, 466p.

Bearman, P.W. (1971), "An investigation of the forces on flat plates in turbulent flow", NPL Aero Report 1296.1969. See also Jnl Fluid Mechanics, Vol. 46, 177p.

Cherry, N.J., Hillier, R. and LaTour, M.M.E.P. (1984), "Unsteady measurements in a separated and reattaching flow", Journal of Fluid Mechanics, Vol. 144, pp 123-146.

Gartshore, I.S. (1973), "The effects of freestream turbulence on the drag of rectangular two dimensional prism", University of Western Ontario, Canada BLWT-4-73.

Hillier, R. and Cherry, N.J. (1981), "The effects of stream turbulence on separation bubbles", Journal of Wind Engineering and Industrial Aerodynamics, Vol.8, pp. 49-58.

Jensen, M. (1958), "The model law for structures in the natural wind", Ingenioren, International Edition, Vol. 2.

Kiya, M., Sasaki, K. and Arie, M. (1983), "Discrete-vortex simulation of a turbulence separation bubble", Journal Wind Engineering and Industrial Aerodynamics, Vol. 14, pp. 375-386.

Kiya, M., Sasaki, K. (1983), "Structure of a turbulent separation bubble, Journal Fluid Mechanics", Vol. 137, pp. 83-113.

Kiya, M. and Sasaki, K. (1985), "Structure of large-scale vortices and unsteady reverse flow in the reattaching zone of a turbulent separation bubble", Journal Fluid Mechanics, Vol. 154, pp. 463-491.

Lee, B.E. (1975), "The effect of turbulence on the surface pressure field of a square prism", Journal Fluid Mechanics, Vol. 69, pp. 263-282.

Melbourne, W.H. (1993),"Turbulence and the leading edge phenomenon", Journal of Wind Engineering and Industrial Aerodynamics, Vol. 49, pp. 45-64.

Melbourne, W.H. (1993), "The response of large roofs to wind action", Proc. 3rd Asia-Pacific Symposium on Wind Engineering, Hong Kong, pp. 27-38.

Melbourne, W.H. (1979), "Turbulence effects on maximum surface pressures, a mechanism and possibility of reduction", Proc. 5th Intl. Conf. on Wind Engineering, Colorado, USA, Pergamon Press, pp. 541-552.

Melbourne, W.H. (1975), "Crosswind response of structures to wind action", Proc. 4th Int. Conf. on Wind Effects on Buildings and Structures, London, Cambridge Press.

Melbourne, W.H. (1982), "Comparison of model and full scale tests of a bridge and a chimney stack", Proc. Int. Workshop on Wind Tunnel Modelling, Gathersburg, MD, pp. 637-653.

Nakamura, Y. and Ohya, Y. (1984), "The effects of turbulence on the mean flow past two-dimensional rectangular cylinders", Journal Fluid Mechanics, Vol. 149, pp. 255-273.

Nakamura, Y. and Ozono, S. (1987), "The effects of turbulence on a separated and reattaching flow", Journal Fluid Mechanics, Vol. 178, pp. 477-490.

Nakamura, Y., Ohya, Y. and Ozono, S. (1988), "The effects of turbulence on bluff-body mean flow", Journal of Wind Engineering and Industrial Aerodynamics, Vol. 28, pp. 251-259.

Saathoff, P.J. and Melbourne, W.H. (1987), "Freestream turbulence and wind tunnel blockage effects on streamwise surface pressures", Journal of Wind Engineering and Industrial Aerodynamics, Vol. 26, pp. 353-370.

Saathoff, P.J. and Melbourne, W.H. (1989), "The generation of peak pressures in separated/reattaching flows", Journal of Wind Engineering and Industrial Aerodynamics, Vol. 32, pp.121-134.

Vickery, B.J. (1966), "Fluctuating lift and drag on a large cylinder of square cross-section in a smooth and in a turbulent stream", Journal Fluid Mechanics, Vol. 25, pp. 481-494.

APPLICATIONS OF CFD TO BLUFF BODY AERODYNAMICS

S. MURAKAMI and A. MOCHIDA

*Institute of Industrial Science, University of Tokyo,
7-22-1, Roppongi, Minato-ku, Tokyo 106, Japan*

SUMMARY

This paper reviews the recent developments in Computational Wind Engineering and their applications to bluff body aerodynamics conducted by us. Here, turbulent vortex shedding flow past a long square cylinder, which is two-dimensional (2D) in the mean, is discussed. The accuracy of the predicted results by CFD is examined by comparing with many experiments from various view points. At the first part of the paper, the results of Large Eddy Simulation (LES) for 2D and 3D computations are compared. The LES results given from 3D computation correspond with the experimental results very well, whereas the results based on 2D computation are different from those based on 3D computation as well as from those given from experiments. The second part of the paper concerns detailed testing of LES (3D) computations for the analysis of periodic and stochastic fluctuations, for which detailed experimental data are available. Results of LES computations are compared with those from various turbulence models based on Reynolds-averaged Navier-Stokes equations (RANS models, i.e., k-ε model, second-moment closure model) as well as with those from experiments. Lastly, results of LES (3D) of unsteady pressure field and wind force acting on an oscillating square cylinder are presented for both cases of the forced oscillation and the wind-induced free oscillation.

INTRODUCTION

In the field of fluid dynamics, computational methods have been developed rapidly in recent years as a result of the burgeoning growth of computer technology. This paper addresses some aspects of the current status and future prospects of Computational Fluid Dynamics (CFD) in the field of wind engineering.

Fig. 1 Instantaneous streamlines past 2D square cylinder

Distributions of the fluctuating surface pressure and the wind force acting on bluff-shaped bodies are of great practical interest in wind engineering. A great number of wind tunnel experiments and numerical studies concerning wind-structure interaction problems have been conducted in order to clarify these complex phenomena. Nonetheless, there are a lot of unresolved issues concerning our understanding of wind-structure interactions.

CFD originated in the field of aeronautics and now have a history of more than 30 years. Numerical methods for simulationg turbulent flows have been developed mainly in the fields of mechanical and aeronautical engineering during the past several decades. Historically CFD has been applied to simple flows such as a channel flow or a flow around an aerofoil in these fields. However, airflow around a bluff body is far more complicated. In many cases, the current CFD technology borrowed from other fields is thus inadequate. A critical evaluation of numerical methods concerning their accuracy for simulating complex turbulent flowfields around bluff bodies needs to be undertaken.

In these years, we have examined the relative performance of various turbulence models, i.e., the standard k-ε model, Algebraic Second-moment closure Model (ASM), Differential Second-moment closure Model (DSM) and Large Eddy Simulation (LES) when applied to complex turbulent flowfields around a bluff body (Murakami et al.(1990b, 1992, 1993a, 1993c, 1993d), Sakamoto et al.(1993b)).

This paper focuses on the numerical prediction of turbulent vortex shedding flow past a fixed or an oscillating long square cylinder. This flowfield is two-dimentional (2D) in the mean, but the instantaneous flowfield is highly three-dimensional around the cylinder (Fig. 1). The accuracy of predicted results by CFD is examined from various view points.

BRIEF REVIEW OF TURBULENCE MODELS

Kolmogolov Microscale and Grid Discretization

The flowfields treated in wind engineering are usually expressed by the incompressible Navier-Stokes equations (hereafter N-S equations). This set of nonlinear partial differential equations is infamous for the difficulty with which they are solved. Furthermore, the flows treated in this field are always highly turbulent, and we must thus rely on the theory of turbulence statistics when solving these N-S equations by means of numerical methods.

Turbulent flowfields are characterized by various scales of turbulent eddies. Hence, the most important decision in numerical simulation is how finely we should resolve these eddies in numerical analysis. The accuracy of the prediction and the amount of CPU time are completely dependent on this judgment. Resolving small eddies requires much CPU time. The smallest eddy in turbulent flow, the so-called "Kolmogorov microscale η", is very small around 0.1~1mm for natural wind. It will be evident that discretization of the flowfield corresponding to the Kolmogorov microscale η is very difficult in practice.

Numerical analysis of a flowfield, whether the numerical method be finite difference or finite element, begins with the discretization of the flowfield; computation is carried out using these discretized expressions in the N-S equations etc. While the amount of CPU time increases with greater numbers of these discretizations of space and time, the accuracy of the prediction is governed by the fineness of discretization. As was stated above, turbulent flow can be reproduced with complete accuracy only if the flowfield is discretized corresponding to the Kolmogorov microscale. However, such a fine discretization is almost impossible to obtain in practice. The size of the microscale(η) decreases as the Reynolds number ($=ul/\nu$, hereafter Re, u:characteristic velocity scale, l : integral scale) increases. Consequently the number of grid points must be increased in proportion to the value of $Re^{9/4}$, according to the theory of turbulence statistics.

The numerical method based on this concept of microscale discretization is called "Direct Numerical Simulation" (hereafter DNS, Kim et al.(1987), Le et al.(1993)). While this simple and clear method uses no turbulence modelling, it requires a huge number of grid points if the Reynolds number of the flowfield is large. The Reynolds number of the flowfields treated in wind engineering is about 10^5~10^7, so DNS of such a flowfield requires around 10^{11}~10^{16} grid points. On the other hand, the maximum number of grid points ever used is around 10^7~10^8. Therefore, it can easily be understood why it is practically impossible to solve a fully turbulent flow by means of DNS even if computer performance is advanced greatly in the future.

Turbulence Models — Current Status and Future Potential

The numerical analysis of turbulent flow faces a serious obstacle of the Kolmogorov microscale, the concept of which comes from the theory of turbulence statistics. In order to overcome this obstacle, various types of turbulence models have been invented. Turbulence modelling is designed to simulate the averaged flowfield (coarse graining, cf. Fig. 2) rather than the original flowfield. Such small scale eddies, which are difficult to resolve, are thus neglected in the process of coarse graining. The small eddies neglected by coarse graining can be incorporated in the simulation through the turbulence modelling.

There are two types of coarse graining ; ensemble (or time, Reynolds) averaging and space averaging. These types of averaging and turbulence models are illustrated in Fig. 3, including DNS which uses no turbulence model. By averaging a flowfield with steep fluctuations, the averaged fluctuations of the flow variables become less steep and the relatively coarse grid discretization can resolve the flowfield with a certain accuracy. Hereafter, the concept of turbulence modelling is described briefly.

Turbulence models based on Reynolds-averaged Navier-Stokes equations (RANS models).

The ensemble (or time) averaged forms of the N-S equations are called the Reynolds-averaged Navier-Stokes Equations. These equations express only the movements of large scale eddies. Since the flowfield contains only larger eddies, nu-

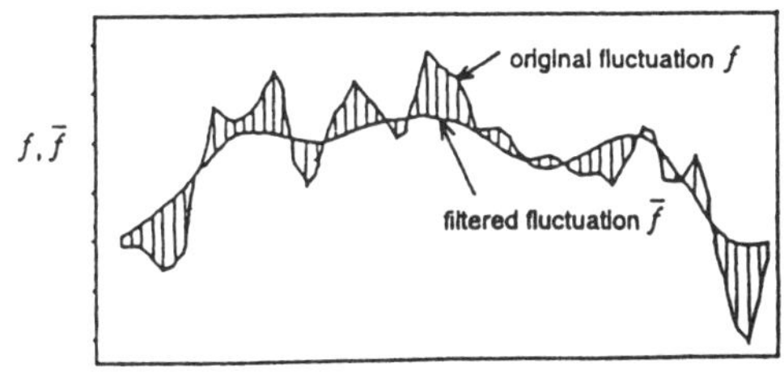

Fig. 2 Filtering of flowfield

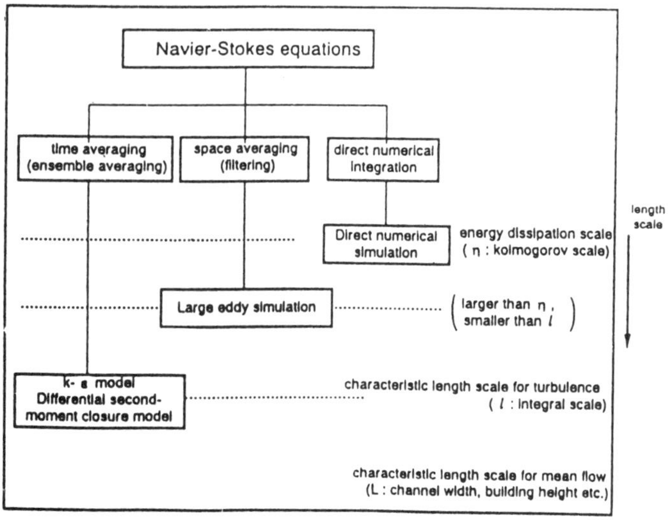

Fig. 3 Methods of numerical simulation and turbulence models

merical analysis can be conducted with a rather coarse grid discretization. Most CFD simulations in the field of engineering have been based on this type of modelling partly because of inadequate computer performance. The most famous and most widely-used is "the k-ε two equation turbulence model" (hereafter k-ε model, Launder and Spalding(1972)). The model equations of the standard type of k-ε model is shown in Table 1. However, it has become clear that the k-ε model possesses some shortcomings (Murakami et al. (1990b, 1992, 1993a)). Some new models of k-ε have been proposed for improving the accuracy of the standard k-ε model for predicting the flowfield past a bluff body (Launder and Kato(1993), Przulj and Younis(1993)). The high-precision version of the Reynolds-averaged Navier-Stokes equations (RANS model) is the Differential Second-moment closure Model (hereafter DSM, Launder, Reece and Rodi (1975)). Applications of DSM and its simplified version, ASM (Algebraic Second-moment Closure Model, Rodi (1976)) are increasing year by year (Murakami et al. (1990a, 1991a, 1993b)). The commonly used forms of DSM and ASM are summarised in Table 2.

Turbulence model based on space-filtering (LES)

Another method for coarse graining is "space-filtering", the concept of which is illustrated in Fig. 2. The degree of averaging can be adjusted easily by changing the width of the filter. Since eddies smaller than the width of the filter(eddies of subgrid scale) are not analyzed directly and eddies larger than the filter width(eddies of resolvable or grid scale) are simulated, this type of modelling is called "Large Eddy Simulation" (hereafter LES, Smagorinsky(1963), Deardorff(1970), Schumann(1975), Moin et al.(1982), Murakami et al.(1987), Germano et al.(1991)). In LES, the effect of small eddies neglected by filtering is incorporated in the simulation through the turbulence model of subgrid scale (SGS) stresses. For modelling

Table 1 Model equations for k-ε EVM

$$\frac{\partial \langle u_i\rangle}{\partial t}+\frac{\partial \langle u_i\rangle\langle u_j\rangle}{\partial x_j}=-\frac{\partial \langle p\rangle}{\partial x_i}+\frac{\partial}{\partial x_j}\nu_t\left(\frac{\partial \langle u_i\rangle}{\partial x_j}+\frac{\partial \langle u_j\rangle}{\partial x_i}\right) \quad (1.1)$$

$$\frac{\partial \langle u_i\rangle}{\partial x_i}=0 \quad (1.2)$$

$$\frac{\partial k}{\partial t}+\frac{\partial k\langle u_j\rangle}{\partial x_j}=\frac{\partial}{\partial x_j}\left(\frac{\nu_t}{\sigma_k}\frac{\partial k}{\partial x_j}\right)+\nu_t\left(\frac{\partial \langle u_i\rangle}{\partial x_j}+\frac{\partial \langle u_j\rangle}{\partial x_i}\right)\frac{\partial \langle u_i\rangle}{\partial x_j}-\varepsilon \quad (1.3)$$

$$\frac{\partial \varepsilon}{\partial t}+\frac{\partial \varepsilon\langle u_j\rangle}{\partial x_j}=\frac{\partial}{\partial x_j}\left(\frac{\nu_t}{\sigma_\varepsilon}\frac{\partial \varepsilon}{\partial x_j}\right)+\frac{\varepsilon}{k}\left(C_{\varepsilon 1}\nu_t\left(\frac{\partial \langle u_i\rangle}{\partial x_j}+\frac{\partial \langle u_j\rangle}{\partial x_i}\right)\frac{\partial \langle u_i\rangle}{\partial x_j}-C_{\varepsilon 2}\varepsilon\right) \quad (1.4)$$

$$\nu_t=C_\mu\frac{k^2}{\varepsilon} \quad (1.5)$$

$$C_\mu: 0.09,\ \sigma_k: 1.0,\ \sigma_\varepsilon: 1.3,\ C_{\varepsilon 1}: 1.44,\ C_{\varepsilon 2}: 1.92 \quad (1.6)$$

Table 2 Model equations for DSM (ASM)

$$\frac{\partial \langle u_i\rangle}{\partial t}+\frac{\partial \langle u_i\rangle\langle u_j\rangle}{\partial x_j}=-\frac{\partial \langle p\rangle}{\partial x_i}-\frac{\partial \langle u_i'u_j'\rangle}{\partial x_j} \quad (2.1)$$

$$\frac{\partial \langle u_i\rangle}{\partial x_i}=0 \quad (2.2)$$

1) DSM

$$\frac{\partial \langle u_i'u_j'\rangle}{\partial t}+\frac{\partial \langle u_i'u_j'\rangle\langle u_j\rangle}{\partial x_j}=P_{ij}+D_{ij}+\Phi_{ij}-\varepsilon_{ij} \quad (2.3)$$

$$\frac{\partial \varepsilon}{\partial t}+\frac{\partial \varepsilon\langle u_j\rangle}{\partial x_j}=D_\varepsilon+\frac{\varepsilon}{k}(C_{\varepsilon 1}P_k-C_{\varepsilon 2}\varepsilon) \quad (2.4)$$

2) ASM

$$\frac{\partial k}{\partial t}+\frac{\partial k\langle u_j\rangle}{\partial x_j}=D_k+P_k-\varepsilon \quad (2.5)$$

$$\frac{\partial \varepsilon}{\partial t}+\frac{\partial \varepsilon\langle u_j\rangle}{\partial x_j}=D_\varepsilon+\frac{\varepsilon}{k}(C_{\varepsilon 1}P_k-C_{\varepsilon 2}\varepsilon) \quad (2.6)$$

$$(P_k-\varepsilon)\frac{\langle u_i'u_j'\rangle}{k}=P_{ij}+\Phi_{ij}-\varepsilon_{ij} \quad (2.7)$$

$$D_{ij}=\frac{\partial}{\partial x_m}\left(C_k\langle u_m'u_l'\rangle\cdot\frac{k}{\varepsilon}\cdot\frac{\partial \langle u_i'u_j'\rangle}{\partial x_l}\right) \quad (2.8)$$

$$D_k=\frac{\partial}{\partial x_m}\left(C_k\langle u_m'u_l'\rangle\cdot\frac{k}{\varepsilon}\cdot\frac{\partial k}{\partial x_l}\right) \quad (2.9)$$

$$D_\varepsilon=\frac{\partial}{\partial x_m}\left(C_\varepsilon\langle u_m'u_l'\rangle\cdot\frac{k}{\varepsilon}\cdot\frac{\partial \varepsilon}{\partial x_l}\right) \quad (2.10)$$

$$P_k=-\langle u_i'u_j'\rangle\frac{\partial \langle u_i\rangle}{\partial x_j} \quad (2.11)$$

$$P_{ij}=-\langle u_i'u_k'\rangle\frac{\partial \langle u_j\rangle}{\partial x_k}-\langle u_j'u_k'\rangle\frac{\partial \langle u_i\rangle}{\partial x_k} \quad (2.12)$$

$$\varepsilon_{ij}=\frac{2}{3}\cdot\delta_{ij}\varepsilon \quad (2.13)$$

$$\Phi_{ij}=\Phi_{ij(1)}+\Phi_{ij(2)}+\Phi^w_{ij(1)} \quad (2.14)$$

$$\Phi_{ij(1)}=-C_1\frac{\varepsilon}{k}\left(\langle u_i'u_j'\rangle-\frac{2}{3}\delta_{ij}k\right) \quad (2.15)$$

$$\Phi_{ij(2)}=-C_2\left(P_{ij}-\frac{2}{3}\delta_{ij}P_k\right) \quad (2.16)$$

$$\Phi^w_{ij(1)}=\sum_{w=1}^{w0}C'_1\frac{\varepsilon}{k}\left(\langle u_k'u_m'\rangle\cdot n_k^{(w)}\cdot n_m^{(w)}\cdot\delta_{ij}-\frac{3}{2}\langle u_k'u_i'\rangle\cdot n_k^{(w)}\cdot n_j^{(w)}-\frac{3}{2}\langle u_k'u_j'\rangle\cdot n_k^{(w)}\cdot n_i^{(w)}\right)\cdot\frac{k^{3/2}}{C_l\cdot h_n^{(w)}\varepsilon} \quad (2.17)$$

$$C_\mu: 0.09,\ \sigma_k: 1.0,\ \sigma_\varepsilon: 1.3,\ C_1: 1.8,\ C_2: 0.6,\ C_1': 0.5,\ C_2': 0.5,\ C_k: 0.22,\ C_\varepsilon: 0.16,\ C_{\varepsilon 1}: 1.44,\ C_{\varepsilon 2}: 1.92,\ C_l: 2.5 \quad (2.18)$$

Table 3 Model equations for LES (standard Smagorinsky model)

$$\frac{\partial \bar{u}_i}{\partial t}+\frac{\partial \bar{u}_i\bar{u}_j}{\partial x_j}=-\frac{\partial \bar{p}}{\partial x_i}-\frac{\partial}{\partial x_j}(\nu+\nu_{SGS})\left(\frac{\partial \bar{u}_i}{\partial x_j}+\nu\frac{\partial \bar{u}_j}{\partial x_i}\right) \quad (3.1)$$

$$\frac{\partial \bar{u}_i}{\partial x_i}=0 \quad (3.2)$$

$$\nu_{SGS}=(C_S\bar{\Delta})\left\{\frac{\partial \bar{u}_i}{\partial x_j}\left(\frac{\partial \bar{u}_i}{\partial x_j}+\frac{\partial \bar{u}_j}{\partial x_i}\right)\right\}^{1/2} \quad (3.3)$$

$$\bar{\Delta}=(h_1h_2h_3)^{1/3}) \quad (3.4)$$

$$C_s: 0.1\sim 0.2 \quad (3.5)$$

the SGS stresses, the SGS eddy viscosity model proposed by Smagorinsky has been most commonly used (Smagorinsky(1963), cf. Table 3). Since Reynolds-averaging is not applied to the basic equations used in LES, the scale of resolved eddies in LES is much smaller than those of RANS models, i.e., k-ε model, ASM and DSM. Consequently the effect of modelling in subgrid scale on the turbulence structure of the flowfield is relatively small, compared to those of RANS models. In fact, LES has succeeded in reproducing the properties of a highly anisotropic flowfield in wind engineering problems (Murakami et al. (1990b, 1991a, 1992, 1993a, 1993c)).

In recent years, more refined SGS models for LES have been developed in the field of meteorology and mechanical engineering ; for example

1) extended versions of the standard Smagorinsky model
(Mason (1989), Mason and Derbyshire (1990), Yoshizawa (1991))
2) new versions of the SGS eddy viscosity models (Horiuti (1993b)),
3) the SGS one-equation model (Deadorff(1980), Moeng(1984), Yoshizawa and Horiuti (1985), Nieuwstadt et al. (1991))
4) the SGS second-moment closure model
(Deadorff (1973, 1974), Schmidt and Schumann (1989)) etc.

Recently, Germano et al.(1991) developed a new SGS closure method called dynamic SGS model. In this model, the model coefficient is decided dynamically as a function of time and space in accordance with the progress of the computation.

To sum up these various surveys, LES is most promising for the simulation of the flowfields treated in wind engineering. The weak point of LES is the large amount of CPU time required for computation, which is much larger than that for RANS models. While the RANS models give an ensemble or time-averaged flowfield, LES must reproduce and store a long sequence of time-dependent flowfields. However this weak point will be overcome to a certain degree by future growth in computer hardware technology.

In the following sections, accuracy of LES computations are examined by comparing LES results with those of experiments and RANS computations.

COMPARISON BETWEEN 2D AND 3D LES COMPUTATIONS FOR THE FLOW PAST 2D SQUARE CYLINDER

The first topic is the comparison of the accuracy of 2D and 3D LES computations. In this paper, turbulent vortex shedding flow past a 2D square cylinder, which is 2D in the mean, is discussed. When the time-averaged flowfield is 2D, it is not unusual for engineering problems to be solved by an unsteady 2D computation because CPU time is reduced greatly in comparison with 3D computations. It is clear that 2D LES is fundamentally incorrect from the view point of turbulence dynamics. Here, 2D and 3D LES computations are compared , and the accuracy of 2D LES computation is examined. Here, the standard Smagorinsky type SGS model, shown in Table 3, was used. The value of the Smagorinsky constant Cs was 0.1 for 3D computation. This value was increased by 50 % for 2D computation in comparison with that for 3D computation in order to compensate for the limited diffusion in the spanwise direction. Specifications for 2D and 3D computations are summarized in Table 4. The Reynolds number based on the cylinder width, H, and the velocity at the inflow boundary, U_0, is 1.0×10^5.The details of LES computations are described in Note 1.

Table 4 Specification for LES computations

case	$L_1 \times L_2 \times L_3$	$N_1\times N_2\times N_3$	boundary conditions: inflow	outflow	x_2	x_3
2D LES	32.85×11.0	99×63	$\bar{u}_1$=1.0, $\bar{u}_2$=0.0	$\bar{u}_1, \bar{u}_2$: $\partial/\partial x_1$=0	periodic	------
3D LES	32.85×11.0×2.0	99×63×10	$\bar{u}_1$=1.0, $\bar{u}_2=\bar{u}_3$=0.0	$\bar{u}_1, \bar{u}_2, \bar{u}_3$: $\partial/\partial x_1$=0	periodic	periodic

L_i : computational domain for x_i direction (normalized by H) N_i : number of grid points in the x_i direction
Δt : interval for time advancement = 0.0002 h_w : mesh interval adjacent to the solid wall = 0.05
No velocity fluctuation is imposed at the inflow boundary.

Mean Surface Pressure (Fig. 4)

Fig. 4 shows distributions of mean surface pressure coefficients $<\overline{C}_p>_t$. Here $< >_t$ means time averaging and $\overline{\ }$ indicate grid-filtering (cf. Nomenclature). The distribution of the 3D computation is similar to that of the experiment by Ohtsuki et al.(1978) except for the leeward face, while in the case of the 2D computation, the negative value of $<\overline{C}_p>_t$ becomes too large at the center of the side faces.

RMS Value of Fluctuating Pressure (Fig. 5)

Distributions of the rms pressure coefficients Cp_{rms} given by the two computations are compared with the results of experiments by Bearman et al.(1982) and Lee (1975) in Fig. 5. It can be concluded that the 3D computation reproduces the pattern of Cp_{rms} distributions in the experiments very well. In the 2D computation, the values of Cp_{rms} differ greatly at both the side face and the leeward face. This lack of agreement between the results of the 2D computation and the experiment is not acceptable.

Power Spectrum of Fluctuating Lift and Drag Forces (Fig.6)

Fig. 6 shows the computed and measured spectra of the fluctuating lift force F_L and the drag force F_D acting on the cylinder. As is shown here, the spectrum curves of F_L have a sharp peak in the results of both computations as well as in that of the experiment. The frequency of each peak corresponds to the fluctuation caused by the corresponding the vortex shedding. The spectrum shapes of F_L and F_D given by the 3D computation correspond very well with those given by the experiments. On the other hand, the peak frequencies of the fluctuations of F_L and F_D in the 2D computation do not correspond with those in the experiments, but are located at higher frequencies for both F_L and F_D. As is shown here, the 2D computation cannot reproduce accurately the surface pressure field and the resulting lift and drag forces.

Difference in the Flowfield Structures near the Side Face (Fig. 7)

Fig.7 compares the power spectra of u_1' based on computations near the side face. This figure reveals significant differences in the flow structures predicted by the 2D and

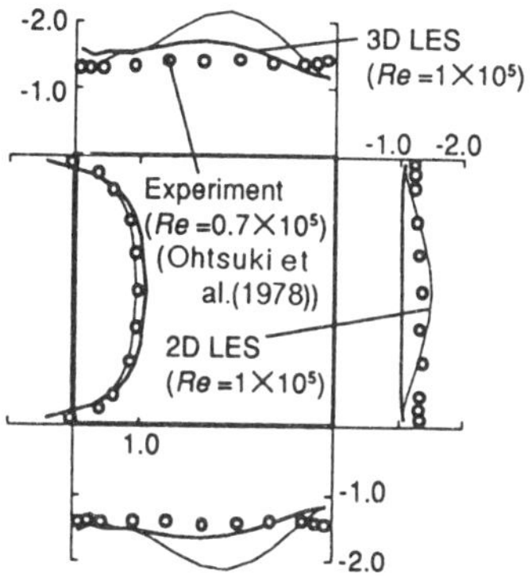

Fig. 4 Distribution of $<\overline{C}_p>_t$

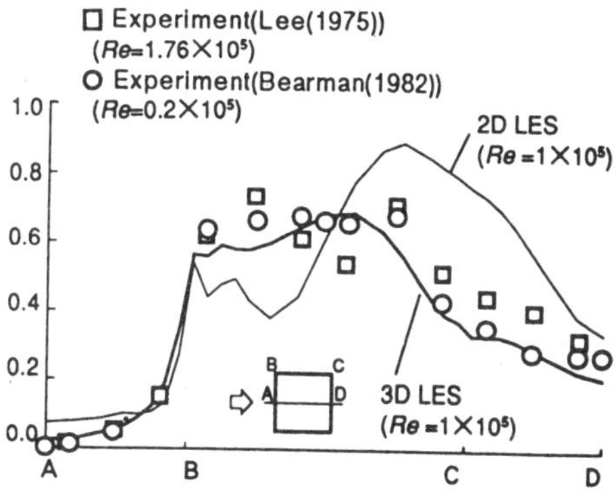

Fig. 5 Distribution of Cp_{rms}

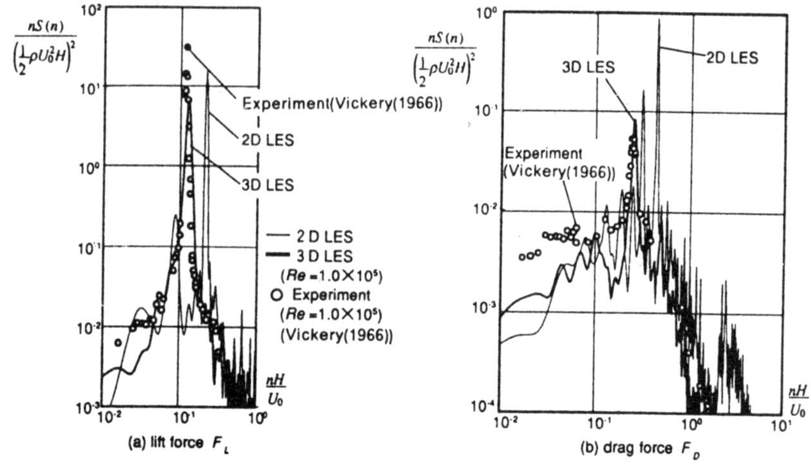

Fig. 6 Power spectrum of fluctuating lift force and drag force

the 3D computations. The spectrum of the 2D computation is confined to a narrow band, and has a sharp peak at the same frequency as that of F_L. On the other hand, the spectrum of the 3D computation extends over a wide band. From this figure, it becomes clear that the energy transfer mechanism throughout a wide spectrum range cannot be reproduced by the 2D computation. This is hardly surprising when we recall that the vortex stretching mechanism, which is essentially three-dimensional and is not reproduced at all in the 2D computation, plays an important role in the energy cascade around this point (Tennekes and Lamlay(1972)). Hereafter, analyses by 3D LES computations are presented.

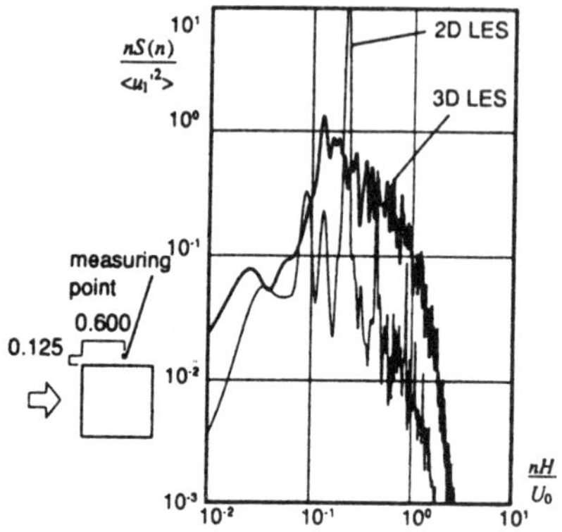

Fig. 7 Power spectrum of velocity fluctuation u_1'

COMPARISON OF LES WITH k-ε / DSM COMPUTATIONS AND MEASUREMENTS FOR THE FLOW PAST 2D SQUARE CYLINDER

In the case of unsteady vortex shedding flows past a 2D square cylinder at high Reynolds numbers, stochastic 3D turbulent fluctuations are superimposed onto the 2D periodic vortex-shedding motion. This is illustrated in Fig. 8 (Franke and Rodi (1991)) where f is the instantaneous value of a qantity, $\tilde{f}$ the periodic fluctuation, f" the stochastic turbulent fluctuation, $\langle f \rangle_t$ the time averaged value and $\langle f \rangle_p$ the phase averaged value.

Proper testing of the accuracy of a numerical simulation for this type of flow is possible only when time-resolved measurements are available. These measurements give information on the temporal variation of the ensemble-averaged quantities and separate the fluctuations into periodic and stochastic turbulent ones. Lyn et al.(1992, 1994) carried out this kind of measurements on the flow past a square cylinder at Re = 2.2×10^4. Here, we analyzed the same flowfield by LES (3D) computations.

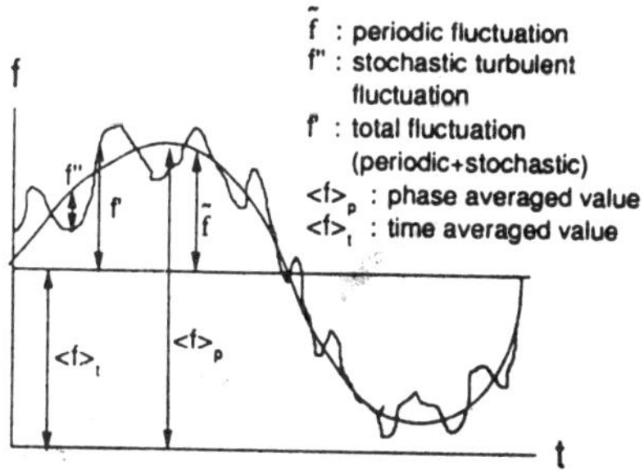

Fig. 8 Periodic and stochastic fluctuations in turbulent vortex shedding flow (Franke and Rodi(1991))

In the computations using RANS models, 3D stochastic turbulent fluctuation f'' is averaged out by ensemble-averaging procedure and its effects are modeled, while 2D periodic fluctuation $\tilde{f}$ can be calculated explicitly. Thus, the only 2D equations need to be solved in the computations using RANS models (Rodi (1993)). Franke and Rodi (1991) carried out the 2D computations using RANS models for the same flowfield. They used the standard k-ε model and the standard Differential Second-moment closure Model (DSM) of Launder, Reece and Rodi (1975), with wall corrections to the pressure-strain terms due to Gibson and Launder (1987). For both models, two types of near-wall treatments are applied ; wall function and the two-layer approach in which the near-wall region is resolved with a one-equation model due to Norris and Reynolds (1975). The results given by LES are compared with those given by the RANS model computations of Franke and Rodi as well as with the experimental results (Lyn et al.(1992, 1994),

Table 5 Computed cases of LES and RANS models and comparison of Integral Parameters

cases		$L_1 \times L_2 \times L_3$	$N_1 \times N_2 \times N_3$	S_t	$\langle C_D \rangle_t$	$\tilde{C}_D$	$\tilde{C}_L$
LES (Smagorinsky model)		20×14×2	104×69×10	0.132	2.09	0.13	1.60
Franke and Rodi (1991)	standard k-ε two layer	20×14	186×156	0.124	1.79	0.0	0.32
	DSM wall funct.	20×14	70×64	0.136	2.15	0.38	2.11
	DSM two layer	20×14	186×156	0.159	2.43	0.08	1.84
Launder and Kato(1993)	standard k-ε wall funct.	20×14	104×70	0.127	1.66		0.1
	modified k-ε wall funct. (eqn.(8) in Note4)	20×14	104×70	0.145	2.05	0.03	1.16
Przulj and Younis(1993)	modified k-ε wall funct. (eqn.(10) in Note5)	35×12	114×92	0.132	2.33	0.92	2.36
Lyn(1992)	experiment (Re=22,000)			0.135	2.05 - 2.23		
Durao(1988)	experiment (Re=14,000)			0.139			

$\tilde{C}_D$, $\tilde{C}_L$: Amplitude of phase averaged drag and lift coefficient,
No velocity fluctuation is imposed at the inflow boundary for the LES computation.

Durao et al.(1988)). Here, the standard Smagorinsky type subgrid model, shown in Table 3, was applied. The outline of LES computations for this case is summarized in Note 2.

Comparison of LES with k-ε / DSM Computations and Experiments

Integral parameters

Values of various integral parameters (e.g., Strouhal number, drag coefficient, etc.) predicted with the various computational methods are compared in Table 5 with experimental values. In the computations by Franke and Rodi(1991), results with the standard k-ε model using wall functions yielded a steady solution and no vortex shedding. Thus this case is not included in Table 5. The two-layer k-ε model yielded too low shedding frequency and drag coefficient $\langle C_D \rangle_t$. The two-layer DSM predicted too high Strouhal number and drag, while the DSM with wall functions predicted these parameters in good agreement with the measurements. The LES computation reproduced slightly lower values, which, however, still agree fairly well with the measurements.

Recently some effort has been made to develop the more refined k-ε models which can be applied to the vortex shedding flows such as the flowfield analyzed here (Launder and Kato(1993), Kato and Launder(1993), Przulj and Younis(1993)). The results of the modified k-ε models, which are also shown in Table 5, are discussed later in section 4.2.

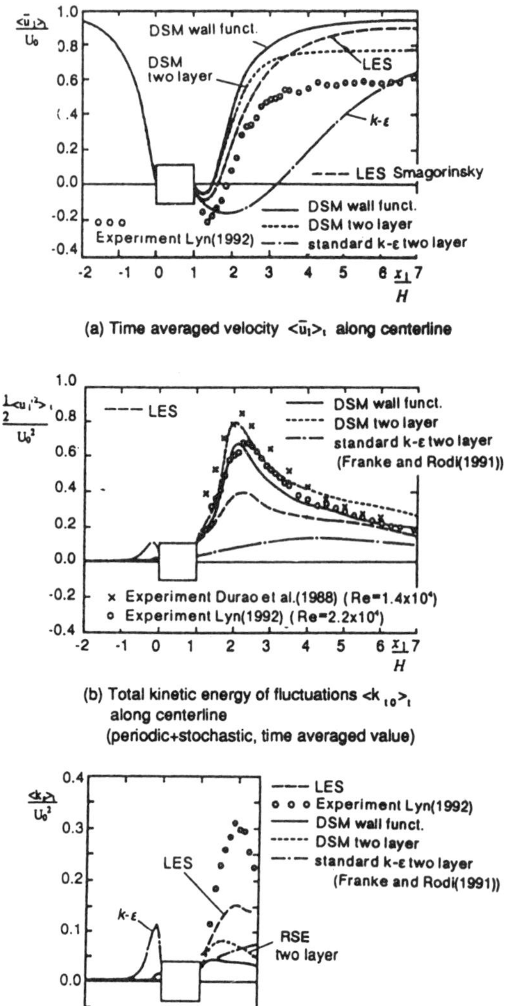

(a) Time averaged velocity $\langle \bar{u}_1 \rangle_t$ along centerline

(b) Total kinetic energy of fluctuations $\langle k_{t0} \rangle_t$ along centerline (periodic+stochastic, time averaged value)

(c) Time averaged kinetic energy of stochastic turbulent fluctuation $\langle k_s \rangle_t$ along centerline (cf.Note 3)

Fig. 9 Comparison of LES with k-ε/DSM and experiments

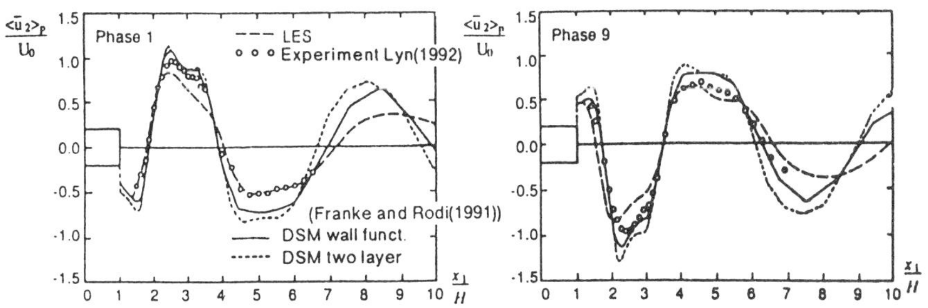

Fig. 10 Phase averaged lateral velocity along centerline at 2 phases (phase 1=1/20T, phase 9=9/20T)

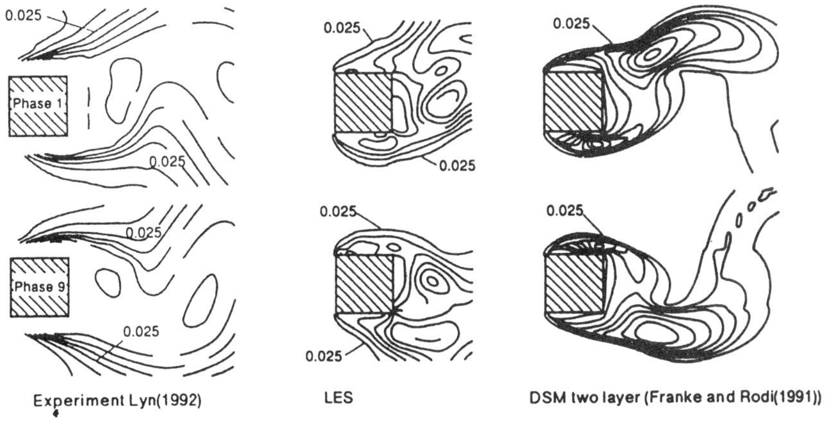

Fig. 11 Contours of phase averaged stochastic turbulent kinetic energy $<k_s>_p$ at 2 phases

Time averaged velocity $<\bar{u}_1>_t$ (Fig. 9 (a))

Experimental data of Lyn (1992) and Durao et al.(1988) are also included in Fig. 9. In front of the cylinder, the results are not much influenced by the computation method used, but there are fairly large differences in the wake region. The k-ε model overestimates the length of the reverse flow zone considerably. Both DSM variants predict too short separation, and there is little difference between the results obtained with the two versions of near wall treatment. The LES shows better agreement with the experiments than the DSM computations, although LES still underestimates the length of the separation region.

Kinetic energy of fluctuations $<k_{to}>_t$, $<k_s>_t$ (Fig. 9 (b), (c))

As shown in Fig. 9 (b), the two-layer k-ε model underestimates severely the total fluctuation level behind the cylinder, while the DSM computations give approximately the correct level of distribution of total fluctuations. In front of the cylinder, the standard k-ε model yields an unrealistically high fluctuation level (here, only turbulent fluctuations exist) due to excessive turbulence production by this model in stagnation regions, as pointed out by Murakami et al. (1990b, 1992, 1993a) and Mochida et al. (1993a). The

DSM does not have this problem. But, Franke and Rodi have shown that all turbulence models tested by them produce too low turbulent fluctuations, $\langle k_s \rangle_t$, in the wake of the cylinder, as shown in Fig. 9(c). This means that the periodic fluctuations are greatly overestimated by the DSM. As for the LES computation, it predicts the correct distribution of the fluctuations, but the level is too low in general for both $\langle k_{to} \rangle_t$ and $\langle k_s \rangle_t$. LES produces a higher turbulent fluctuation level, $\langle k_s \rangle_t$, in downstream region, which would be nearly correct, but is too low by about a factor of 2 in the peak region around $x_1/H=2$. Hence, both total and stochastic turbulent fluctuations are underestimated in the LES simulation. Individually, the LES results are both closer to the measured level than those predicted by the DSM.

Phase averaged velocity $\langle \overline{u_2} \rangle_p$ (Fig.10)

The agreement of wall function DSM and LES computations with the measurements is fairly good in general. On the other hand, the result of two-layer DSM slightly overestimates the value of $\langle \overline{u_2} \rangle_p$, in comparison with the results of LES and wall function DSM.

Phase averaged turbulent kinetic energy $\langle k_s \rangle_p$ (Fig.11)

Here, the LES computation appears to reproduce the observed behavior better than the computations with the DSM. The asymmetry in the contour lines predicted by the DSM is basically the same as in the experiments, but the maxima of $\langle k_s \rangle_p$ in the experimental data are located much closer to the centerline than in the computations. This points to a weakness in the DSM, which appears to predict the centers of the shed vortices considerably farther from the

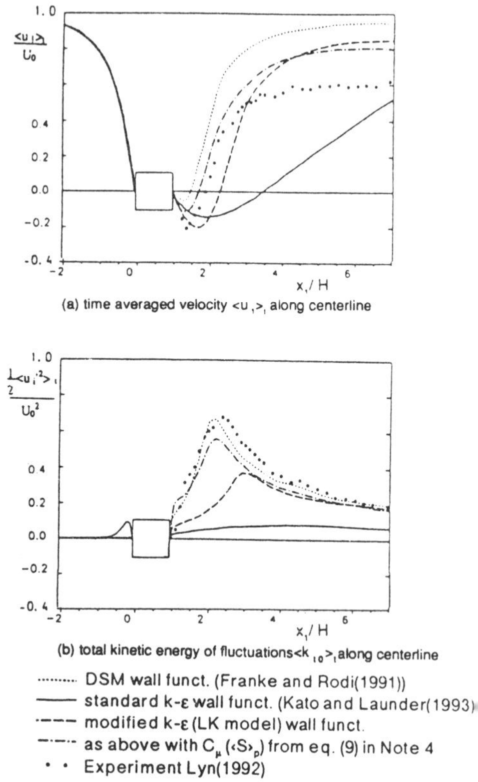

Fig. 12 Comparison of LK model with standard k-ε model

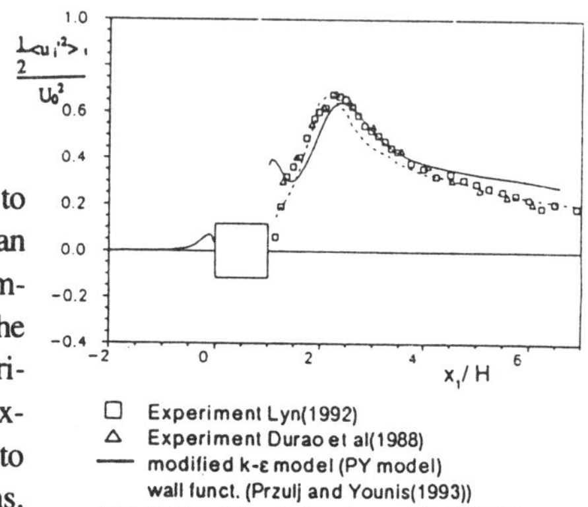

Fig. 13 Comparison of PY model with experiment (total kinetic energy of fluctuations $\langle k_{to} \rangle_t$ along centerline)

centerline than the experiments indicate. The LES computation appears to predict the maxima of $<k_s>_n$ closer to the centerline.

Performance of the Modified k-ε Models

Recently, two different types of modified k-ε models were proposed by Launder and Kato (1993) (hereafter LK model), and Przulj and Younis (1993) (PY model). Details of these models are given in Notes 4 and 5 respectively.

The results of these models are also included in Table 5. The consequence of the modification for P_k in the case of LK model is clear in the values of $\tilde{C}_L$. There is an order of magnitude increase in $\tilde{C}_L$ resulting from the use of modified production as is shown in Table 5. The use of eqn.(8) in Note 4 in place of eqn.(7) gives remarkably closer agreement with the LES. On the other hand, $\tilde{C}_L$ given from PY model becomes too high in comparison with that of LES.

Fig.12 (a) compares the time-averaged velocity $<u_1>_t$ along the centerline. LK model (with $C_\mu(<S>_p)$) reproduces the size of the reverse flow zone rather well..

The total kinetic energy of fluctuations $<k_{to}>_t$ is illustrated in Fig.12 (b). $<k_{to}>_t$ given from the standard k-ε model computation conducted by Launder and Kato (1993) shows excessively large production of turbulent fluctuations in the stagnation region as well as in the result with the standard k-ε model by Franke and Rodi(1991) (Fig.9(b), (c)). Unfortunately this problem still exists in the result of PY model as is indicated in Fig.13, while the accuracy for the prediction of $<k_{to}>_t$ is improved remarkably in the stagnation region by LK model (Fig.12 (b)). On the other hand, $<k_{to}>_t$ is somewhat underestimated in the wake region by LK model when C_μ is treated as constant. Significant improvement is attained by varying the value of C_μ in space using eqn.(9) in Note 4 (Fig.12 (b)). It is rather surprising that distribution of $<k_{to}>_t$ given by PY model shows very close agreement with the experiment in the wake region although $<k_{to}>_t$ is severely overestimated in the stagnation region, by this model (Fig.13).

LES COMPUTATIONS OF UNSTEADY PRESSURE FIELD AROUND AN OSCILLATING SQUARE CYLINDER

The interactive phenomena between fluid and a bluff-shaped body represent one of the most difficult problems in the field of fluid dynamics. This section deals with the interaction between fluid and body, i.e., the flowfields around oscillating square cylinders both in cases of forced oscillation and wind-induced free oscillation. Here, computations based on LES (3D) using the standard Smagorinsky model are presented. A very simple procedure to incorporate the influence of body motion on the flowfield is used. The accuracy of the computations using this procedure is made clear by comparing them with the experimental results of Bearman et al.(1982) and Nakamura et al. (1975).

Outline of Numerical Method for Oscillating Cylinder

Fig.14 illustrates the flow conditions treated here. The cylinder oscillates in two different ways. One way is by forced oscillation and another is by wind-induced free oscillation. In forced oscillation, the cylinder is forced to oscillate in a sinusoidal motion in the x_2 (lateral) direction. In the case of free oscillation, the cylinder is spring-mounted in the x_2 direction and oscillates by the lift force acting on it. Table 6 lists the cases analyzed: one case of the fixed cylinder (case 0), fourteen cases of forced oscillation (cases 10-16, cases 20-26) and four cases of free oscillation (cases 30-33). In the forced oscillation cases, two different values for the amplitude (A/H= 0.10, 0.25), and nine different values for the reduced velocity ($U_0/(NH)$= 7.0-12.0) are imposed. In the experiment, the lock-in phenomena occur for all the cases considered herein. As for the cases of free oscillation, the reduced velocity is varied from 9.0 to 12.0, and the Scruton number($S_c= 2m\delta/\rho H^2 L_3$) is set at 1.36. In this range, the self-induced oscillation of the cylinder was observed in the experiment by Nakamura et al.(1975).

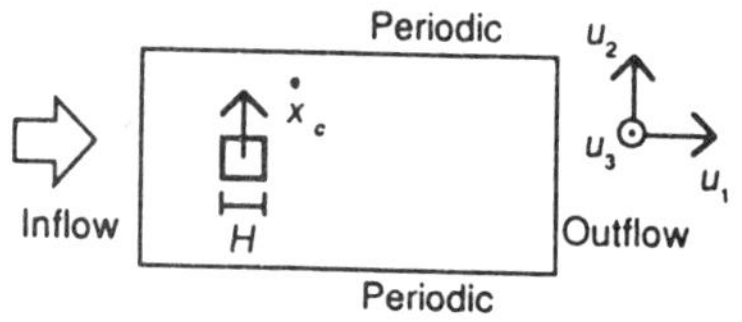

$\dot{x}_c = 2\pi NA\cos(2\pi N t)$
in the case of forced oscillating cylinder

Fig. 14 Flow conditions for simulation

Table 6 Computed cases

Reduced velocity V_r (U_0/NH)	Fixed cylinder	Oscillating cylinder		
		Forced		Free
		Amplitude (A/H)		S_c
		0.10	0.25	1.36
---	case 0			
7.0		case 10	case 20	---
7.5		case 11	case 21	---
8.0		case 12	case 22	---
8.5		case 13	case 23	---
9.0		case 14	---	case 30
9.5		case 15	---	---
10.0		case 16	case 24	case 31
11.0		---	case 25	case 32
12.0		---	case 26	case 33

V_r: reduced velocity ($V_r = U_0/NH$)
(reciprocal of non-dimensional frequency)
N : frequency of the oscillating cylinder
(cf. Nomenclature)
S_c: Scruton number of spring-mounted cylinder
($S_c = 2m\delta/\rho H^2 L_3$)

The governing equations are given in Tables 7 and 8. The standard Smagorinsky model is used for turbulence modelling of sub-grid scale. In the cases of the fixed cylinder and forced oscillating cylinder, the Reynolds number based on H and U_0 is 2.2 x 10^4, which corresponds to that in the experiment by Bearman et al. (1982). As for the case of the free oscillating cylinder, the Reynolds number is 5 x 10^4, which approximately corresponds to that in the experiment by Nakamura et al.(1975). Numerical methods and boundary conditions are summarized in NOTE 6.

Table 7 Model equations for LES
(Standard Smagorinsky model modified for including the effect of oscillation)

$$\frac{\partial \bar{u}_i}{\partial t} + \frac{\partial \bar{u}_i \cdot \bar{u}_j}{\partial x_j} = -\frac{\partial}{\partial x_i}\left(\frac{\bar{p}}{\rho} + \frac{2}{3}k_{SGS}\right) + \frac{\partial}{\partial x_j}(\nu + \nu_{SGS})\cdot\bar{s}_{ij} - \ddot{x}_c\,\delta_{i2}$$

$$\frac{\partial \bar{u}_i}{\partial x_i} = 0$$

$$\nu_{SGS} = 2(C_s\bar{\Delta})^2(\bar{s}_{ij}\,\bar{s}_{ij})^{\frac{1}{2}} \quad \bar{s}_{ij} = \frac{1}{2}\left(\frac{\partial \bar{u}_i}{\partial x_j} + \frac{\partial \bar{u}_j}{\partial x_i}\right) \quad k_{SGS} = \frac{\nu_{SGS}^2}{(C_K\cdot\bar{\Delta})^2}$$

$$\bar{\Delta} = (h_1 h_2 h_3)^{1/3} \times (1 - \exp(x_n^+/25)),\ C_s = 0.10,\ C_K = 0.094$$

Table 8 Equation for cylinder
(One mass system for free oscillation)

$$m\ddot{x}_c + c\dot{x}_c + kx_c = F_L$$

$$c = 2Nm\delta,\quad k = (2\pi N)2m$$

$$\delta = 1.88\times10^{-2},\quad m/(\rho H^2 L_3) = 36.1$$

c : damping coefficient, k : spring stiffness,
δ : logarithmic damping, m: mass of the cylinder

In the present computation, the whole mesh layout (i.e., the

Table 9 Computational domain and boundary conditions

		Boundary conditions		
$L_1 \times L_2 \times L_3$	$N_1 \times N_2 \times N_3$	Inflow	Outflow	x_2, x_3
$20 \times 12 \times 2$	$104 \times 69 \times 10$	$u_1=1.0, \overline{u_2}=-\dot{x}_c, \overline{u_3}=0.0$	$\overline{u_1}, \overline{u_2}, \overline{u_3} : \partial/\partial x_1=0$	periodic

Δt : time interval for time marching = 0.001
h_w : grid interval adjacent to solid wall = 0.025
Wall boundary conditions: power law type two layer model (Werner and Wengle(1991))
No velocity fluctuation is imposed at inflow boundary.

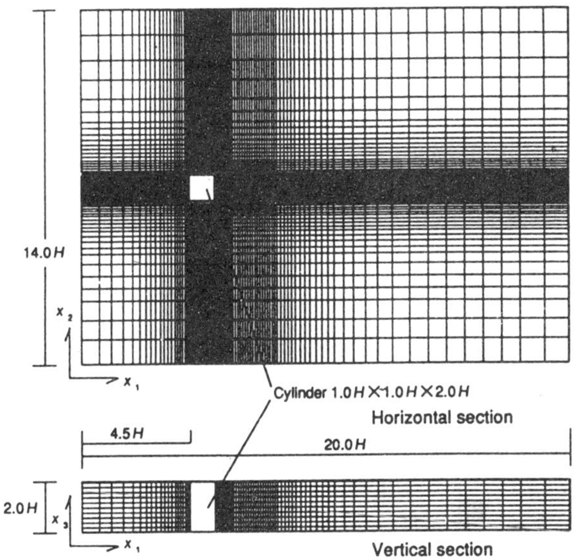

Fig. 15 Computational grid ($104(x_1)$x$69(x_2)$x$10(x_3)$=71,760)

coordinates) is treated to move together with body motion. In order to incorporate the influence of body motion, a technique is applied to simplify the computation. Instead of moving the body, here the fluctuating inertia force term, $-\ddot{x}_C$, whose magnitude is equal to the acceleration of the body but with the opposite sign, is added to the transport equation of $\overline{u_2}$ as an external force at every grid point of the fluid for every time step, including points at the inflow boundary (cf. Tables 7 and 9). Equivalent form is easily obtained, using the ALE (Arbitrary Lagrangian-Eulerian) method proposed by Hirt et al. (1974), by letting the spatial derivative of $\dot{x}_C$ be 0.

Results and Discussion

Forced oscillation

(1) Instantaneous Velocity and Pressure Fields for cases of Positive and Negative Aerodynamic Force Damping(Fig. 16)

The instantaneous velocity and pressure fields around forced oscillating cylinders are illustrated in Fig.16 for cases of 10 (A/H=0.1, NH/U_0=1/7.0) and 14 (A/H=0.1, NH/U_0=1/9.0). The figures show the instantaneous flowfields when moving velocity, $\dot{x}_C$, has its peak value in each case. Cylinders are moving to the left side (i.e., upper side in Figs. 16(a) and 16(b)). In case 10, large negative pressure occurs at the right sideof the cylinder (lower side in Fig.16(a)). On the other hand, it occurs at the left side in case 14 (Fig. 16(b)). This means that the pressure field would damp the cylinder oscillation in case 10 but amplify the oscillation in case 14. The significant difference between the results of both cases is also observed more clearly in the difference of the phase angles between the cylinder displacement and the pressure acting on the side faces of the cylinder for two cases. This point will be discussed later in section 5.2.1 (3).

(2) Frequency of Lift Force Acting on the Cylinder

Fig.17 compares the lift force acting on the fixed cylinder (case 0) and the forced oscillating cylinder (case 23), wherein the imposed body frequency is 0.12 (1/8.5=NH $/U_0$). The Strouhal number $(S_t)_f$, which is evaluated from the time history of the lift force, is 0.14 (1/7.3) for the case of the fixed cylinder

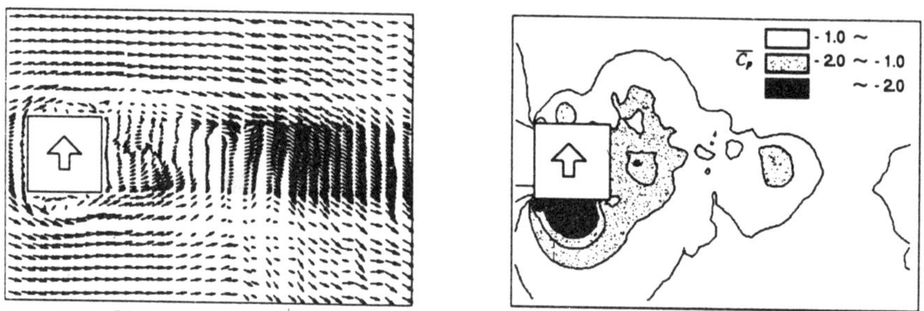

Velocity vector field

Pressure field

(a) case 10 (A/H= 0.10, NH/U_0= 1/7.0)

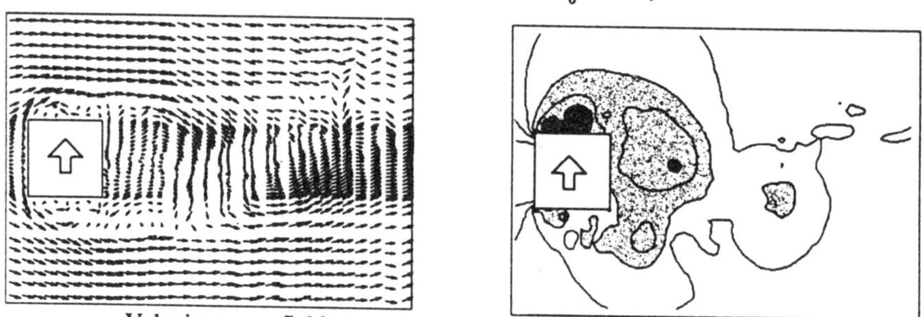

Velocity vector field

Pressure field

(b) case 14 (A/H = 0.10, NH/U_0= 1/9.0)

Fig. 16 Instantaneous velocity and pressure fields around forced oscillating cylinder for the cases of two different reduced velocities
(case 10: aerodynamic force damping is positive,
case 14: aerodynamic force damping is negative.)

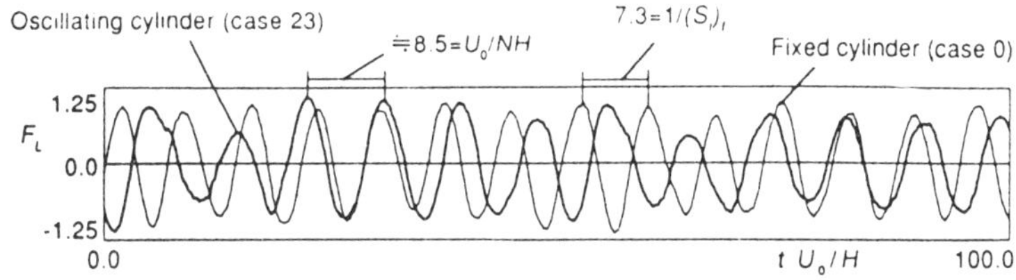

Fig. 17 Time history of lift force

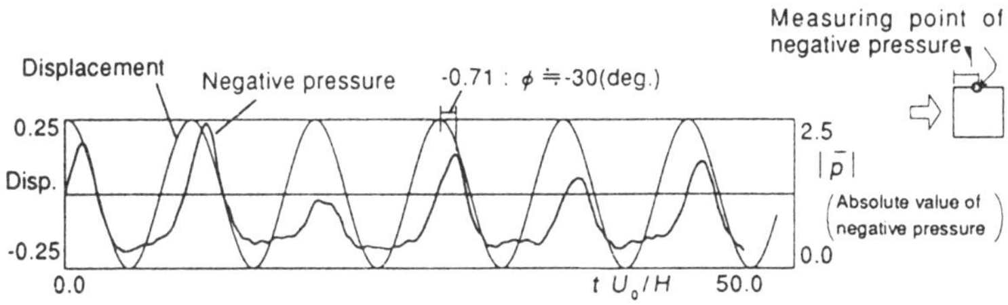

Fig. 18 Time history of negative pressure on side and body displacement (Computation : case23 ; A/H=0.25, NH/U_0=1/8.5)

while it decreases to 0.12 (1/8.5), the value of the imposed body frequency for the case of the oscillating cylinder. This change in the vortex shedding frequency by body motion is known as the lock-in phenomenon, which is reproduced successfully in the present computation. The lock-in phenomenon is also confirmed in all other cases conducted here.

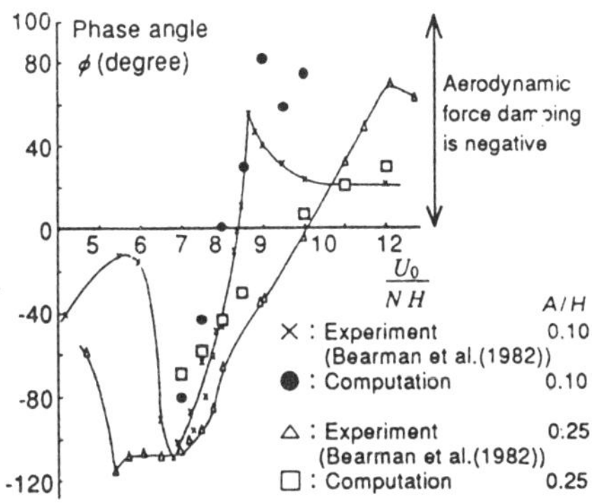

Fig. 19 Phase angle of negative pressure on side and body displacement

(3) Phase Angle between Body Motion and Fluid Force

The time histories of the displacement of the forced oscillating cylinder and the negative pressure acting at the center of the side face of the cylinder for case 23 are shown in Fig. 18. The phase angle between the two is evaluated as about -30° by Fourier analysis. In Fig. 19, the phase angles for the other thirteen cases at A/H = 0.10 and 0.25 are also plotted. The predicted phase angles for these cases correspond rather well with the experimental values. When the phase angle is within the range $0° < \phi < 180°$, the aerodynamic force damping is negative. This means that the lift force would amplify the cylinder oscillation. As shown in Fig. 19, the phase angle for case 10 (A/H = 0.10, U_0/(NH) =7.0) is about -80° which would damp the oscillation, while the angle for case 14 (A/H = 0.10, U_0/(NH) =9.0) is about 80° and the aerodynamic force damping is negative. Thus, the oscillation would be amplified by the aerodynamic force in case 14. The aerodynamic force damping of positive and negative observed in the experiment is reproduced well in the computation.

Wind-induced free oscillation

Fig. 20 shows the oscillation amplitudes of the cylinders in the cases of free oscillation (cases 30-33), where the cylinder is spring-mounted and oscillates within x_2 direction by the lift force (cf. Table 8). In the experiment, self (vortex)-induced oscillation starts when the reduced velocity, U_0/NH, is about 8 and the oscillation amplitude of the cylinder increases rapidly according to the increase of reduced velocity. The computational results reproduce this tendency very well.

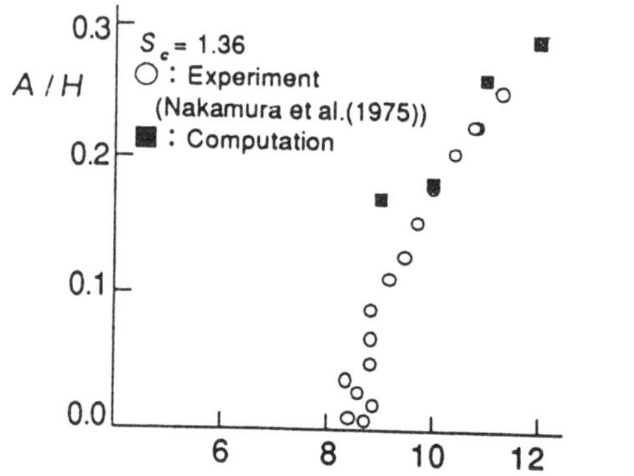

Fig. 20 Amplitudes of free oscillating cylinder

CONCLUDING REMARKS

The numerical prediction of turbulent flowfield around a square cylinder is one of the most difficult problems in the field of CFD at this stage, since the flowfield is very complicated, characterized by the impinging, the separation at the frontal corner with steep streamline curvature, the recirculation behind the cylinder and the vortex shedding, etc.

The numerical analyses of vortex shedding flow with various turbulence models are presented here. The standard k-ε model did not succeed in reproducing the vortex shedding at all. However, the modified k-ε model succeeded in reproducing it well, with an accurate prediction of some integral parameters. DSM reproduced the vortex shedding rather well, but the results show many discrepancies between the experiments, e.g., the overestimation of periodic fluctuation. In general, LES shows best agreement with the experimental results in many respects, even though the standard Smagorinsky SGS model is used. However, recent progress in wind engineering require more accurate SGS models for LES. The dynamic SGS model is expected to become a powerful tool for flow analysis in wind engineering problems in near future.

The improvement of accuracy usually requires the increase of CPU resources and computation instability. Thus a strategy for selecting a turbulence model may be proposed according to the purpose of the analysis and the accuracy required. The relative abilities of various turbulence models are compared in Table 10. LES is most suited to analyze the complicated flow situations usually treated in the applications of wind engineering, but LES needs a great deal of CPU time. Further efforts should be made to develop more reliable turbulence models and more simple and less time-consuming CFD methods.

Table 10 Relative comparison of various turbulence models for practical modelling of flowfields related building engineering (Murakami(1993a))

Turbulence model	Standard k − ε	Low-Re. No. k − ε	Standard DSM (ASM)	Low-Re. No. DSM (ASM)	LES	LES
wall boundary condition	wall function	non-slip	wall function	non-slip	wall function	non-slip
1. Simple flows (channel flow, pipe flow, etc) (local equilibrium is valid)	○	○	○	○	○	○
2. Flow with streamline curvature						
1) weak curvature	○	○	○	○	○	○
2) strong curvature (flow around bluff body)	X , △	X , △	○ (△)	○ (△)	○	○
3. Impinging flow	X , △	X , △	△ , ○	△ , ○	○	○
4. Flowfield with low Reynolds No.	X , △	△ , ○	X , △	△ , ○	△ , ○	○
5. Non − isothermal flow						
1) weak stratification	○	○	○	○	○	○
2) strong stratification	X , △	X , △	○	○	○	○
6. Convective heat transfer at wall	X , △	○	X , △	○	X , △	○
7. Unsteady flow, unsteady diffusion						
1) highly unsteady	X	X	X	X	○	○
2) vortex shedding	X	△	○	○	○	○
8. Jet						
1) normal	○	○	○	○	○	○
2) swirl	X	X	○	○	○	○

○ : functions well △ : insufficiently functional X : functions poorly

ACKNOWLEDGMENTS

The LES computations presented in this paper were carried out in cooperation with Mr.S.Sakamoto, Taisei Corporation and Mr. Y. Tominaga, I.I.S., Univ. of Tokyo. This paper was written with the assistance of Mr. Y. Tominaga. The authors would like to express their sincere gratitude for their valuable contributions.

NOTE 1 Outline of 2D and 3D LES computations of flow past fixed square cylinder at Re=1.0x10^5

No velocity fluctuation is imposed at the inflow boundary. The turbulence levels are 0.04 % in the experiment of Bearman (1982) and 0.2 % in the experiment of Ohtsuki(1978), respectively. A staggered grid was adopted and a second-order centered difference scheme was applied for the calculation of spatial derivatives. For time advancement, the Adams-Bashforth scheme was used for both convection and diffusion terms. For boundary conditions at solid walls, the generalized logarithmic law (Launder and Spalding (1972)) expressed by

$$\frac{\langle(\bar{u}_t)_w\rangle_t}{\langle\tau_w/\rho\rangle_t}\,(C_\mu^{1/2}k_w)^{1/2}=\frac{1}{\kappa}\left(\frac{E\frac{1}{2}h_w(C_\mu^{1/2}k_w)^{1/2}}{\nu}\right) \tag{1}$$

was employed in order to estimate the time-averaged wall shear stress, $\langle\tau_w\rangle_t$. Furthermore, the following relation was used to specify the instantaneous wall shear stress τ_w (Grötzbach(1987), Piomelli et al. (1989), Murakami et al. (1990b)).

$$\tau_w=\langle\tau_w\rangle_t\times\frac{(\bar{u}_t)_w}{\langle(\bar{u}_t)_w\rangle_t} \tag{2}$$

In the 3D computation, the computational domain covers 32.85H x 11.0H x 2.0H in the streamwise (x_1), lateral (x_2) and spanwise (x_3) directions, respectively. 99 x 63 x 10 grid points are placed in each direction. For the 2D computation, the computational domain covers the same length of streamwise and lateral directions, and the same number of grid points are placed in both directions, but the computation has no dimension in the spanwise direction.

Nomenclature

x_i : three components of spatial coordinate (i = 1,2,3:streamwise,lateral,spanwise)
L_i : computational domain for x_i direction
h_i : mesh interval in x_i direction
u_i : three components of relative velocity vector to the cylinder velocity
f : instantaneous value of f
$\overline{f}$: grid-filtered value of f
$\langle f \rangle_t$: time-averaged value of f
$\langle f \rangle_p$: phase-averaged value of $\overline{f}$
f' : deviation from $\langle f \rangle_t$, $f' = \overline{f} - \langle \overline{f} \rangle_t$
$\tilde{f}$: periodic fluctuation, $\tilde{f} = \langle f \rangle_p - \langle f \rangle_t$
f'' : stochastic fluctuation, $f'' = f - \langle f \rangle_p$
t : time, n : frequency
H : width of square cylinder
U_0 : u_1 value at inflow of computational domain
p : pressure,
$\langle P_0 \rangle$: reference static pressure
C_p : instantaneous pressure coefficient ($C_p = (p - \langle P_0 \rangle)/(1/2 \rho U_0^2)$)
C_{prms} : rms pressure coefficient ($C_{prms} = (\langle p'^2 \rangle_t)^{1/2}/(1/2 \rho U_0^2)$)
$S(n)$: power spectrum
F_L : instantaneous lift force
F_D : instantaneous drag force
$\langle k_{t0} \rangle_t$: total kinetic energy of fluctuation, (periodic+stochastic, time averaged value),
k_s : kinetic energy of stochastic turbulent fluctuation, ($\langle k_{t0} \rangle_t = 1/2 \langle u_i' u_i' \rangle_t$, $k_s = 1/2 \langle u_i'' u_i'' \rangle$)
C_s : Smagorinsky constant
ν_{SGS} : subgrid scale eddy viscosity
k_{SGS} : subgrid component of k
h_w : grid interval adjacent to the solid wall
$(u_t)_w$: tangential velocity component at the near wall node,
k_w : total kinetic energy of fluctuations at the near wall node,
m : mass of the cylinder
δ : logarithmic damping of the cylinder
S_c : Scruton number ($S_c = 2m\delta / \rho H^2 L_3$)
S_t : Strouhal number
$(S_t)_f$: S_t of the fixed cylinder
x_n^+ : $u_* x_n / \nu$
x_n : distance from solid wall
u_* : friction velocity

$\ddot{x}_c, \dot{x}_c, x_c$: cylinder acceleration, velocity and displacement in x_2 direction ($x_c = A \sin(2\pi N t)$ in the case of forced oscillation)
A : displacement of the oscillating cylinder
N : frequency of the oscillating cylinder
(forced oscillation: N = imposed oscillation frequency
free oscillation: N = cylinder natural frequency)

When values are made dimensionless, representative length scale H, velocity scale U_0 and air density ρ are used.

The interval of time advancement is 2×10^{-4} in non-dimensional time scale, which is the scale based on H and U_0. Starting from the initial conditions, the computations required 5.0×10^5 time steps (100 in non-dimensional time scale) to reach stable periodic conditions. The computations were then continued further for 2.05×10^6 steps (410 in non-dimensional time scale). Thus statistically reliable averaged values were obtained.

NOTE 2 Outline of LES computations of flow past fixed square cylinder at Re=2.2x10⁴

The value of the Smagorinsky constant C_s was 0.13 for this case. Because the Reynolds number, which is 2.2×10^4, is lower than that in the previous section, the grid scales are multiplied by the Van Driest type wall damping function, $1-\exp(-x_n^+/25)$, in order to account for the near wall effect. For the computations conducted here, a second-order centered difference scheme proposed by Piacsek and Williams (1970), which insures spatial conservation of quadratic values, was adopted for the spatial derivatives. For time advancement, the Adams-Bashforth scheme was used for the convection terms and the Crank-Nicolson scheme for the diffusion terms. The interval for time advancement is 1.0×10^{-3} in non-dimensional time scale based on U_0 and H.

The computational domain covers 4.5H upstream of the cylinder, 14.5H downstream and 6.5H on either side of the cylinder. The size of the domain in the plane perpendicular to the cylinder is exactly the same as that in Franke and Rodi(1991) 's computation . 104 x 69 grid points were placed in the streamwise and lateral directions, respectively. The grid interval h_w adjacent to the cylinder wall is 0.022. This value is the same as that for Franke and Rodi's wall function case. Franke and Rodi set h_w = 0.00125 in the two-layer approach. At the inflow boundary, the approach flow was set to be constant and uniform and no velocity fluctuations were prescribed in the LES computations, while the turbulence level of the approach flow was 2% in Lyn's experiments and Franke and Rodi's computations. At the outflow boundary, zero gradient conditions were imposed. Symmetry conditions were employed for the lateral boundaries, and periodicity conditions were imposed for the boundary planes perpendicular to the cylinder axis. For the boundary condition at the solid walls, Werner and Wengle(1991) 's approach assuming a linear or 1/7 power law distribution of the instantaneous velocity is adopted:

$$\frac{\bar{u}_t}{u^*} = x_n^+ \qquad (x_n^+ \leq 11.81) \tag{3}$$

$$\frac{\bar{u}_t}{u^*} = 8.3 x_n^{+\,1/7} \quad (x_n^+ \geq 11.81) \tag{4}$$

This approach was chosen because the Reynolds number, which is 2.2 x 10^4, is lower than that for the previous case.

The computations required 1.5 x 10^5 time steps (150 in non-dimensional time scale) to reach stable periodic conditions. The computations were continued further for 1.0 x 10^5 steps (100 in non-dimensional time scale) covering 13 shedding periods. The time averaged values were determined by time averaging over these 13 periods as well as by averaging over the spanwise direction.

NOTE 3 We could sample data necessary for the $\langle k_s \rangle_t$ computation only in the area close to the cylinder. This was due to the limitation of the computer memory.

NOTE 4 Modified k-ε model proposed by Launder and Kato (1993) (LK model)

Launder and Kato defined dimensionless strain and vorticity parameters:

$$\langle S \rangle_p = \frac{k_s}{\varepsilon}\left\{\frac{1}{2}\left(\frac{\partial \langle u_i \rangle_p}{\partial x_j} + \frac{\partial \langle u_j \rangle_p}{\partial x_i}\right)^2\right\}^{1/2} \tag{5}$$

$$\langle \Omega \rangle_p = \frac{k_s}{\varepsilon}\left\{\frac{1}{2}\left(\frac{\partial \langle u_i \rangle_p}{\partial x_j} - \frac{\partial \langle u_j \rangle_p}{\partial x_i}\right)^2\right\}^{1/2} \tag{6}$$

The production term of turbulent kinetic energy can be written as follows using the definition of $\langle S \rangle_p$ in eqn.(5).

$$P_k = C_\mu \varepsilon \langle S \rangle_p^2 \tag{7}$$

In LK model eqn.(7) is replaced by the following equation.

$$P_k = C_\mu \varepsilon \langle S \rangle_p \langle \Omega \rangle_p \tag{8}$$

The values of $\langle S \rangle_p$ become very large in the stagnation region and the very large values of $\langle S \rangle_p$ lead to the excessive level of P_k when the original form of eqn.(7) is used. However, the deformation near a stagnation point is nearly irrotational, i.e. $\langle \Omega \rangle_p$=0. Thus the replacement

$P_k = C_\mu \varepsilon \langle S\rangle_p \langle \Omega\rangle_p$ leads to a marked reduction in turbulence production P_k near the stagnation point while having no effect in a simple shear flow.

Further modifications have been proposed by Craft et al. (1993) involving a dependence of C_μ on $\langle S\rangle_p$. Kato and Launder (1993) carried out a computation using the model by Craft et al.(1993) in which C_μ is treated as a function of $\langle S\rangle_p$:

$$C_\mu = \min\left[0.09,\ 0.3/(1+0.35\times S^{*\,1.5})\right] : S^* = \min\left[20,\ \langle S\rangle_p\right] \tag{9}$$

NOTE 5 Modified k-ε model proposed by Przulj and Younis (1993) (PY model)

Przulj and Younis (1993)added the time-dependent source term to the ε equation as follows.

$$C_{\varepsilon 1} P_k \frac{\varepsilon}{k_s}\left(1 + C_t \frac{k_s}{\varepsilon}\frac{|\partial q/\partial t|}{q}\right), \tag{10}$$

where q is the mean kinetic energy $1/2\langle u_i\rangle_p \langle u_i\rangle_p$. A new coefficient C_t was introduced; its value was set as 0.26, assigned by computer optimization.

NOTE 6 Numerical methods and boundary conditions for LES computations of flow past oscillating square cylinder at Re=2.2x10⁴

The size of the computational domain and the boundary conditions are summarized in Table 9. The number of grid points is also indicated in the Table. Grid arrangements are illustrated in Fig. 15. The minimum grid interval adjacent to the solid wall h_w was set at 0.025. For the boundary conditions at the solid wall, Werner and Wengle(1991) 's approach is adopted, as well as in the computations in the previous section.

The interval of time advancement is 1×10^{-3} in the time scale which is made dimensionless by H and U_0. In the case of the fixed cylinder, the computation required 1.0×10^5 time steps (100 in non-dimensional time scale) to reach stable periodic conditions, starting from the initial conditions. As for the oscillating cylinder, starting from the result of the fixed cylinder, 1.5×10^5 steps in the case of forced oscillation and 2.5×10^5 steps in the case of free oscillation are required to reach stable periodic conditions respectively. The computation is then continued further for 1.0×10^5 steps in the case of the fixed cylinder and for 0.7×10^5 - 1.2×10^5 steps (70-120 in non-dimensional time scale) covering 10 times of period of the cylinder oscillation in the case of the oscillating cylinder. Thus, statistically reliable averaged values are obtained. The computation for one case takes 20-60 CPU hours on a Fujitsu VP2600 machine.

REFERENCE

Bearman, P.W. and Obasaju,E.D. (1982) "An experimental study of pressure fluctuations on fixed and oscillating square-section cylinders", J.Fluid Mech., vol.119,pp.297-321.

Craft, T.J., Launder, B.E. and Suga, K. (1993) "Extending the Applicability of Eddy Viscosity Models through the Use of Deformation Invariants ", Proc. of 5th IAHR Conference on Refined Flow Modelling and Measurements

Deardorff,J.W. (1970) "A numerical study of three-dimensional turbulent channel flow at large Reynolds numbers", J.Fluid Mech., 41,453-480

Deardorff, J. W. (1973) "The use of subgrid transport equations in a three-dimensional model of atmospheric turbulence", J. Fluids Eng., 429-438.

Deardorff, J. W. (1974) "Three-dimensional numerical study of height and mean structure of a heated planetary boundary layer. Boundary-layer Meteor., vol.7, 81-106.

Deardorff, J. W.(1980) "Stratocumulus-capped mixed layers derived from a three -dimensional model". Boundary-layer Meteor., vol.18,495-527.

Durao, D.F.G., Heitor, M.V. and Pereira, J.C.F. (1988) "Measurements of turbulent and periodic flows around a square cross-section cylinder, Experiments in Fluids, vol.6, pp.298-304

Franke, R. and Rodi, W. (1991) "Calculation of vortex shedding past a square cylinder with various turbulence models", Proc. 8th Symp. on Turbulent Shear Flows, p.189.

Germano, M., U. Piomelli, P. Moin and W. H. Cabot,(1991) "A Dynamic subgrid scale eddy viscosity model", Phys. Fluids, A3(7),1760-1765.

Gibson, M.M. and Launder, B.E. (1987) "Ground Effects on Pressure Fluctuations in the Atmospheric Boundary Layer", J.Fluid Mech., Vol.86, p.491,

Grözbach, G. (1987) "Direct numerical and large eddy simulation of turbulent channel flows", Encyclopedia of Fluid Mechanics, Vol.6, Complex Flow Phenomena and Modeling, pp.1337-1391,

Hirt, C.W. et al.(1974) "An Arbitrary Lagrangian-Eulerian Computing Method for All Flow Speeds", J.Comput. Phys. 14, pp.227-253

Horiuti, K. (1993a) "A proper velocity scale for modeling subgrid-scale eddy viscosities in large eddy simulation", Phys. Fluids A5(1),,pp.146-157.

Horiuti, K. and Mansour, N. (1993b) "A normal stress subgrid-scale eddy viscosity model in large eddy simulation", Ninth Symp. on Turbulent Shear Flows,13-2

Inamuro, T. Adachi,T.and Sakata, H. (1993) "Simulation of aerodynamic instability of bluff body", J. of Wind Eng. and Ind. Aerodyn., vol.46-47

Kato,M. and Launder,B.E.(1993) "The Modelling of Turbulent Flow around Stationary and Vibrating Square Cylinders", 9th Symp. on Turbulent Shear Flow, Kyoto

Karanth, D., Rankin, G.W. and Sridhar, K.(1993) "Numerical simulation of flow past an oscillating cylinder, Proc. of Forum on Unsteady Flows, ASME Fluid Engineering Conference (FED-Vol.157), pp.167-175

Kawai, H.(1990) "A discrete vortex analysis of flow around a vibrating cylinder with a splitter plate", pp.259-273, J. of Wind Eng. and Ind. Aerodyn., vol.35

Kim, J., Moin, P. and Moser, R. (1987) "Turbulence statistics in fully developed channel flow at low Reynolds number", J.Fluid Mech., 177

Launder,B.E. and Spalding, J.L. (1972) "Mathematical models of turbulence, Academic Press, New York

Launder,B.E.,Reece, G.J. and Rodi, W.(1975) "Progress in the development of a Reynolds stress turbulence closure", J.Fluid Mech., 68, 537-566

Launder, B.E. and Kato, M. (1993) "Modelling Flow-induced Oscillations in Turbulent Flow around Square Cylinder", Proc. of Forum on Unsteady Flows (FED-vol.157), ASME 1993, pp.189

Le, H. , P. Moin and J. Kim (1993)," Direct numerical simulation of turbulent flow over a backward-facing step", Ninth Symp. on Turbulent Shear Flows, 13-2

Lee, B.E. (1975) "The effect of turbulence on the surface pressure field of a square prism", J.Fluid Mech., vol.69, pp.263-282

Lyn, D.A. (1992) "Ensemble-averaged measurements in the turbulent near wake of a square cylinder", Private communication

Lyn, D.A. and Rodi, W. (1994) "The flapping shear layer formed by flow separation from the forward corner of square cylinder", J. Fluid Mech., vol. 267, PP.353-376

Mason, P. J.(1989) "Large eddy simulation of the convective atomospheric boundary layer". J. Atomos. Sci., vol.46,1492-1516.

Mason, P. J. and Derbyshire,S. H. (1990) "Large-eddy simulation of the stable-stratified atomospheric boundary layer", Bounbary-Layer Meteor., 53, 117-162.

Mochida,A. Murakami,S. Shoji,I and Ishida,Y. (1993a) "Numerical Simulation of Flowfield around Texas Tech Building by LES, J.Wind Eng.Ind.Aerodyn.,46-47, pp.455-460

Mochida,A., Ishida,Y. and Murakami,S. (1993b) "Numerical Study on Flowfield around Structures with Oblique Wind Angle based Composite Grid System", Proc. 7th U.S.Conf. Wind Eng., pp. 463-472.

Moeng, C. H. (1984) "A large eddy simulation model for the study of planetary boundary layer turbulence". J. Atomos. Sci., vol.41,2052-2062.

Moin,P. and Kim, J. (1982) "Numerical investigation of turbulent channel flow", J.Fluid, Mech., 18, 341-377

Murakami,S.,Mochida, A. and Hibi,K. (1987) "Three-dimensional numerical simulation of airflow around a cubic model by means of large eddy simulation, J.Wind Eng.Ind.Aerodyn., 25, 291-305

Murakami,S.,Kato, S. and Kondo, Y. (1990a) "Examining k-ε EVM by means of ASM for a 3-D horizontal buoyant jet in enclosed space", Engineering Turbulence Modelling and Experiments, Elsevier Science Publishing,205-214

Murakami,S., Mochida,A. and Hayashi, Y. (1990b) "Examining the k-ε model by means of a wind tunnel test and Large-eddy simulation of the turbulence structure around a cube",.J.Wind Eng.Ind.Aerodyn., 35, 87-100.

Murakami,S. (1990c) "Computational Wind Engineering", J.Wind Eng. Ind. Aerodyn., 36, pp. 517-538

Murakami,S., A.Mochida and Y.Hayashi (1991a) "Scrutinizing k-ε EVM and ASM by means of LES and wind tunnel for flowfield around cube, 8th Symp.on Turbulent Shear Flows,

Murakami,S., Mochida,A., Hayashi, Y. and Sakamoto, S. (1992) "Numerical study on velocity-pressurefield and wind forces for bluff bodies by k-ε,ASM and LES", J.Wind Eng.Ind.Aerodyn., 41-44,2841-2852.

Murakami,S. (1993a) "Comparison of various turbulence models applied to a bluff body" J.Wind Eng.Ind.Aerodyn., 46-47, pp.21-36

Murakami,S., Mochida, A. and Ooka, R. (1993b) "Numerical simulation of flowfield over surface-mounted cube with various second-moment closure models, 9th Symp on Turbulent Shear Flows, 13-5-1~15-5-6

Murakami,S., Rodi,W., Mochida,A. and Sakamoto,S.(1993c) "Large Eddy Simulation of Turbulent Vortex Shedding Flow past 2D Square Cylinders", Proc. of Symp. on Eng. Applications of Large Eddy Simulations (FED-vol. 162), ASME 1993,pp.113

Murakami, S. and Mochida, A. (1993d) "On Turbulent Vortex Snedding Flow Past 2D Square Cylinder Predicted by CFD", Third Asia-Pacific Symp. on Wind Eng., 1993.

Nakamura,Y. and Mizota,T. (1975) "Unsteady lifts and Wakes of Oscillating Rectangular Prisms", ASCE., J. of Eng. Mech. Div., vol.101, No. EM6, p.855

Nieuwstadt, F. T. M. , Mason,P. J. , Moeng, C. H. and Schumann, U. (1991) "Large-eddy simulation of the convective boundary layer : A comparison of four computer codes", 8th Symp.on Turbulent Shear Flows

Nomura, T.(1993) "A numerical study on vortex-excited oscillations of bluff cylinders", J. of Wind Eng. and Ind. Aerodyn., vol.50

Norris,H.L. and Reynolds,W.C. (1975) "Turbulent Channel Flow with Moving Wavy

Boundary", Department Mech. Eng. Rept. FM-10, Stanford Univ.
Ohtsuki, Y. et al. (1978) "Wind tunnel experiments on aerodynamic Forces and pressure distributions of rectangular cylinders in a uniform flow", Proc. the 5th Symp. on Wind Effects on Structures, pp.169-175, Tokyo Japan
Okajima, A., Kitajima, K. and Ueno, H. (1993) "Numerical study on wake patterns and aerodynamic forces of an oscillating cylinder with a circular and rectangular cross-section", J. Wind Eng. and Industrial Aerodyn., vol.50
Piacsek S. A. and Williams,S.A. (1970) "Conservation properties of convection difference schemes", J. of Comput. Phys. , 6, pp.392-405.
Piomelli, U., Ferziger, J.H. and Moin, P. (1989) "New Approximate Boundary Conditions for Large Eddy Simulations of Wall-Bounded Flows", Phys. Fluids, A1 (6), 1061
Przulj, V. and Younis,B.A. (1993) "Some Aspects of the Prediction of Turbulent Vortex Shedding from Bluff Bodies", [FED-vol.149], ASME, pp.75
Rodi,W.,(1976) ZAMM, 58,T219-T221
Rodi, W. (1993) "On the Simulation of Turbulent Flow Past Bluff Bodies", J.Wind Eng. Ind. Aerodyn., 46-47, pp.3-20.
Sakamoto, S., Murakami,S., Kato, S. and Mochida, A. (1993a) "Unsteady Pressure Field around Oscillating Prism prediceted by LES", J. of Wind Eng. and Ind. Aerodyn.,vol.46-47,pp.551-556
Sakamoto, S., Murakami, S. and Mochida, A. (1993b) "Numerical study on flow past 2D square cylinder by Large Eddy Simulation : Comparison between 2D and 3D computations", J. of Wind Eng. and Ind. Aerodyn., vol.50, pp.61-68
Schmidt, H. and Schumann U. (1989) "Coherent structure of the convective boundary layer derived from large eddy simulation". J. Fluid Mech., vol.200, 511-562.
Schumann,U. (1975) "Subgrid scale model for finite difference simulation of turbulent flows in plane channels and annuli", J. Comput. Phys., 18, 376-404
Shimura, M. and Zienkiewicz, O.C. (1991) Interation Analysis between Structure and FluidFlow Using the Direct Laplacian Method, Proc. of 4th ICCCBE, pp.267-274
Shirato, H., Matsumoto, M. and Shiraishi, N. (1993) "Unsteady aerodynamic force characteristics on 2-D oscillating bluff body", J. of Wind Eng. and Ind. Aerodyn., vol.46-47
Smagorinsky,J.S. (1963) "General circulation experiments with the primitive equations, Part1, Basic experiments", Mon.Weather Rev., 91, 99-164
Tamura, T.and Kuwahara, K. (1992) "Numerical Study on Aerodynamic Instability of Oscillating Rectangular Cylinders", J. of Wind Eng. and Ind Aerodyn.,vol.41-44, pp.253-254
Tennekes, H. and Lumley, J.L. (1972) "A First Course in Turbulence", the MIT Press
Vickery, B.J. (1966) "Fluctuating lift and drag on a long cylinder of square cross-section in a smooth and in a turbulent stream", J.Fluid Mech., vol.25, pp.481-494
Werner, H and Wengle, H. (1991) "Large-eddy simulation of turbulent flow over and around a cube in a plate channel", 8th symposium on turbulent shear flows, 19-4
Yoshizawa, A. and Horiuti, K. (1985) "Statistically-derived subgrid-scale kinetic energy model for the large-eddy simulation of turbulent flows". J. Phys. Soc. Japan. vol.54, No. 8, pp.2834-2839.
Yoshizawa, A. (1991) "Eddy-viscosity-type subgrid-scale model with a variable Smagorinsky coefficient and its relationship with the one-equation model in large eddy simulation",Phys. Fluid A3(8),pp2007-2009.

STATE OF THE ART VOLUME

FIELD MEASUREMENTS OF HIGH WINDS AND THEIR INTERPRETATION

ALAN P. JEARY
City University of Hong Kong, Hong Kong

SUMMARY

A review of measurements of the characteristics of strong winds is presented. Measurements and analysis techniques are discussed for velocities over 20 m/s. Major investigations of strong wind characteristics conducted since 1960, and their significance, are discussed.

INTRODUCTION

In this paper consideration is given to the occurrence of strong winds, to their measurement and to the interpretation of those measurements in the context of the way such data are used for the purpose of the design of structures.

In this paper high wind speeds are considered to be those that, at a height of 10 metres have a sustained mean in excess of 20 metres per second.

The number of full-scale measurements made of wind in the planetary boundary layer is relatively small, and those that are made fall into two broad categories. Firstly, most countries maintain a database of wind measurements, with anemometers placed at strategic locations. This generates the largest number of data about boundary layer winds. Secondly, and much more rarely, full-scale measurements of wind data in the field for specific purposes are sometimes mounted. These experiments are often for the purpose of investigating specific situations so as to ascertain the characteristics of the wind in such circumstances.

The distinction between light winds and strong winds has an important bearing on measurements as we shall see later. In particular the statistical techniques currently used, assume that stability exists in the atmosphere, both in terms of random fluctuations achieving stationarity, and of the atmosphere itself being neutrally stable. If the adiabatic lapse rate is -1°C/100 metres then the atmosphere is considered to be neutrally stable. This temperature gradient fundamentally affects the structure of wind in the boundary layer. In light winds, fluctuations in the flow may be supplemented by convective action or vertical transportation may be blocked by thermal inversion. These factors may have considerable influence on the turbulent fluctuations in the atmosphere under light winds.

There is considerable difficulty associated with the measurements themselves. Not only are there problems for the physical integrity of the anemometers, but also the precision of the measuring devices has been called into question. The recent attempts to use novel transducers have shown that the measurement of short term gusts (say of one second duration) has been impossible until the latter half of the last decade.

HISTORICAL PERSPECTIVE

The rotating cup anemometer was invented by John Robinson in 1846, and consisted of three or four rotating cups. This is still the most common form for the measurement of wind speeds in the boundary layer. The use of such devices was researched by W.H. Dines at the end of the 19th Century. His experiments with the simultaneous measurement of wind speeds close to the top of his house showed that the different anemometers did not indicate the same speeds, and so he progressively raised them until at a height of 4.5 metres above roof level they were all reading similar values.

During the last 50 years a revolution in data handling and collection has made possible enormous strides in field measurements of the behaviour of wind. A number of technological developments were made which have subsequently become common-place in the allied fields of meteorology and atmospheric physics. These developments have centred on the use of various types of laser, sonic or microwave devices for monitoring the movement of air. Many measurements are now made from orbiting satellites.

The major problem for measurements of the boundary layer using remote sensing is one of resolution. It is quite possible to use a microwave source and to monitor winds at different heights through the measurement of reflected images backscattered from different levels. The problem is that at present the minimum height at which such measurements can be made is 500 metres, and with a resolution of 500 metre intervals. At the time of writing the most secure forms of data capture inside the boundary layer are still those associated with readings that are made from a device that is supported from the ground. This effectively limits such readings to a height of 300 to 400 metres. Even so mounting experiments in which the structure of the wind is monitored throughout this height range is sufficiently difficult and expensive that the number of attempts is exceedingly small.

HIGH WIND SPEEDS

The highest speed ever measured was that registered at the USAF airforce base at Thule Greenland at which a peak gust of 92.5 m/s was monitored. It is generally reckoned that the windiest location on Earth is the Commonwealth Bay George V Coast of Antarctica. At this location a daily mean wind speed of 48.3 m/s was monitored on 21st/22nd March 1951.

Most countries maintain records for their local area. The original purpose of monitoring the wind speeds was to try to predict the forthcoming occurrence of very strong winds that could be dangerous to shipping. This situation is quite typical of the early impetus for the monitoring of wind activity and in turn reflects the fact that winds over the oceans are stronger than those over land areas. Early measurements then were more likely to be instituted by seafaring nations.

High surface winds are themselves a function of entire weather systems tracking over the surface of the planet or of locally induced phenomena. These mechanisms may include cyclonic weather systems, fronts and depressions, tornados, and thunderstorms and may be modified by such effects as Lee waves, or mechanisms associated with gravity or thermally induced movements of air in mountainous regions.

DESCRIPTION OF THE ATMOSPHERIC BOUNDARY LAYER

For the purposes of modern design it is necessary to describe the nature of the wind in terms of its structure at a location, and of its probability of recurrence. Perhaps the most important factor though, is that in keeping with the spirit of scientific inquiry, such models must be tested against full scale measurements and must earn their status through verification.

An impetus to the measurement of the structure of strong winds was given by the quantum leap in the conceptualisation of the processes at work through the work of Davenport in the early 1960's. The description of the nature of wind in terms of its structure close to the surface was the single most important concept in wind engineering during this century. Several experiments were quickly mounted to test the theory and the outline was quickly substantiated and enlarged by other significant contributions to the field.

These methods require a description of the variation of the mean wind profile with height, the turbulence associated with strong winds, the intensity of that turbulence, the gust maximum velocity, the relationship between the mean windspeed and the peak gust and the spectrum of wind gusts with respect to their frequency content. In addition, the size of gust eddies was of particular concern because there was an early understanding that the interaction of structures with the gusts was crucially dependent on the ratio of the structure to gust size. Ground friction forces, turbulent eddies and shear forces at the lowest levels are affected by increases in the mean wind velocity and measurements at high speeds are therefore essential.

THE STRUCTURE OF WINDS AND ITS MEASUREMENT

The atmospheric boundary layer in the first 2000 metres is now discussed and the parameters are considered in a little more detail together with comments on their measurement:

Velocity and direction

Measurements of the velocity of the wind and its direction of attack are seemingly straightforward. Unfortunately this is not the case. Despite the design of several different types of anemometer the rotating cup type mentioned above and introduced through the work of John Robinson, or the mechanically similar Gill-type propeller and vane type are easily the most robust types for use in rugged field environments.

Of other types introduced recently sonic devices that effectively measure wind velocity by measuring the time taken for a sonic pulse to travel between transmitter and receiver, are perhaps the most common. The harsh outdoor environment is particularly hard on researchers who do not respect the contrariness of nature.

The rotating cup or Gill type propeller anemometer is itself a mechanical device with mechanical properties. A force acting on such devices may be in error by as much as 2 m/s. Also reported (Maribo Pedersen et al 1992) is an error of as large as 6% in the reading obtained from a rotating cup anemometer if the mounting is within 15 diameters of the boom.

Similarly a fin behind a propeller or underneath a group of rotating cups also takes a finite time to respond to a change of direction. In this case a reducing force as the vane acquires the "correct" direction leads to a dead zone that increases the apparent variability of the direction of attack.

Radar type devices have recently been used to study the effect of winds higher in the atmosphere. Very recently the different techniques have resulted in devices that are close to being usable in the wind engineering field rather than purely for meteorological research purposes. An example is given of German experience (Vogt and Thomas 1993) in which a SODAR device has been developed and is usable to resolve wind speeds at 10 metre intervals from ground level upwards rather than from the 1 Km upwards previously attainable (Meek et al 1985). Unfortunately the work of Vogt and Thomas shows that the use of the acoustic Doppler SODAR device whilst promising is still prone to differences from more conventional measuring techniques of 30%.

Such devices can be used to generate an electrical analogue of both speed and velocity that can be converted into digital form and stored in a computer. Wind records of this type can then be scanned for both mean and peak data with appropriate averaging times.

If the wind speed and direction data are sampled at a rate of once a second, then for a single anemometer over 63 million data must be archived for one year's worth of readings. For this reason hourly mean data are normally archived, and the majority of high quality data is discarded.

Davenport's work of the early 60's opened the study of dynamic behaviour in which the gustiness of the wind must be defined if the effect on structures is to be predicted rationally. This description must, of necessity, be statistical in nature.

From the basic measurement of windspeed and of direction can be deduced, through computer analysis several important parameters, such as:

Intensity of Turbulence (I_u)
For a wind record the mean velocity is calculated appropriate to a preselected averaging period. The standard deviation of the fluctuations of the wind velocity about this mean value is then computed.

Gust Factor (K) is the ratio of the peak gust to the mean wind speed.

Peak Factor (g) is a related quantity and is expressed as $g = (K - 1)/ I_u$.
The provision of a record of wind data in a digital computer allows further analysis to be conducted and a series of other measures of the statistical properties are possible:

Spectrum of horizontal gustiness

Since the energy in the wind impinges on a structure through the mechanism of an interaction of gusts of wind with the structure then it is important to know how much energy is present as a function of frequency. A spectrum may be generated directly from wind speed data. Alternatively the spectrum of pressures may be presented through a linear transformation, or the ordinates may be weighted by the frequency to produce a nS(n) spectrum in which the power in the signal is represented as a function of frequency or of reduced frequency. The reason for presenting the data in this fashion is that in the nS(n) form the value of the energy in the wind can be fed directly into equations for the response of a structure to wind.

This implies that the response of a structure to a wind load can be thought of in terms of two parts. The first is essentially quasi static and may be thought of as the response to the mean wind. The second part is dynamic and is the structure's response to the sub inertia part of the spectrum at higher frequencies. There are two implications. As previously mentioned the measurement of the wind speed and its representation in the frequency domain requires the standard assumptions for spectral analysis to be true.

The data set forming the record of fluctuating wind speeds must therefore form a stationary sample if the spectrum of wind fluctuations as a function of frequency is to be valid. Secondly the separation of the sub inertial range of the spectrum from the inertial range must be valid. The spectral gap must exist in order to remove trends from the data sets. This spectral gap is itself assumed to exist largely as a result of investigations such as that at Brookhaven (Van Der Hoven 1957). The amount of decoupling between the microclimatological sub inertia regime, and the macroclimatological low frequency regime is exaggerated in such plots since the ordinate is multiplied by frequency.

In wind tunnel tests the separation of the micro and macro climatological parts of the spectrum is induced. This in turn creates a condition in which stationarity of the remaining turbulent spectrum is much easier to achieve. The investigator in the field has no such luxury. He must attempt to remove the non-stationary data before beginning the analysis. It is a salutary warning that in reviewing many of the largest studies of wind observations the mention of tests for stationarity is one of the rarest events reported. Some authors have attempted to circumvent the problem by using the autoregressive moving average method. However, it is difficult to determine variance errors in this way.

Vertical profiles

Measurements made at a series of vertically separated heights can provide a direct measurement of the vertical profile of the boundary layer either in terms of means or of gusts. Averaging over periods such as one hour will produce (normally) well behaved vertical velocity profiles. Turbulence intensity variation with height can also be presented in this manner. If the averaging process is reduced to the region of a few seconds then a gust profile may be constructed. In the field, vertical transportation of wind gusts tends to make the gust velocity profile and the short term turbulence intensity profile rather variable quantities. Nevertheless they are such important parameters that several attempts have been made to measure them.

In addition to the direct measurements noted above there are some inferences that can be made:

Integral length scales $^{x}L_{i}$, $^{y}L_{i}$, $^{z}L_{i}$

Integral length scales are a measure of the average size of a gust and are made in the x,y and z senses. There are nine integral scales, and they are of particular interest in scaling wind tunnel measurements.

The measurement of the integral length scale in the u direction can be achieved by integrating the autocorrelation function from a single point (or anemometer). The autocorrelation is itself the inverse Fourier transform of the spectrum. Alternatively, it can be determined by measuring the maximum in an nS(n) spectrum.

This is because the autocorrelation function effectively plots the correlation of a shaft of air moving past that point with itself. A zero value of the autocorrelation function would indicate that a coherent mass of air was exhausted at that distance and would therefore effectively define the limit of the coherent gust. Unfortunately the autocorrelation function correlates noise signals as well as the useful information and the function often does not decay to zero. Under such circumstances it is common practice to invoke Taylor's hypothesis to convert from an integral time to an integral length. The peak in a spectrum merely indicates the most likely wavelength associated with gusts. Energy at other wavelengths is also present.

Many difficulties have been experienced with the measurement of integral lengths. This is perhaps not surprising, given the fact that it is rare to test for stationarity, but additionally the autocorrelation function averages information at all frequencies present. Considering that gust sizes are associated with different frequencies in the microclimatological peak, and these are mixed with trends from low frequency energy, it is perhaps surprising that no attempts to measure integral lengths as a function of frequency appear in the literature.

Values of ${}^{y}L_i$ and ${}^{z}L_i$ may be obtained in a similar manner with the exception that cross correlations between anemometers in the vertical or the lateral sense must be performed.

The roughness length Z_o, The friction velocity U_*, Surface drag C_D can all be deduced from the measurement of mean and gust speeds, via simple relationships.

The parameters described above can be determined only if the atmosphere is well behaved, the data are stationary and a stable terrain exists. Clearly there are many situations when this is not the case yet measurements from such situations must still be sought.

DESCRIPTION OF STATISTICS FOR STRONG WINDS

Moving from the meteorological to the wind engineering domains, we encounter a field that draws heavily from both. The description of statistics about the return of strong winds normally gives as an end product, a description of a wind in terms of its probability of occurrence. To obtain this basic windspeed suitable for any particular location analysis of the wind data collected over a number of years is normally effected. Such techniques have developed and are epitomised by the description by Cook and Prior (1987). Recording stations are normally composed of a series of anemometers that are run not only by government agencies. The data must be corrected to an "effective height" before incorporating normally as measures of mean wind speeds averaged over a suitable period such as one hour. The statistics of extremes may then be used to provide a description of the likelihood of the occurrence of a particular windspeed at some location.

This basic technique can be supplemented to produce more efficient estimates drawn from fewer data. Firstly the data relevant to storms can be accumulated rather than for instance yearly data. Secondly, the Best Linear Unbiased Estimator (Lieblein BLUE) can be used to give more emphasis to the data for shorter periods thereby estimating the extremes more efficiently.

In general terms there seem to be some general principles that reduce confusion in the estimation of extremes. Firstly it is better not to mix mechanisms. If strong winds are generated by cyclonic weather systems and by thunderstorms, separation of the two types of data reduces errors. A study of thunderstorm activity (Twisdale and Vickery 1992) showed that when the mechanisms were separated then the winds generated by thunderstorms at the stations that are located far from the coast are underestimated if the other mechanisms are not separated before the analysis. The use of Fisher Tippett type I distributions are theoretically applicable to only one mechanism at a time. The use of a Fisher Tippett type II distribution to simulate mixed mechanisms is merely pragmatic.

There is also another very useful method for establishing windspeeds when a very strong storm has passed through an area causing damage or severe damage. This method that may be termed "reductionism" is epitomised by the approach after Cyclone Tracy (Walker 1975) or after Hurricane Hugo (Sparks 1992). In the latter Sparks assembled data from several military bases and airports mostly with anemometers at nonstandard heights to establish the probable area in which the strongest winds occurred. He then used the diaries of some of the local inhabitants to establish the most probable wind speeds. Sparks and his coauthors, commented that the observation of sea height was of particular use since its rise was correlated with the occurrence of the strongest winds. He was able to ascertain that at Bulls Bay in South Carolina the maximum gust was approximately 65 m/s, and the mean at the height of the storm was 38 m/s. Sparks and his coauthors have produced a list of types of damage correlated with gusts of various speeds in unprotected areas.

SOME SIGNIFICANT MEASUREMENTS

Davenport's contributions (1961, 1965) to the understanding of the structure of wind and in recommending an efficient way in which to represent this structure, were based on a review of measurements made in many different conditions and in several different countries and terrains. Accordingly he recommended values for the power law index, , the gradient height and the surface drag coefficient from his review of these measurements. He described three terrain categories with rising values for the three parameters in each case.

The early theoretical input coupled with references to many measurements led to several investigators mounting experiments to verify or modify the original descriptions.

Harris (1971) was able to use the Rugby radio station that is located in level terrain with little obstruction anywhere nearby to make measurements at four different heights. He then went on to extend the areas of investigation. Deaves and Harris (1980) extended this work with measurements conducted at Cardington and were to present considerable information about profiles, spectra, the variation of turbulence intensity with height and integral length scales. They showed integral length scales to be a function of the terrain roughness, although the difficulties with analysis of the autocorrelation functions was apparent.

Duchene Marullaz (1975) conducted a series of experiments at Nantes. There were three independent masts each of which had anemometers at 10, 20, 40 and 60 metres height. Measurements of Gust factors, Integral lengths, spectra, U_*, z_o, and were made.

Results were presented for two gales. The velocity profile was found to give a value of in the range of 0.3 to 0.36, z_o in the range 1.15 m to 1.65 metres and U_* in the range 1.6 to 2.2 m/s. Values of turbulence intensity were noted to reduce with increasing height dropping from 0.3 at 10 metres to 0.2 at 60 metres.

Integral length scales showed a consistent variation with height. However, the claim that the values obtained were similar to those measured by Harris (1971) was only true in that they were of the same order of magnitude. When the reported value of 0.3 to 0.36 was taken into consideration then the value of ${}^xL_u$ would be 190 metres rather than 300 metres at a height of 60 metres. Spectra had a similar form to that proposed originally by Davenport.

A large effort had, in the meantime been mounted in Hong Kong at Cape D'Aguilar (Mackey 1975). This consisted of a ten story experimental station located on the foreshore at the south easterly tip of Hong Kong. Velocity and direction measurements were obtained from five vertical lines of anemometers at heights of 13, 28, 43 and 61 m. Mackey and subsequently Choi (1983) reported on measurements made during five different typhoons. The mean velocity profile provided a steady value of 0.19 for wind speeds of 10 m/s up to nearly 40 m/s (Mackey 1975). However, in the case of the gust profile it was found that a time lag generally existed. A gust at the upper levels would be followed after a delay by a maximum at the lower levels. Plots of the vertical profile during typhoons when maxima are reached at each recording level were made. Instantaneous values of fitted to such profiles vary considerably (from 0.08 to 0.4). The implication is that is that during very strong winds there is considerable vertical transportation, and that if, over the duration of a storm, the maximum gusts as a function of height are presented then the gust velocity profile reaches an asymptote soon after a height of 10 metres. Despite problems with the analysis of autocorrelation functions, integral length scales were proposed which were claimed to agree very well with the Rugby measurements. However, these values are 3 to 4 times larger than those established later.

Ishizaki (1983) summarised a considerable amount of work conducted in Japan and which had until that time been published only in Japanese. The work of Arai et al conducted form 1964 to 1967 monitored six typhoons at six different heights near the coast, using aerovane anemometers. Of particular concern was an attempt to establish a relationship between the mean windspeed and (n) the inverse of , the power law exponent. There was so much variation in the relationship that the authors were forced to dismiss a rational relationship. Subsequently the work coordinated by the Architectural Institute of Japan at Tarama Island, Okinawa, between 1975 and 1978 was reported. Rotating Cup anemometers were used, and a calibration established that they followed a 4 second periodicity with about 90% accuracy. Once again an attempt to study the relationship between (n) and mean wind speed for typhoons was made, and on this occasion a tentative relationship was established. When the hourly mean values were taken then there is a linear increase in the value of (n) as the mean wind speed increases. There is, however, a good deal of scatter. Turbulence intensity showed considerable scatter at low wind speeds. However above 20 m/s the variation was limited to 8 to 15% and approached an asymptote of about 10%. This result was obtained from the tests at Tarama and was confirmed by other tests at Satoura from measurements made at a height of 40 metres. These data established a good linear relationship between the gust factor and I_u.

These data led Ishizaki to comment that turbulence is an inherent property of an individual typhoon. It is modified by the terrain roughness, but this is not essential to produce the turbulence. This was a very clear indication that each typhoon has its own character. Notwithstanding this the data were used as the basis for standard values for typhoon wind profiles and for values of , Z_o, and I_u for winds of mean speed up to 60 m/s at a height of 10 metres.

By the late 70's meteorologists established the velocity profile to a height of 10,000 metres (Powell 1979), through a combination of buoys, ships, aircraft and radar measuring devices. These measurements were generally able to discriminate only to 500 or 1000 metres height resolution, and in the absence of measurements in the first 500 metres the assumption was that the power law held for a neutral atmosphere.

The data from Wallops island (Tielman and Mullins 1979) were obtained at 15 metre intervals at 5 levels. The usual cup and vane anemometers were supplemented with the use of split film devices for the accurate measurement of turbulence properties. The measurement of temperature profiles was of particular significance in that the authors were able to identify low level inversions that created a separated regime. This in turn led, at certain times to the blocking of the low frequency wind excitation. The experiment addressed the problem of the measurement of integral length scales. A comparison of the measurement of integral lengths through the integration of the autocorrelation function and through the establishing of the wave number at the peak of the Von Karman spectrum led the authors to identify that the two methods only gave similar results when the blocking of the low frequency range winds occurred. Since the spectral method represents measurements only in the sub inertial higher frequency range and the autocorrelation method includes the low frequency data as well, the clear implication of this result is that much of the variation in previous (and subsequent) measurements is caused by a lack of checking for stationarity of data, since data recorded for only a short period and having significant low frequency components in them must test positive for trend-induced non-stationarity.

Measurements were made at Nakagawa and at Satowa (Shiotani and Iwatani 1979) at 5 levels, 130m to 150m from the beach and in an array of 5 towers with anemometers all located at 40 metres height in a 190 metre line along the coast respectively. Above 20 m/s at both sites the value of I_u closely approached 10% as an asymptote. The trend continued to 60 m/s.

Teunissen (1983) reported on experiments conducted under the aegis of the Atmospheric Environment Service into flow over three dimensional hills in neutrally stable conditions. This started with the Kettles hill experiment (Taylor et al 1983) and continued with the Askervein experiment. At Kettles hill ten ten metre towers were placed on the hill in a straight line array, and were supplemented by ten movable 3 metre tall towers. A sonic anemometer at the summit of the hill was used for turbulence measurements. Mean wind speeds in the range 15 to 20 m/s were measured. These experiments were essentially to underwrite the basis for design using either wind tunnels or the Jackson and Hunt formulation. Both of the modelling techniques were found to overestimate the full scale measurements, the wind tunnel by more than the mathematical model.

Maeda and Makino (1988) reported on measurements made in Fukuoka City and included measurements from the time of a typhoon. In this case the Gill-type anemometers were supplemented with the use of ultra sonic anemometers for the measurement of short term fluctuations. The measurements were made at a height of 38 metres and the typhoon maximum gust was 51 m/s with an hourly mean of 24.7

m/s. Measurements of integral scales and of the turbulence intensity were made using an Autoregressive technique to establish the spectral ordinates. The spectra for the U, V and W directions were compared with theoretical curves based on the Von Karman description of a spectrum for high wind speeds. The authors reported that in the u direction the agreement was good, but that in the v and w directions agreement was only obtained if values of ${}^{v}L_x$, and ${}^{w}L_x$ were substituted for ${}^{u}L_x$.

Naito (1988) reported on measurements taken from an offshore tower located 1 Km offshore with two sonic anemometers located 6.85m and 23 metres above sea level. The measurements were made for a large number of lower velocity winds and for a typhoon. During this period of stronger winds the characteristics of gust factor were very similar to the low velocity regime until there was large velocity excursion up to 50 m/s. The gust factor then rose from a range of 1.2 to 1.5 up to 2.0 to 3.2 for a period of about 50 minutes. Turbulence intensity measurements for high wind speeds showed small values and measurements of spectra showed the peak value to be considerably lower than for those made over land. The implication is that over the ocean large scale turbulence dominates.

Flay and Stevenson used eight towers over a 315 metre fetch to make measurements of integral lengths at windspeeds of up to 20 m/s. They observed that the use of the spectral peak method always gave values that were less than those obtained using the autocorrelation method, and went on to suggest ways in which the autocorrelation could be used to obtain similar results. The main method of measuring the integral time over the period up to a reduction to 1/e effectively excludes the very low frequency data, although there would also be an element of very high frequency as well. This high frequency noise comes from the electronics and correlates at all time delays leaving an unwanted constant offset in the autocorrelation function. The authors went on to show comparisons of results for ${}^{x}L_u$, ${}^{x}L_v$, ${}^{x}L_w$, ${}^{y}L_u$, ${}^{y}L_v$, ${}^{y}L_w$, with measurements quoted by Counihan, ESDU, Shiotani and Teunissen. Their claim that the comparison was reasonable must be seen within the context of 100% differences in some values.

Two towers in Tokyo at 31 metres and 55.7 metres height, and one in Yokohama at 86 metres were reported (Kato et al 1992). Ultrasonic transducers were used and the authors criticised the older mechanically based devices as not being capable of resolving short duration gusts. The authors show measurements of gust factors for 1 second averages although there is a great variability to these measurements at lower windspeeds. However, as the speed rises then the values of I_u and gust factor reach stable values. Additionally the U direction spectrum is quoted and agrees well with the Von Karman formulation.

Snaebjornsson et al (1992) presented measurements made using a 38 metre high tower with anemometers of the Gill-type located at 10 metres and 37 metres. The lava field that surrounded the site gave surprisingly low values for z_o despite the fact that the elements themselves were up to 2 metres in diameter. The maximum velocity measured was 29 m/s and this was interesting because many of the measured parameters settled down to relatively constant values after a velocity of 20 m/s was passed. I_u settled to 18% and I_v settled to 10%. The authors reported on integral length scales for u, v and w that settled to 250 metres, 125 metres and 37.5 metres respectively although the authors note that the variability in the measurements may be caused by non-stationarity of the data.

Miyashita et al (1993) reported on measurements from a communications tower in suburban Tokyo. Their measurements came from four heights (45, 80, 115 and 150 metres) at each of which were located three cup anemometers and a sonic anemometer. The authors selected a typhoon (typhoon 9011 August 8 1990) and a depression (October 11 1991) to make comparisons of the measurement of parameters in the two different regimes. The maximum mean windspeed at the 80 metre level in the typhoon was measured as 23 m/s whilst for the depression it was 15 m/s. The comparison showed that the value for remained unchanged, the mean value of I_u was similar whilst measurements of ${}^xL_u$ was variable in both cases. Power spectra for the u direction were similar in both cases and agreed well with the Von Karman formulation. The values measured for ${}^xL_u$ actually agree well with measurements made during the Askervein experiment.

There are many other measurements that have been made on winds and some of these have been notable (particularly Aylesbury, Askervein and at Texas Tech) however, all experiments are at the mercy of the random process associated with the generation of winds near the surface. The artificial selection of a cutoff point at 20 m/s in this consideration has excluded some of these measurements however they serve to show standards against which to compare measurements at higher wind speeds.

There are other types of strong winds that are not considered here. For instance tornados, with maximum windspeeds of 150 m/s and thunderstorms with maxima of

75 m/s present significant loading to structures. However, the study of such local phenomena is relatively young and details of structure are less well developed.

IN CONCLUSION

Much effort has gone into the measurement of wind data from full scale experiments so as to underpin theoretical models. Unfortunately it is clear that the atmosphere presents many difficulties to the investigator. In particular very low frequency activity often introduces bias and non stationarity into data records. More modern techniques to avoid such problems are available and are being used in isolated cases. Whilst the major components of parametric models are reasonably well defined for many different situations there still exist major problems to establishing consistent integral length scales. This problem exists not only because of problems with stationarity but also because the averaging process takes in different gust sizes on different occasions.

REFERENCES

Arai H., Hayashida C., and Tsuchiya J. (1962) "Study on Vertical distributions of wind velocities and evaluation of angles of sea wind", Railway technical Research Report no. 315 (In Japanese)

Cook N.J. and Prior M.J. (1987), "Extreme Wind Climate of the United Kingdom", JWEIA, 26, pp 371-389.

Choi E.C.C (1983), "Gradient height and velocity profile during typhoons", 6th ICWE, Australia.

Davenport A.G. (April 1961), "The spectrum of horizontal gustiness near the ground in high winds", Quarterly Jnl. of the Royal Met Society, 87, pp 194-211

Davenport A.G. (August 1961), "The application of statistical concepts to the wind loading of structures", Proc Civ. Engnrs., 19, 449.

Duchene-Marullaz P. (1975), "Full-scale measurements of atmospheric turbulence in a suburban area", Proc. of the 4th int'l Conf. on Wind Effects on Buildings and Structures. pp 23-32.

Deaves D.M. and Harris R.I. (1978), "A Mathematical Model Of the Structure of Strong Winds", CIRIA Report No. 76.

Davenport A.G.(1965), "The relationship of wind structure to wind loading", Paper No. 2 Proc. Conf on Wind Effects on Buildings and Structures. N.P.L. 1963

Flay R.G.J. and Stevenson D.C. (1988), "Integral Length Scales in Strong Winds below 20m", JWEIA, 28, pp 21-30.

Harris R.I. (October 1968), "On the spectrum and auto-correlation function of gustiness in high winds", ERA report 5273.

Harris R.I. (1971), "The nature of the wind". Paper 3. pp29-56. The Modern Design of Wind-sensitive structures. CIRIA.

Ishizaki H.(1983), "Wind Profiles, Turbulence Intensity and Gust Factors for Design in Typhoon-Prone Regions", 6ICWE, Australia.

Kato N., Okhuma T., Kim J.R., Marukawa H. and Niihori Y (1992), "Full scale Measurements of Wind Velocity in Two Urban Areas using an Ultrasonic Anemometer", 8ICWE, WEIA. 41-44 pp 67-78.

Miyashita K., Tamura Y., Asami Y. and Naito S. (1993), hird Asia/Pacific Symposium on Wind Engineering Dec 1993. Hong Kong. pp 423-428.

Mackey S. and Ko P.K.L (1975), "Spatial Configuation of Gusts", Proc. 4th Int'l Conference on Wind Effects on Buildings and Structures. pp 41-52.

Maeda J. and Makino M. (1988), "Power Spectra of Longitudinal and Lateral Wind Speed near the Ground in Strong Winds", JWEIA. pp 31-40.

Maribo Pedersen B., Hansen K.H., Oye S., Brinch M. and Fabian O (1992), "Some experimental investigations on the influence of the mounting arrangements on the accuracy of cup-anemometer measurements", JWEIA, 39 (1992) 373-383.

Meek C.E., Reid I.M., and Manson A.H. (1985), "Observation of mesosophic wind velocities", Radio Science V 20, No. 6, Nov-Dec. pp 1383-1402.

Naito G. (1988), "Turbulent Properties and Spectral Behaviours of Ocean Winds Observed at an offshore Tower", JWEIA, 28 pp 51-60.

Powell M.D. (1979), "An Evaluation of Diagnostic Marine Boundary Layer Models Applied to Tropical Cyclones", 5ICWE, Colorado. pp 133-144.

Shiotani M. and Iwatani Y. (1979), "Gust Factors over Flat Terrains and their Modification by a Barrier", 5ICWE, Colorado. pp 203-214.

Sparks P.R., Reid G.T., Reid W.D., Welsh S. and Welsh N. (1992), "Wind Conditions in Hurricane Hugo by Measurement, Inference, and Experience", JWEIA, 41-44 pp 55-66.

Snaebjornsson J.T. and Sigbjornsson R. (1992), "Wind Structure over Postglacial Lava Surface" , JWEIA 41-44. pp 305-315.

Teunissen H.W. (1983), "Wind-tunnel and Full-scale comparisons of mean wind flow over an isolated low hill", 6ICWE, Gold Coast. Australia. 1983.

Taylor P.A., Mickle R.E., Salmon J.R. and Teunissen H.W. (1983), "The Kettles Hill experiment: Site description and mean flow results", Report AQRB-83-001-L. AES, Ontario, Canada.

Tielman H.W. and Mullins S.E. (1979), "The Structure of Moderately Strong Winds at a Mid Atlantic Coastal Site (below 75m). 5ICWE, Fort Collins Colorado. pp 145-160.

Twisdale L.A and Vickery P.J. (1992), "Research on Thunderstorm Wind Design Parameters", 7th US Nat. Conf. on Wind Eng. & Int. Jnl. of Wind Engineering and Industrial Aerodynamics.

Van Der Hoven I. (1957), "Power Spectrum of horizontal wind speed in the frequency range from 0.0007 to 900 cycles per hour", Jnl. of Meteorology, 1957, Vol 14, pp 160-164.

Vogt S. and Thomas P. (1993), "SODAR A Useful Remote Sounder to Measure Wind and Turbulence", 3rd Asia Pacific Symp on Wind Engineering. pp 435-440.

Von Karman T: Proc Natl Acad Sci. USA (1948) **34** 530.

Walker G.R. (1975),"Report on Cyclone Tracy - Effects on Buildings - December 1974", Vol 3 Australian Dept. of Housing and Construction.

ESTIMATION OF EXTREME WIND SPEEDS

EMIL SIMIU

Building and Fire Research Laboratory,
National Institute of Standards and Technology,
Gaithersburg, MD 20899, USA

SUMMARY

Extreme wind loads used in design include nominal design wind loads (e.g., the 50-yr wind load) and ultimate wind loads. This paper briefly reviews the relationship between extreme wind loads and extreme wind speeds, assessments of epochal versus 'peak-over-threshold' approaches for estimating extreme non-tornadic winds in areas not subjected to tropical storms, and methods for estimating extremes from short records. Also reviewed are wind direction effects, and the estimation of extreme winds due to tropical cyclones (hurricanes) and tornadoes. We point out uncertainties due to model shortcomings and insufficient data, safety concerns due to current inconsistent uses of reliability concepts, and the implications of these concerns for code writing.

INTRODUCTION

A nominal design wind load is an extreme load with specified probability of being exceeded during a given time interval. In the United States that interval is usually 50 years. For example, for the inland Miami, Florida area, the ASCE Standard 7-93 (1993) specifies a nominal 50-year load based on a specified 50-year nominal design wind speed of 110 mph (49.17 m/s).

A structure or element thereof is expected to withstand loads substantially in excess of a 50- or 100-year wind load without loss of integrity. The wind load beyond which loss of integrity can be expected is referred to as ultimate wind load. The nominal ultimate strength provided for by the designer is based on an assumed ultimate wind load equal to the design wind load times a wind load factor. This statement is valid for the simple case where wind is the dominant load. It needs to be modified if load combinations are considered, but for clarity we refer here only to this case.

The load factor should be selected so that the probability of occurrence of the ultimate load is acceptably small. This probabilistic concept is important from an economic or insurance point of view. To the extent that evacuation or similar measures cannot be counted on to prevent loss of life, it is also important from a safety point of view.

A probabilistic approach has proven helpful in a number of cases, particularly for relative assessments of alternative design provisions, e.g., for mobile homes. However, in most cases the difficulties of obtaining wind load factors by probabilistic methods have proven to be substantial if not prohibitive. For this reason code writers have largely relied on wind load factors implicit in traditional codes and standards. For example, the ASCE Standard 7-93 specifies a wind load factor of 1.3 (e.g., the nominal ultimate wind load for the inland Miami area would correspond to a wind speed of $(1.3)^{1/2}110 = 125.4$ mph (56.06 m/s)).

Reliance on traditional code values is sometimes referred to as "calibration against existing practice." Traditional codes were generally adequate for many types of structures, but questions remain on whether safety margins implicit in those codes may be applied to modern structures, which can differ substantially from their predecessors in their materials and design/construction techniques. We discuss later an example involving wind direction effects.

Much effort has been and is being expended in an attempt to develop a practical probabilistic methodology for design. Probabilistic studies aimed at improving estimates of extreme wind loads are part of this effort. In this paper we review the relation between extreme wind loads and extreme wind speeds, and briefly discuss turbulence and wind direction effects on loads that don't entail significant dynamic amplification or aeroelastic effects. We also review the estimation by epochal and 'peaks-over threshold' methods of non-tornadic extreme winds in areas not subjected to tropical storms. To our knowledge this last topic has not yet been the object of a wind engineering state-of-the-art review. Its treatment in this paper is therefore more extensive than for such topics as the estimation of extreme winds from short records or the estimation of extreme winds due to tropical cyclones and tornadoes, that have been covered in some detail elsewhere (e.g., Simiu and Scanlan (1986), ASCE Committee on Wind Effects state-of-the art review (1987), Sill and Sparks (1991), and Marshall (1993), which contain numerous relevant references)

RELATION BETWEEN EXTREME WIND SPEEDS AND EXTREME WIND LOADS

Assume that the wind velocity is known at a particular location and elevation near a structure, where it is unaffected by any obstructions and is therefore indicative of the ambient wind environment. The wind pressure at a point on the building surface, or the wind force on a member, is a function of that wind speed and can be determined from results of wind tunnel or full-scale tests.

Turbulence and Flow Separation Effects.

The wind load on a particular member is obtained by integrating the pressures over the member's tributary area. Since, owing to turbulence and flow separation effects, pressures are time-dependent and imperfectly correlated spatially, with generally unknown correlation, the integration cannot be performed analytically, except for a very few simple cases, and is performed instead by a variety of techniques in wind tunnel or full-scale tests. When the tributary area is small correlation effects are relatively small, and the fluctuating force excursions may be many times larger than the standard deviation of the fluctuations. The magnitude of the force fluctuation to be specified for design purposes is an issue for which no clear and consistent reliability-based solution appears to be available at this time. This issue is complicated by questions on the extent to which loads measured on small scale models can provide satisfactory indications on the magnitude of their prototype counterparts, particularly if the results of interest involve large fluctuating pressure excursions.

Wind Direction Effects.

Pressures (and therefore forces) depend on both wind speed and direction. The dependence is of the form

$$p(\theta)=\tfrac{1}{2}\rho c(\theta)v(\theta)^2 \qquad (1)$$

where ρ, c, p, v and θ denote air density, the aerodynamic coefficient, pressure (or force), wind velocity, and direction, respectively. Two methods have been proposed for obtaining extremes of the vector $p(\theta)$. The first method relies on techniques for estimating the rate of upcrossing by $p(\theta)$ of a limit state $s(\theta)$ (Davenport, 1977). The second method is based on the creation of a set of i ($i=1,2,..,8$ or $i=1,2,..,16$) time series

$$p_j(\theta_i)=\tfrac{1}{2}\rho c(\theta_i)v_j(\theta_i)^2, \qquad j=1,2,..,N \qquad (2)$$

based on a set of i recorded time series $v_j(\theta_i)$, where θ_i are the eight or sixteen directions for which directional wind speeds are measured. From these sets of time series the single time series

$$w_j=\{\max_i[p_j(\theta_i)]\}^{1/2} \qquad (3)$$

is extracted. In Eq. 3 $\max_i$ denotes the maximum over all i's. To within a dimensional constant, w_j may be interpreted as an equivalent wind speed. The time series w_j consists of the largest equivalent wind speeds affecting the structure during the intervals $[t_{j-1}, t_j]$ ($j=1,2,..N$). It is subjected to a statistical analysis and yields the extreme values w, and therefore the extreme pressures (forces) acting on the structure, w^2, for the mean recurrence intervals of interest. For a numerical example, see Simiu and Scanlan (1986). For brevity we refer to the pressures $p=w^2$ as **actual** extreme pressures.

Extreme Loads Calculated Without Regard for Wind Direction Effects.

Extreme load estimation is simplified if the idealized time series

$$W_j = \tfrac{1}{2}\rho\{\max_i[c(\theta_i)]\ \max_i[v_j(\theta_i)^2]\}^{1/2}, \tag{4}$$

obtained by ignoring direction effects, is considered in lieu of the time series w_j. For any given mean recurrence interval, depending upon the directional dependence of the aerodynamic coefficient and the wind climate, the extreme values of the variable $P=W^2$ are in general larger — in many instances much larger — than the actual extreme pressures $p=w^2$. We refer here to the pressures P as **idealized** extreme pressures. Past experience shows that idealized extreme pressures P (based, say, on a 50- or 100-year extreme wind estimated without regard for direction), used in conjunction with a load factor of 1.3, normally result in acceptably small failure probabilities. However, use of the smaller actual extreme pressures p (with a 50- or 100-year mean return period) in conjunction with a load factor of 1.3 will result in higher failure probabilities that could well prove to be unacceptable and cannot be justified by invoking past experience.

This issue has not yet been adequately addressed by standard-writing bodies. For example, the ASCE Standard 7-93 as well as the draft ASCE Standard 7-95 allow nominal wind loads to be estimated on the basis of ad-hoc wind tunnel tests. For many special structures estimates of nominal wind loads based on such tests account for wind direction effects. However, the standard fails to indicate that the use of those estimates in conjunction with the wind load factor specified by the standard generally results in higher failure probabilities than those implicit in the provisions for ordinary structures. In this writer's opinion, which was duly communicated to the ASCE Subcommittee on the ASCE 7-95 Standard, this omission could have serious safety repercussions and deserves careful scrutiny.

A similar failure to address the reliability problem in a consistent fashion led recently to a strong increase in effective safety margins for window glass design, which in the writer's view is largely unwarranted. This issue is discussed in detail by Simiu and Hendrickson (1987).

ESTIMATION OF EXTREME WIND SPEEDS WITHOUT REGARD FOR DIRECTION

To within a constant dimensional factor, the time series W_j is the same as the time series $\max_i[v_j(\theta_i)]=V_j$ $(j=1,2,..,N)$. Extreme idealized wind loads can therefore be obtained from estimates of the extreme variate V inferred from this time series.

The vast majority of structural engineering calculations for wind are based on idealized extreme pressures, rather than actual extreme pressures. This state of affairs is due to: (1) the difficulty of codifying the estimation of actual extreme wind loads for most ordinary structures, (2) the generally inadequate availability of directional aerodynamic and wind climatological data, and (3) computational inconvenience. This last factor carries less weight in the age of personal computers, and it may be that in the near future expert systems with adequate data bases will increasingly allow directional effects to be accounted for in the estimation of wind loads (Simiu et al. 1993). Nevertheless, for the time being, estimating extremes wind speeds without regard for direction remains an important structural engineering problem.

The estimation of nominal design wind speeds (e.g., wind speeds with, say, a 50-year return period) is in general not unduly sensitive to the choice, within reasonable limits, of the statistical estimation procedure and the distributional form assumed to underlie the data. For example, the method of moments is inferior to the probability plot correlation coefficient (ppcc), but using it, instead of the ppcc, to estimate 50-year wind speeds entails errors of about 3 to 5 percent. Similar errors are inherent in the use of the assumption that a Fréchet distribution with tail length parameter $\gamma=9$, rather than a Gumbel distribution, best fits the data. However, if ultimate loads (or load factors) are of interest, the results can be sensitive to the choice of estimation procedure and distribution.

Extreme largest value distributions are, strictly speaking, valid only in the asymptotic limit of large extremes. It is nevertheless reasonable to assume that extreme winds are described probabilistically — at least approximately — by extreme largest value distributions. There are exactly three such distributions. In order of increasing tail lengths, they are the reverse Weibull distribution, the Gumbel distribution, and the Fréchet distribution. (The reverse Weibull and Fréchet are more properly referred to as families of distributions, each distribution being characterized by a particular value of the tail length parameter.) A remarkable feature of the reverse Weibull distribution is its finite upper tail.

The American National Standard A58.1-1972 (a predecessor of the current ASCE Standard 7-1993) was based on the assumption that extreme wind speeds are described by a Fréchet distribution with tail length parameter $\gamma=9$. As shown by subsequent studies, it may be confidently assumed that the Gumbel distribution — which is shorter-tailed than the Fréchet distribution with $\gamma=9$ — is a better probabilistic model of the extreme speeds (Simiu and Scanlan, 1986). However, even studies based on the Gumbel model result in apparently unrealistically high estimates of failure probabilities (Ellingwood et al., 1980). This may be explained in part by the fact that those studies do not adequately account for wind direction effects. However, an additional explanation may be that the extreme speeds are best fitted not by Gumbel distributions, which have infinite upper tails, but rather by reverse Weibull distributions which — like wind speeds in nature — have finite upper tails.

Recent substantial advances in extreme value theory appear to justify efforts to develop more realistic probabilistic models of extreme wind speeds and, consequently, more realistic wind load factors. This is likely to be true in spite of difficulties such as the limited availability of long-term data, the current insufficiency of comprehensive meteorological models available to the extreme wind analyst, and limitations inherent in statistical procedures. We describe some recent contributions to these efforts.

Classical Extreme Value Theory and 'Peaks over Threshold Methods.'

Classical extreme value theory is based on the analysis of data consisting of the largest value in each of a number of basic comparable sets called epochs (a set consisting, e.g., of a year of record, or of a sample of data of given size; in wind engineering, it has been customary to define epochs by calendar years). For independent, identically-distributed variates with cumulative distribution function F, the distribution of the largest of a set of n values is simply F^n. With proper choice of the constants a_n and b_n, and for reasonable F's, $F^n(a_n + b_n x)$ converges to a limiting distribution, known as the asymptotic distribution. As mentioned earlier, a notable result of the theory is that there exist only three types of asymptotic extreme largest value distributions, known, in order of decreasing tail length, as the Fréchet (or Fisher-Tippett Type II), Gumbel (Type I), and reverse Weibull (Type III) distributions (Lechner et al., 1993· Gross et al., 1994).

In contrast to classical theory, the theory developed in recent years makes it possible to analyze all data exceeding a specified threshold, regardless of whether they are the largest in the respective sets or not. An asymptotic distribution — the Generalized Pareto Distribution (GPD) — has been developed using the fact that exceedances of a sufficiently high threshold are rare events to which the Poisson distribution applies. The expression for the GPD is

$$G(y) = \text{Prob}[Y \leq y] = 1-\{[1+(cy/a)]^{-1/c}\} \qquad a>0,\ (1+(cy/a))>0 \qquad (5)$$

Equation 5 can be used to represent the conditional cumulative distribution of the excess Y = X - u of the variate X over the threshold u, given X > u for u sufficiently large (Pickands, 1975). c>0, c=0 and c<0 correspond respectively to Fréchet, Gumbel, and reverse Weibull (right tail-limited) limiting distributions. For c=0 the expression between braces is understood in a limiting sense as the exponential exp(-y/a) (Castillo, 1988, p. 215).

The peaks over threshold approach reflected in Eq. 5 can extend the size of the sample being analyzed. Consider, for example, two successive years in which the respective largest wind speeds were 30 m/s and 45 m/s, and assume that in the second year winds with speeds of 31 m/s, 37 m/s, 41 m/s and 44 m/s were also recorded, at dates separated by sufficiently long intervals (i.e., longer than a week, say) to view the data as independent. For the purposes of threshold theory the two years would supply six data points. The classical theory would make use of only two data points. In fact it may be argued that, by choosing a somewhat lower threshold, the number of data points used to estimate the parameters of the GPD could be considerably larger than six in our example.

Description of CME, Pickands and Dekkers-Einmahl-De Haan Methods.

Several methods have been proposed for estimating GPD parameters: the Conditional Mean Exceedance method (CME), the Pickands method, and the Dekkers–Einmahl–de Haan method (or, for brevity, the de Haan method).

Conditional Mean Exceedance (CME) Method. The CME (or mean residual life — MRL — as it it usually termed in biometric or reliability contexts) is the expectation of the amount by which a value exceeds a threshold u, conditional on that threshold being attained. If the exceedance data are fitted by the GPD model and $c < 1$, $u > 0$, and $a+uc > 0$, then the CME plot (i.e., CME vs. u) should follow a line with intercept $a/(1-c)$ and slope $c/(1-c)$ (Davisson and Smith, 1990). The linearity of the CME plot can thus be used as an indicator of the appropriateness of the GPD model, and both c and a can be estimated from the CME plot.

Pickands Method. Following Pickands' (1975) notation, let $X_{(1)} \geq \ldots \geq X_{(n)}$ denote the order statistics (ordered sample values) of a sample of size n. For s = 1, 2, .., [n/4] ([] denoting largest integer part of), one computes $F_s(x)$, the empirical estimate of the exceedance CDF

$$F(x;s) = \mathrm{Prob}(X-X_{(4s)} < x \mid X > X_{(4s)}) \tag{6}$$

and $G_s(x)$, the Generalized Pareto distribution, with a and c estimated by

$$\hat{c} = \frac{\log\{(X_{(s)} - X_{(2s)})/(X_{(2s)} - X_{(4s)})\}}{\log(2)} \tag{7}$$

$$\hat{a} = \frac{\hat{c}(X_{(2s)} - X_{(4s)})}{2^{\hat{c}} - 1} \tag{8}$$

One takes for Pickands estimators of c and a those values which minimize (for $1 < s < [n/4]$) the maximum distance between the empirical exceedance CDF and the GPD model.

Following a critique of an earlier implementation of the Pickands method (Pickands, 1975, Castillo, 1988), an alternative implementation was developed (Lechner et al., 1991), which entailed the following steps: (1) choose as threshold u an order statistic of the sample; (2) compute the empirical exceedance CDF for the data above u; (3) nonlinear least–squares fit the GPD model for the parameters c and a; (4) plot the resulting c estimates against u for each order statistic. If the plot of $\hat{c}$ is stable around some horizontal level for most of the order statistic thresholds plotted, then the plot is presumptive evidence for the GPD model being applicable and can be used to yield numerical estimates of c; the distribution is Weibull, Fréchet or Gumbel according as c is negative, positive, or fluctuates around zero. The approach just described was suggested by Bingham (1990).

De Haan Method. Recent work by de Haan (1994) and coworkers provides a moment-based estimator which, like Pickands' estimator, is asymptotically unbiased for the true tail parameter and, in addition, is asymptotically normal. We now describe this estimator, using the order-statistic notation introduced above.

Let n denote the total number of data and k the number of data above the threshold u. (Note that u is then the (k+1)-th highest data point.) Compute, for r=1 and r=2, the quantities

$$M_n^{(r)} = \frac{1}{k} \sum_{i=0}^{k-1} [\log X_{n-i,n} - \log X_{n-k,n}]^r \tag{9}$$

where $X_{n-i,n}$ denotes the (i+1)-th highest value in the set (Note that $X_{n-k,n}=u$.) The estimators of a and c are then

$$\hat{a} = uM_n^{(1)}/\rho_1 \qquad (\rho_1=1 \text{ for } \hat{c}\geq 0;\ \rho_1=1/(1-\hat{c}) \text{ for } \hat{c}<0) \tag{10a}$$

$$\hat{c} = M_n^{(1)} + 1 - \frac{1}{2\{1 - (M_n^{(1)})^2/M_n^{(2)}\}} \tag{10b}$$

The standard deviation of the asymptotically normal estimator of $\hat{c}$ is

$$\text{s.d.}(\hat{c}) = [(1+\hat{c}^2)/k]^{1/2} \qquad \hat{c}\geq 0 \tag{11a}$$

$$\text{s.d.}(\hat{c}) = \{1/k[(1-\hat{c})^2(1-2\hat{c})(4-8(1-2\hat{c})/(1-3\hat{c})+(5-11\hat{c})(1-2\hat{c})/(1-3\hat{c})/ (1-4\hat{c}))]\}^{1/2} \qquad \hat{c}<0 \tag{11b}$$

Estimation of Variates with Specified Mean Recurrence Intervals.

For wind engineering purposes the estimates of the wind speeds corresponding to various mean recurrence intervals are of interest. We give expressions that allow the estimation from the GPD of the value of the variate corresponding to any percentage point $1 - 1/(\lambda R)$, where λ is the mean crossing rate of the threshold u per year (i.e., the average number of data points above the threshold u per year), and R is the mean recurrence interval in years. Set

$$\text{Prob}(Y < y) = 1 - 1/(\lambda R) \tag{12}$$

From Eqs. 5 and 12, we have

$$1 - [1 + cy/a]^{-1/c} = 1 - 1/(\lambda R) \tag{13}$$

Therefore

$$y = -a[1 - (\lambda R)^c]/c \tag{13}$$

The value being sought is

$$x_R = y + u \quad (14)$$

where u is the threshold used in the estimation of c and a.

Relations Between Distribution Parameters and Expected Value and Standard Deviation.

Relations between distribution parameters and the expectation E(X) and the standard deviation s(X) for the Gumbel and reverse Weibull distribution are given below. (Subscripts G and W refer to the Gumbel and reverse Weibull distributions, $F_G(x)$ and $F_W(x)$, respectively.)

$$F_G(x) = \exp\{-\exp[-(x-\mu_G)/\sigma_G]\} \quad (15)$$

$$F_W(x) = \exp\{-[(\mu_W - x)/\sigma_W]^\gamma\}, \quad x < \mu_W \quad (16)$$

$$\sigma_G = (6^{1/2}/\pi)s(X) \quad (17)$$

$$\mu_G = E(X) - 0.57722\sigma_G \quad (18)$$

$$E[(X-\mu_W)/\sigma_W] = -\Gamma(1 + 1/\gamma) \quad (19)$$

$$s[(X-\mu_W)/\sigma_W] = \{\Gamma(1+2/\gamma)-[\Gamma(1+1/\gamma)]^2\}^{1/2} \quad (20)$$

where Γ is the gamma function (Johnson and Kotz, 1972). For the GPD,

$$E(X) = a/(1-c) \quad (21)$$

$$s(X) = a/\{(1-c)(1-2c)^{1/2}\} \quad (22)$$

(Hosking and Wallis, 1987).

Results of Monte Carlo Simulations.

Preliminary Monte Carlo studies reported by Gross et al. (1994) led to the following tentative conclusions:

Comparison of Estimation Methods. The CME and the de Haan methods are competitive. Both methods are superior to the Pickands method. The de Haan method gives better estimates than the CME method for extremes with large mean recurrence intervals. Note, however, that the de Haan method as described in Gross et al. (1994) was based on the de Haan estimator of the parameter c (Eq. 10b), and on an estimator of the parameter a less precise than Eq. 10b. For this reason it is reasonable to expect that the de Haan method that makes use of both Eqs. 10a and 10b performs better than the CME method.

Optimal Crossing Rate. A high threshold reduces the bias since it conforms best with the asymptotic assumption on which the GPD distribution is based; however, because it results in a small number of data, it increases the sampling error. It appears that, with no significant error, an approximately optimal threshold corresponds to a mean exceedance rate of 5/yr to 15/yr.

(We note here a typographical error in Gross et al. (1994): in Table 5 the population value for c should be -0.275, instead of -0.5 -- cf. p. 142, line 6 of Gross et al. (1994).)

Results of Extreme Wind Speed Analyses.

Results of analyses performed on sets of about 20 to 45 yearly maximum wind speeds recorded at various U.S. sites were reported by Lechner et al. (1992). About one hundred data samples of size 20 to 45 years recorded at stations not affected by hurricanes were analyzed by the CME procedure. For more than two-thirds of the samples the c values estimated by the modified Pickands method were negative. The same data were recently analyzed by Simiu and Heckert (1995) by using the de Haan method. These analyses confirmed the results of Lechner et al (1992). However, because the number of data available in these samples is small, especially for large thresholds, the confidence bands for the estimates tend to be relatively wide.

Analyses were also done for 48 sets of daily data records with lengths 15 to 24 years. As explained in Gross et al. (1995), the number of data in the sets was reduced by a factor of four to decrease the effect of correlation due to wind speeds recorded in the same storm. Results based on de Haan's method (Eq. 10a) -- as opposed to the more inconclusive results based on the CME method reported by Gross et al. (1995) -- showed an unmistakable tendency of the estimated values of c to be negative Simiu and Heckert, 1995). These results are significant. They provide evidence that extreme value statistics reflect the physical fact that wind speeds are bounded. However, it appears that dependable quantitative information for use in structural reliability estimates and the development of wind load factors for building standards would require larger data sets than are presently available. We note that, using a different approach, Kanda (1994) also showed that extreme winds are best fitted by distributions with limited tails. See also Walshaw (1994).

Sampling Errors in Estimation of Extreme Wind Speeds.

Estimates of sampling errors are available under the assumption that the extreme annual wind speeds have a Gumbel distribution -- see Simiu and Scanlan (1986, p. 87). Based on that assumption, the standard deviation of the sampling errors was estimated to be about 5 to 10 percent of the wind speeds obtained from an approximately 30-year long sample of maximum yearly data. Sampling errors so estimated are acceptable approximations for use in reliability calculations.

Gust Wind Speeds versus Fastest Mile Speeds.

Peterka (1992) reported results of extreme wind analyses based on peak gust, as opposed to fastest-mile, records, and used a technique to reduce variability due to sampling error by combining stations with short records into "superstations" with long records. The acceptability of this technique is a function of the degree of mutual dependence of the storm occurrences at the various stations being consolidated.

ESTIMATES OF EXTREME WIND SPEEDS FROM SHORT RECORDS

A procedure for estimating extreme wind speeds without regard for direction at locations where long-term data are not available was reported by Simiu et al. (1982) and Grigoriu (1984). The method, whose applicability was tested for 36 U.S. stations and a total of 67 three-year records, makes it possible to infer the approximate probabilistic behavior of extreme winds from data consisting of the largest monthly wind speeds recorded over a period of three years or longer. Estimators of the wind speed with an N-year mean recurrence interval and of the corresponding standard deviation of the sampling errors are given in Simiu and Scanlan (1986, p. 91).

Inferences concerning the probabilistic model of the extreme wind climate have also been attempted from data consisting of largest daily or largest hourly wind speeds by the authors just quoted and by Gusella (1991). Using such data raises two questions. First, what is the effect on the analysis of the mutual correlation among daily or hourly data? According to an estimate by Grigoriu (1982) that effect is tolerably small. Second, what is the effect of basing the inferences on data that are overwhelmingly representative of weak winds having little in common metorologically with the extreme winds of interest? According to preliminary results by Gross et al. (1995), weak winds may be viewed as noise obscuring the process of interest, rather than providing useful information on wind extremes.

ESTIMATES OF HURRICANE/TROPICAL CYCLONE WIND SPEEDS

In tropical-cyclone-prone regions the winds of interest to the structural engineer are primarily those associated with hurricanes (strong tropical cyclones). Statistical analyses of hurricane winds would therefore be necessary. However, the number of hurricane wind speed data at any one location is in most cases small. The confidence limits for predictions based on hurricane wind speed data at one location would, in general, be uncceptably wide.

For this reason estimates of hurricane wind speeds at a site are obtained indirectly from statistical information on the climatological characteristics of hurricanes, used in conjunction with a physical model of the hurricane wind field. Such a model allows the estimation of maximum wind speeds induced at any given location by a hurricane for which the following climatological characteristics are specified:

-- difference between atmospheric pressures at the center and the periphery of the storm
-- radius of maximum wind speeds

-- speed of storm motion
-- coordinate of crossing point along the coast or on a line normal to the coast.

The probability distribution of the hurricane wind speeds is then estimated as follows:

(1) A region is defined such that hurricanes occurring outside that region have a negligible effect at the site of concern. (In the United States such a region includes 750 km of coastline, say, and a 450 km segment over the ocean, normal to the coast.)

(2) The climatological characteristics of the hurricane, including the frequency of hurricane occurrences in this region, are modeled probabilistically from statistical data obtained in the region under consideration.

(3) The values of the climatological characteristics for a number, n, of hurricanes are obtained by Monte Carlo simulation from these probabilistic models.

(4) The maximum wind speeds, v_i, ($i=1,2,\ldots,n$) induced by each of these hurricanes at the location of concern are calculated on the basis of the climatological characteristics thus obtained and of the physical model of the hurricane wind field, including a model for the decay of the storm as it travels over land. Thus, a set of n hurricane wind speeds is calculated, which is consistent with the statistical data on climatological characteristics of hurricanes in the region of interest.

(5) A statistical estimation procedure is applied to the calculated hurricane wind speeds in order to estimate the probability distribution of the hurricane wind speeds at the location being considered.

The procedure just oulined was first developed by Russell (1971), and was applied with various modifications by, among others, Batts et al. (1980a) and Georgiou et al. (1983), whose respective estimates for the Gulf coast and the East coast of the United States are compared in Simiu and Scanlan (1986). Recent work by Vickery and Twisdale (1994) was aimed at improving certain aspects of the estimates by Batts et al. (1980), and resulted in lower estimates of hurricane wind speeds inland. Models of the ratio between peak gusts and sustained winds in hurricanes used by Batts et al. (1980) were based on data for extratropical storms. Improved models based on hurricane wind speed data have been proposed by Krayer and Marshall (1992). The issue of sampling errors in the estimation of hurricane wind speeds was examined by means of Monte Carlo simulations (Batts et al., 1980b), which showed that the standard deviation of the errors was typically 5 to 10 percent of the estimated speeds. A computer program for estimating direction effects in hurricanes is described in Hendrickson and Simiu (1986).

Lessons drawn from observations of characteristics and effects of hurricanes Hugo and Andrew are the object of studies by Sill and Sparks (1991), Reinhold et al. (1993), and Marshall (1993). The latter study reviews in detail, and contains recommendations for the improvement of, wind load provisions for manufactured home construction.

TORNADO WINDS

The approach to the estimation of hurricane winds discussed in the preceding section could be described, roughly, as the monitoring of a billiard game played by nature with balls in the form of hurricanes having complex physical and probabilistic characteristics, against various targets such as towns. A similar approach is used for tornado winds.

Let A_0 be some large reference area, and let n be the estimated number of tornado occurrences per year in area A_0. The estimated probability that a tornado will strike a particular location within A_0 in any one year is asssumed to be $P(H) = n\ a/A_0$, where a = estimated mean area of individual tornado path. The probability that a tornado with maximum wind speeds higher than some specified value V_0 will strike a location in any one year can be written as $P(H,V_0)=P(V_0)P(H)$, where $P(V_0)$ is the probability that the maximum wind speed is at least V_0, given that a tornado has occurred. The estimation of $P(H,V_0)$ relies on relatively few and uncertain data — inferred mostly from observations of damage — on tornado occurrences, path areas, and wind speeds, and on largely subjective extrapolations from these data to small probability levels. According to estimates presented by Markee et al. (1974), maximum tornado wind speeds corresponding to a probability $P(H,V_0)=10^{-7}$ in any one year vary between 400 mph (179 m/s) in Oklahoma and Nebraska to 240 mph (107 m/s) in Northern California and Oregon. More refined estimates have subsequently been proposed (see Simiu and Scanlan (1986) for a summary of codified estimated tornado winds, pressure drops and pressure drop rates in the United States, and information on estimated tornado-borne missile speeds, and for additional references).

REFERENCES

American Society of Civil Engineers (1987), **Wind Loading and Wind-Induced Structural Response**, A state-of-the-art report prepared by the Committee on Wind Effects.

ASCE 7-93 (1993), **Minimum Design Loads for Buildings and Other Structures**, American Society of Civil Engineers, New York.

Batts, M.E., Russell, L.R., and Simiu, E. (1980), "Hurricane Wind Speeds in the United States," **Journal of the Structural Div., ASCE, 106**, 2001-2015.

Batts, M.E., Simiu, E. and Cordes, M.R., (1980), "Sampling Errors in Estimation of Extreme Hurricane Winds," **Journal of the Structural Division, ASCE, 106**, 2111-2115.

Bingham, N.H., "Discussion of the Paper by Davisson and Smith,"

Journal of the Royal Statistical Society , series B, 52, p. 431.
Castillo, E. (1988), **Extreme Value Theory in Engineering**, Academic Press, New York.
Davenport, A.G. (1977), "The Prediction of Risk Under Wind Loading," **Proceedings 2nd Intern. Conf. on Stroctural Safety and Reliability**, Munich, 511-538.
Davisson, A.C., and Smith, R.L. (1990), "Models of Exceedances Over High Thresholds," **Journal of the Royal Statistical Society**, Series B, **52**, pp. 339-442.
De Haan, L. (1994), "Extreme Value Statistics," **Extreme Value Theory and Applications**, Galambos, J., Lechner, J. and Simiu, E., (eds.), Kluwer, Boston.
Ellingwood, B.R. et al. (1980), **Development of a Probability-Based Load Criterion for American National Standard A58**, NBS Special Publication 577, National Bureau of Standards, Washington, DC.
Georgiou, P.N., Davenport, A.G. and Vickery, B.J. (1983), "Design Wind Speeds in Regions Dominated by Tropical Cyclones," **Journal of Wind Engineering and Ind. Aerodynamics**, 13, 139-152.
Grigoriu, M. (1982), "Estimates of Design Winds from Short Records," **Journal of the Structural Div., ASCE**, **108**, 1034-1048.
Grigoriu, M. (1984), "Estimation of Extreme Winds from Short Records," **Journal of Structural Engineering**, **110**, 1467-1484.
Gross, J., Heckert, A., Lechner, J. and Simiu, E. (1994), "Novel Extreme Value Estimation Procedures: Application to Extreme Wind Data," **Extreme Value Theory and Applications**, Galambos, J., Lechner, J. and Simiu, E., (eds.), Kluwer, Boston.
Gross, J., Simiu, E., Heckert, N.A. and Lechner, J.A. (1995), "A Study of Optimal Extreme Wind Estimation Procedures," **Proceedings, Ninth International Wind Engineering Conference**, New Delhi.
Gusella, V. (1991), "Estimation of Extreme Wind Speeds from Short-term Records," **Journal of Structural Engineering, 117**, 375-390.
Hendrickson, E.M. and Simiu, E. (1986), **Directional Hurricane Wind Effects**, NBSIR 86-3317, National Bureau of Standards, Gaithersburg, MD.
Hosking, J.R.M. and Wallis, J.R. (1987), "Parameter and Quantile Estimation for the Generalized Pareto Distribution," **Technometrics**, **29**, pp. 339-349.
Kanda, J. (1994), "Application of an Empirical Extreme Value Distribution to Load Models," **J. Res. Natl. Inst. Stand. Technol.**, **99**, 413-420.
Krayer, W.J. and Marshall, R.D. (1992), "Gust Factors Applied to Hurricane Winds," **Bulletin of the American Meteorological Society**, **73**, 613-617.
Lechner, J.A., Leigh, S.D., and Simiu, E. (1991), "Recent Approaches to Extreme Value Estimation with Application to Extreme Wind Speeds..," **Journal of Wind Engineering and Industrial Aerodynamics**, 41-44, 509-519.
Lechner, J.A., Simiu, E. and Heckert, N.A. (1993), 'Assessment of 'Peaks over Threshold' Methods for Estimating Extreme Value Distribution Tails," **Structural Safety**, **12**, 305-314.
Markee, E.H., Beckerley, J.G., and Sanders, K.E. (1974), **Technical Basis for Interim Regional Tornado Criteria**, WASH 1300, U.S. Atomic Energy Commission, Washington, DC.

Marshall, R.D. (1993), **Wind Load Provisions of the Manufactured Home Construction and Safety Standards -- A Review and Recommendations for Improvement**, NISTIR 5189, NationalInstitute of Standards and Technology, Gaithersburg, MD.

Peterka, J.A. (1992), "Improved Extreme Wind Prediction for the United States," **Journal of Wind Engineering and Industrial Aerodynamics**, 41-44, 533-541.

Pickands, J. (1975), "Statistical Inference Using Order Statistics," **Annals of Statistics**, 3, 119-131.

Reinhold, T.A., Sill, B.L., Vickery, P.J., and Powell, M., (1993), Wind Speeds in Hurricane Andrew: Myths and Reality, **Conference Proceedings, The 7th U.S. National Conference on Wind Engineering**, Vol. 2, (G.C. Hart, ed.), University of California at Los Angeles.

Russell, L.R. (1971), "Probability Distributions for Hurricane Effects," **Journal of Waterways, Harbors, and Coastal Engineering Division**, **97**, 139-154.

Sill, B.L.and Sparks, P.R. (eds.) (1991), **Hurricane Hugo One Year Later**, American Society of Civil Engineers, New York, N.Y.

Simiu, E., Filliben, J.J. and Shaver, J.R. (1982), "Short-term Records and Extreme Wind Speeds," **Journal of the Structural Division, ASCE**, **108**, 2571-2577.

Simiu, E., and Scanlan, R.H. (1986), **Wind Effects on Structures**, Second Edition, Wiley-Interscience, New York.

Simiu, E., and Hendrickson, E. (1987), "Design Criteria for Glass Cladding Subjected to Wind Loads," **J. Str. Eng.**, **113**, 1199-1203.

Simiu, E., Garrett, J.H., and Reed, K.A. (1993), "Development of Computer-based Models of Standards and Attendant Knowledge-base and Procedural Systems," in **Structural Engineering in Natural Hazards Mitigation**, ASCE Struct. Congr., Irvine, CA., 841-946.

Simiu, E., and Heckert, N.A. (1995), **Extreme Wind Distribution Tails**, Building and Fire Research Laboratory, National Institute of Standards and Technology (in preparation).

Vickery, P.J. and Twisdale, L.A. (1994), "Windfield and Filling Models for Hurricane Windspeed Predictions," submitted to Journal of Structural Engineering.

Walshaw, D. (1994), "Getting the most from your extreme wind data: a step-by-step guide," **J. Res. Natl. Inst. Stand. Technol.**, **99**, 399-412.

INSTRUMENTATION OF PROTOTYPE STRUCTURES

H.P. RUSCHEWEYH
Prof. Dr.-Ing., Department of Wind Engineering
Institut for Steel Structures,
Technical University Aachen, Germany

SUMMARY

Full-scale measurements are described as an important step for the progress in wind engineering knowledge. The whole field of application is indicated. The paper is then concentrated on the instrumentation of prototype structures which has two aspects:

With the modern equipments the researcher is able to design full-scale tests for reliable results, but the handling on site is mostly difficult.

A list of advise based on experiences of many full-scale tests on different structures is given. This list indicates the most important points for the installation and the handling of a full-scale measuring device.

INTRODUCTION

The development of Wind Engineering in the past 30 to 40 years has been enhanced by many groups of scientists and engineers round the world.

Numerous successful conferences (Davenport (1974), IAWE (1963)) indicate the progress of knowledge in that field. Calculation procedures have been developed to predict the response of the structure under wind actions and to describe environmental condition in the wind field.

Nevertheless, all of these high sophisticated procedures and theoretical studies have to be verified by full-scale observations. Only the real structure affected by the wind shows whether the prediction procedure gives realistic values or not. The question arises: Is the calculation procedure on the safe side or is it unsafe? Is it too safe and does it produce an uneconomically design?

Therefore, from time to time full-scale measurements on real structures have to be made. It should be done after theoretical and physical progress but they are also necessary to give a starting impulse for new investigations. An example for the first point is the verification of vortex-induced vibrations of stacks which have shown that some of theoretical predictions developed in the past are far away from reality while other procedures are very close to the full-scale data (Ruscheweyh (1974, 1988), Vickery (1990)).

An example for a starting point for investigations is given by the full-scale measurements on natural draught cooling towers (Niemann (1971), Farell (1974), Ruscheweyh (1974)) which gave results for better Reynolds-number simulation in wind tunnel tests.

Natural wind field studies cannot be done without full-scale measurements. The very famous wind measurements of van der Hoven (1957) became a fundamental source for the understanding of the dynamic characteristics of the wind flow.

Waste gas distribution behind buildings has been simulated by numerical calculations and by wind tunnel tests. High reliable full-scale verifications are not available yet, so that a relatively great unreliability exists in that field.

The full-scale measurements may be devided into the following categories:

- field measurements of wind characteristics
- field measurements of air pollution
- field measurements of wind comfort in passenger zones and for city ventilation
- wind pressure measurements on buildings and structures
- wind force measurements on buildings and structures and on structural elements
- vibration measurements including damping measurements.

This paper is restricted to the instrumentation of prototype structures, i.e. field measurements of the wind, air pollution and wind comfort will not be explained.

SOME HISTORICAL ASPECTS UP TO MODERN TECHNICS

The possibility and accuracy of full-scale measurements strongly depends on the availability and sensibility of the measuring devices. In earlier times, no electric pressure transducers were available, i.e. wind pressure measurements could not be done or the result was limited to only few measuring points with pour accuracy. To overcome this problem, (Mayne (1970) used self-made pressure transducers for his measurements of a building and on the TV tower London, UK. This type of pressure transducer has been modified by Ruschẹweyh (1974) for his full-scale measurements on the TV tower Hamburg and on the cooling towers in Scholven, Germany (Ebner (1974)) and Gösgen, Switzerland (Ruscheweyh (1983)). Today several commercial pressure transducers are available even for low costs, which have a sufficient sensitivity for wind pressure measurements.

Parallel to this development the computer capacity has increased in such an enormous way that the handling of the large number of datas is no more a problem.

In a similar way the measurement technique for vibration measurements has been developed. In the early 50th and in the 60th high sensitive seismic accelerometers have been used for building vibrations. Later on electronic correction circuits have been designed for the transducers, accelerometers became smaller and the frequency range has been expanded.

A problem is still the very low natural frequency of very tall buildings or long span bridges because the sensitivity of the seismic-type accelerometers decreases with the square of decreasing frequency:

$$\ddot{y} = \omega^2 y \qquad (1)$$

A new type of measuring device, a Laser beam device, may solve this problem. A modern type of Laser beam device is applied in a current project for the static and dynamic deflection of a guyed mast in the test field of the University in Aachen, Germany (Ruscheweyh 1994).

For the measurement of the reference wind speed and wind direction traditional cup or fan anemometers have been used and are still in operation. For most cases, these instruments are sufficient.

For the analysis of higher frequencies, higher than 1 Hz, the inertia of such anemometers is too large. For the requirement of high frequency analysis sonic anemometers are now available with a frequency range up to 10 Hz and more and for three components of the wind.

Summarizing the development of instruments we can say that for a modern full-scale measurement a high sophisticated technique is available to solve most of the problems. The only problem which stands against this modern technique in many cases is the financial cost level. Unfortunately for wind engineering measurements the budgets are often extremely limited, so that a possible progress is restricted.

APPLICATION TECHNIQUE TO BUILDINGS AND STRUCTURES

The one side of the full-scale measurements is the instrument-technique the otherside is the application to the structure. The last one very often produces problems which seem to be simple but they are not simple. The following list indicates the most important points which have to be considered for the preparation and during the execution of full-scale tests:

the measurement of the reference wind speed requires an undisturbed location for the anemometer nearby the structure. "Nearby" means on the other hand, that there is always an interference effect with the structure. A calibration of the anemometer output is necessary. In most a second or third anemometer at a different location is necessary.

For pressure measurements an undisturbed reference pressure has to be available. It is very often difficult or impossible to realize this requirement. For a free-standing building one can install a protected hole on the ground surface and connect this point with the measuring level in the height above ground. The reference pressure then is correct, if the temperature distributions within the tube and the outside atmosphere are equal. If not, a correction has to be made by measuring the temperature along the connecting tube.

If the building is situated within a built up surrounding, the reference pressure can be taken from a measuring point above the top of the building. The distance to the top of the building must be large enough in order to be "undisturbed" by the flow around the building.

Pressure measurements on a circular cylindrical structure (towers, chimneys) can be referred to a reference pressure which can be calculated from the characteristic of the pressure distribution assuming the "zero crossing angle" which depends on the Reynold-number. This procedure has been successfully applied to the full scale measurements on the TV tower Hamburg (Ruscheweyh, 1974).

If no reference pressure is available only pressure differences on the building wall can be analysed.

For pressure measurements the pressure holes in the surface of the structure have to be large enough. Otherwise water and/or dust can obstract the measuring point. A diameter of 10 mm should be used as a minimum.

One has to take care that water which has been pressed into the hole by the wind, can flow out easily.

Water should not effect the electrical circuit.

Icing can be a problem depending on the site of the structure.

Protection against lightning should be made at towers and masts, if the measurements include the top of the tower.

For long term measurements the zero drift of the measurement equipment has to be taken into account. This is of minor importance for dynamic measurements but it is important for static or quasi-static measurements of pressure, deflection or forces. Strain gauges have to be temperature compensated and the amplifire must be very stable in its zero points. Nevertheless, calibrations have to be possible from time to time.

This calibration control is also important to check the whole equipment. In a long term measurement failures of single elements occur again and again.

The power supply should be installed in a fail safe way. Have in mind, that within a strong wind storm the local power supply may fail, but your wind load measurement must go on!

During installation of the measurement equipment in a building it has to be taken into account that a building or structure plot is a rough surrounding. The instruments, cables and other elements must be carefully protected.

Also vandalism can endanger the equipment.

During all the activities it is very important, that the responsible researcher controls personally the installation, the calibration and the measurements. Otherwise he cannot be sure that the results are O.k. Have in mind, that a full-scale measurement series cannot be repeated.

Before starting the installation check the safety requirements and the regulations for the safety of the personnel.

REFERENCES

Davenport, A.G. (Edit.) (1974), "Symposium on Full Scale Measurements of Wind Effects on Tall Buildings and other Structures", BLWT University of Western Ontario, London, Canada, June 1974.

Ebner, H. and Ruscheweyh, H. (1974), "Contribution to the Stationary Wind Load on Rotational Symmetric Structures in the Transcritical Reynolds Number Range", PhD Thesis 1971, Ruhr-University Bochum (in German).

Farell, C. and Maisch, F.F. (1974), "External Roughness Effects on the Mean Wind Pressure Distribution on Hyperbolic Cooling Towers", IIHR Rep. No. 164, The University of Iowa, August 1974.

IAWE (1963), " International Conference on Wind Effects on Buildings and Structures and Wind Engineering", Series of Conferences, Teddington (1963), Toronto (1967), Tokyo (1971), London, UK (1974), Fort Collins (1979), Australia (1983), Aachen (1987), London, Canada (1991).

Mayne, J.R. (1970), "A Wind-Pressure Transducer", Current-Paper 17/70, Building Res. Station Carston, Watford, WD2 7JR (1970).

Niemann, H.-J. (1971), "Contribution to the Stationary Wind Load on Rotational Symmetric Structures in the Transcritical Reynolds Number Range", PhD Thesis 1971, Ruhr-University Bochum (in German).

Ruscheweyh, H. (1974), "Static and Dynamic Wind Forces on Circular Cylindrical Structures-Full-Scale Measurements at the TV Tower Hamburg", PhD Thesis 1974, RWTH Aachen (in German).

Ruscheweyh, H. (1983), "Full-Scale Measurements on a Large Natural Draught Cooling Tower", the 6th Internat. Conf. on Wind Engineering, Australia 1983, Postersession, Report Institute for Steel Structures, RWTH Aachen, Germany.

Ruscheweyh, H. and Sedlacek, G. (1988), "Crosswind Vibrations of Steel Stacks, - A Critical Comparison between some recently proposed Codes -, Journ. of Wind Eng. and Industrial Aerodynamics, 30 (1988) 173-183.

Ruscheweyh, H. and Verwiebe C. (1994), "Full-Scale Measurements on a Small Guyed Mast for Nonlinear Dynamic Response", Current Research Project, Windengineering, Tech. University Aachen, Germany.

Van der Hoven, I. (1957), "Power Spectrum of Horizontal Wind Speed in the Frequency Range from 0,0007 - 900 cycles per hour", Journal of Meteorology, Vol. 14 (1957), p. 160-194.

Vickery, B.J. (1990), " Progress and Problems in the Prediction of the Response of Prototype Structures to Vortex-Induced Excitation", BBAA Kyoto, Japan 1988, Elsevier 1990, edit. by M. Ito.

MEASURING THE DYNAMIC CHARACTERISTICS OF PROTOTYPE STRUCTURES

J.D. LITTLER and B.R. ELLIS

Building Research Establishment, Garston, Watford, United Kingdom

SUMMARY

This paper is concerned with measuring the dynamic properties of prototype structures, especially those characteristics which are important for calculating structural response to wind loading. Two different types of test are considered in detail. The first is the forced vibration test which is used when accurate measurements of all of the characteristics of the fundamental modes of vibration are required. The second is the spectral analysis of the response of structures to wind loading, which can yield both structural characteristics and overall modal response. In both cases the test methodology and the potential problems involved are discussed. Measurements made by BRE on a number of prototype structures are used to illustrate the test techniques and the quality of the results which can be obtained.

INTRODUCTION

In the design of a tall building or a large bridge, the evaluation of structural response to the design wind load may be important both for serviceability and safety considerations. Therefore it is important that the designer has reliable methods for assessing both the wind loads and the structural characteristics in order to enable the structural response to be calculated. One way of checking that the methods used for these calculations are adequate, is to measure the characteristics and response of existing structures, and compare the behaviour with that assumed in design. This feedback mechanism will certainly be beneficial to individual designers, and will eventually improve the accuracy of predictions made by the wind engineering profession in general. However, measuring the characteristics or response of an actual structure may not be a simple task, and the purpose of this paper is to consider what measurements can be made and their likely accuracy. Details of how these characteristics are used in analysis and design will be dealt with in other papers in this volume.

The Building Research Establishment (BRE) has been involved with measuring the dynamic characteristics of prototype structures since the mid 1970's, and most of the methods described in the paper are those used by BRE. There are a number of different tests which can be used for various purposes, and these can be divided into two main types of test. First, there is the forced vibration test which is used to determine the characteristics of the lower frequency modes of vibration of a structure, i.e. frequency, damping, stiffness and mode shape. Second, there is the ambient test, or monitoring the response of the structure to wind loading, and this can be used to determine some of the characteristics of the structure (frequency, damping and mode shape), and also to determine the structural response to a given wind speed. However, before describing these tests methods in detail, a number of related points are discussed. These are the accuracy of estimating characteristics at the design stage, the measurement of structural response and simple tests to measure frequency.

ESTIMATING CHARACTERISTICS AT THE DESIGN STAGE

In order to evaluate the dynamic response of a structure to wind loads, it is necessary to estimate the modal characteristics of the structure. For each mode of vibration these characteristics are: frequency, damping, stiffness and mode shape. The lower frequency modes of vibration will have a significant effect on the calculated response of the structure to the wind; hence, it is useful to consider the expected accuracy in estimating these characteristics at the design stage.

A paper by Ellis (1980) examined the accuracy of estimating the fundamental frequency of buildings. It used frequencies measured on a large number of buildings and also made use of published results where both measured and computer based predicted frequencies were given. To demonstrate the accuracy of empirical predictors an optimum formula was derived for the data. This showed that for a simple formula $f = 46/H$, (where f is the fundamental translational frequency and H is the height of the building in metres), errors of ±50%, between measured and predicted frequencies, were not uncommon. However, when the computer based predictions were examined, their correlation with the measured values was worse than that for the simple empirical method. Intuitively, it would seem that detailed computer models should provide better estimates of the fundamental frequencies of buildings than simple empirical predictors, but since 1980 the authors have seen no real evidence that the accuracy of computer based models has improved significantly. Hence unless there is definite evidence to suggest otherwise, errors of ±50% should be anticipated with estimates of natural frequencies.

The importance of obtaining the correct value of natural frequency can be illustrated by examining the case of the response of a tall building to wind loading. The r.m.s. resonant acceleration is inversely proportional to the natural frequency raised to the power of about 1.35. The response of a tall building is also inversely proportional to the mass of the building and is inversely proportional to the square root of the damping. Littler and Murphy (1992) suggest ±20% and -50%, +100% (i.e. a factor of two) as typical bounds of uncertainty for these parameters respectively.

In the previous paragraphs, the accuracy of predicting the characteristics of buildings is discussed, and similar errors are to be expected for other types of structure. However, if a structure is relatively simple and its boundary conditions are well defined, then this is likely to lead to a more accurate structural model. There is no reliable way of calculating damping values and the best estimates of damping are usually obtained by adopting measured values from similar types of structure.

THE MEASUREMENT OF STRUCTURAL RESPONSE

In the following sections a variety of tests are described, but each type of test requires the measurement of the response of the structure. The equipment used for the measurements must be appropriate for the task and used correctly. To monitor response a variety of transducers, e.g. accelerometers or geophones, can be used, but they need to be selected to cover the frequency and amplitude range examined. The signals from the transducers may need to be filtered and amplified before being displayed or recorded. Oscilloscopes are normally used for displaying the signal, with the signal either being recorded on a tape recorder, or onto a computer based data acquisition system. If the data are recorded digitally then careful consideration should be given to the sampling rate used and the potential problem of aliasing. This topic is dealt with in Bendat and Piersol (1986). Digital recording is becoming increasingly common, but analogue recording usually has one advantage in that the same signal can be reanalysed with different parameters if required. Finally it is important to note that each item in the measurement chain (transducers, signal conditioning equipment and the recording device) needs to be calibrated.

SIMPLE TESTS TO MEASURE FREQUENCY

In some circumstances all that may be required from a test is to measure the fundamental frequency of a structure, and this can be a relatively easy task. Most of these tests could be performed by engineers given the right equipment and should yield reasonable values of natural frequency.

Perhaps the simplest way to measure the natural frequency of a structure subjected to wind excitation is to display the signal from a measurement transducer on an oscilloscope and, if the signal resembles a single sine wave, determine the time between successive peaks in the time history. This gives the period of the motion which is the inverse of the frequency. This method works well when the response is primarily in one mode. Careful positioning of the transducer and possible filtering of the signal may be required in order to minimise interference between translational and torsional modes.

Another fairly simple way of measuring frequency is to make a short recording of the response of the structure to wind loading (a response-time history) and then transform this record from the time to the frequency domain using a fast Fourier transform (FFT) routine to produce an autospectrum, or a power spectrum. The short recording could be of several minutes duration. BRE often use this type of analysis when taking remote measurements using their Laser interferometer. This Laser system provides a method of taking measurements without direct access to the structure, which can be useful in certain circumstances.

Alternatively, for relatively small structures, a simple form of forced vibration could be used, which may only take a few seconds and involves monitoring the structural response to a given impact. Such an impact test could be someone jumping on a structure such as a footbridge or a weight being released from a grandstand roof. The frequency can be derived either from the response-time history, or an autospectrum of that record. For a number of structures the decay of vibration may also yield a reasonable estimate of the damping.

FORCED VIBRATION TESTING

Steady-state forced vibration testing is expensive and a test on a single building may take several days; however, if a detailed evaluation of all the characteristics of the lower frequency modes of vibration is desired, it is unrivalled in accuracy by any other test method. In the following sections the test procedure used by BRE is described in detail.

The BRE Test Procedure

BRE has carried out forced vibration tests on many tall buildings and other structures, primarily to evaluate the characteristics of their fundamental modes of vibration. These tests have a major advantage over ambient tests, because the modal stiffness and modal mass can be measured. BRE has a system of four exciters, which use eccentric contra-rotating masses to produce a horizontal sinusoidal force, and these have been used for testing buildings and dams. For vertical motion BRE has a much smaller system, which has been used for testing floors and grandstands. For this type of exciter the force is proportional to the eccentric mass used, its radius from the centre of rotation and the square of the excitation frequency ($mr\omega^2$). The large system can generate a maximum force of 4.1 kN (single peak) at 1 Hz, and the smaller system 167 N at 1 Hz.

Basic test methodology - Frequench sweeps

The exciter(s) are placed in a suitable location and orientation for exciting the structure in the mode to be investigated. The exciters are set to produce a known force and then the forcing frequency is incremented over the required frequency range whilst the response is monitored. The response is usually monitored using an accelerometer aligned to measure motion in the appropriate direction. The data obtained are normalised by converting the measured accelerations to the equivalent displacement and then dividing this displacement by the applied force. The plot of normalised response against frequency is termed the response spectrum. A best-fit curve based on a visco-elastic model is then determined for the response spectrum. The parameters used to determine the best-fit model for each mode are frequency, damping and stiffness. The accuracy of these values can be judged by a visual comparison of the experimental data and the fitted curve.

With this type of forced vibration test, it is important that the exciter can provide sufficient force to produce a measurable response in the structure over the required frequency range, and it is equally important that the system can be controlled accurately and that the frequency control is stable. The BRE system which is used to test buildings works from 0.3 Hz to 20.0 Hz in increments of 0.001 Hz. The phase of each exciter relative to the frequency of the reference signal of the control is checked, and adjusted as necessary, several hundred times per revolution of the weights. The phase lag of each exciter is displayed for the operator to show when a stable situation has been obtained, which is especially important when the required frequency has just been altered.

If the testing is conducted when the wind is blowing, then the ambient movement of the structure will affect the tests. This is usually only a problem for very large low-frequency systems, and in these cases testing has to be undertaken in relatively calm conditions.

The final requirement for the testing is that the increments of frequency used are sufficiently small when the response is changing rapidly in amplitude, so that the spectrum will be resolved adequately. This means that a large number of points are required around resonance in order to define the fitted curve. Fig. 1 shows a response spectrum obtained for a fundamental translational mode from a test on one tall building, Habershon Court.

With the facility for changing the applied force, the force vibration testing procedure is also used to examine the characteristics at various amplitudes of vibration. This has shown that the structural characteristics vary with amplitude, and for the range which is of interest to wind engineers, these changes are a small decrease in frequency and a small increase in damping with increasing amplitude of motion. The visco-elastic model also shows a slight skew fit with the measured data, which is another indication of non-linear response, but for wind engineering purposes this is not important and can be disregarded.

Mode shapes can also be determined by setting the exciters to provide a steady state motion at a natural frequency and monitoring the response at various locations throughout the structure relative to a reference accelerometer. This can be used to examine support conditions which may be important for calibrating any detailed mathematical model of the system.

Damping from decays

The other method of obtaining damping from steady-state testing is by analysing a decay of oscillation. The method now used by BRE is as follows. The exciters are set to produce steady-state motion at a natural frequency and then switched off simultaneously. The decay of oscillation is recorded, usually at a sampling frequency of 1000 Hz for tests on a tall building. If necessary, filtering of the signal can be carried out prior to recording. The upper part of Fig. 2 shows the decay obtained for the same mode and at the same force level as for the response spectrum given in Fig. 1. The analysis is then carried out on a chosen part of this decay. The centre part of Fig. 2 shows the portion of the decay from 95% to 35% of the original steady-state amplitude.

A best-fit decay, based on a visco-elastic model, is then derived for the selected part of the decay using a least squares fitting procedure. All the points in the selected part of the decay are used, not just the peak amplitudes. The program uses frequency, damping and initial amplitude values (based on the selected number of cycles and the initial and final values from the chosen part of the decay) as the initial values for

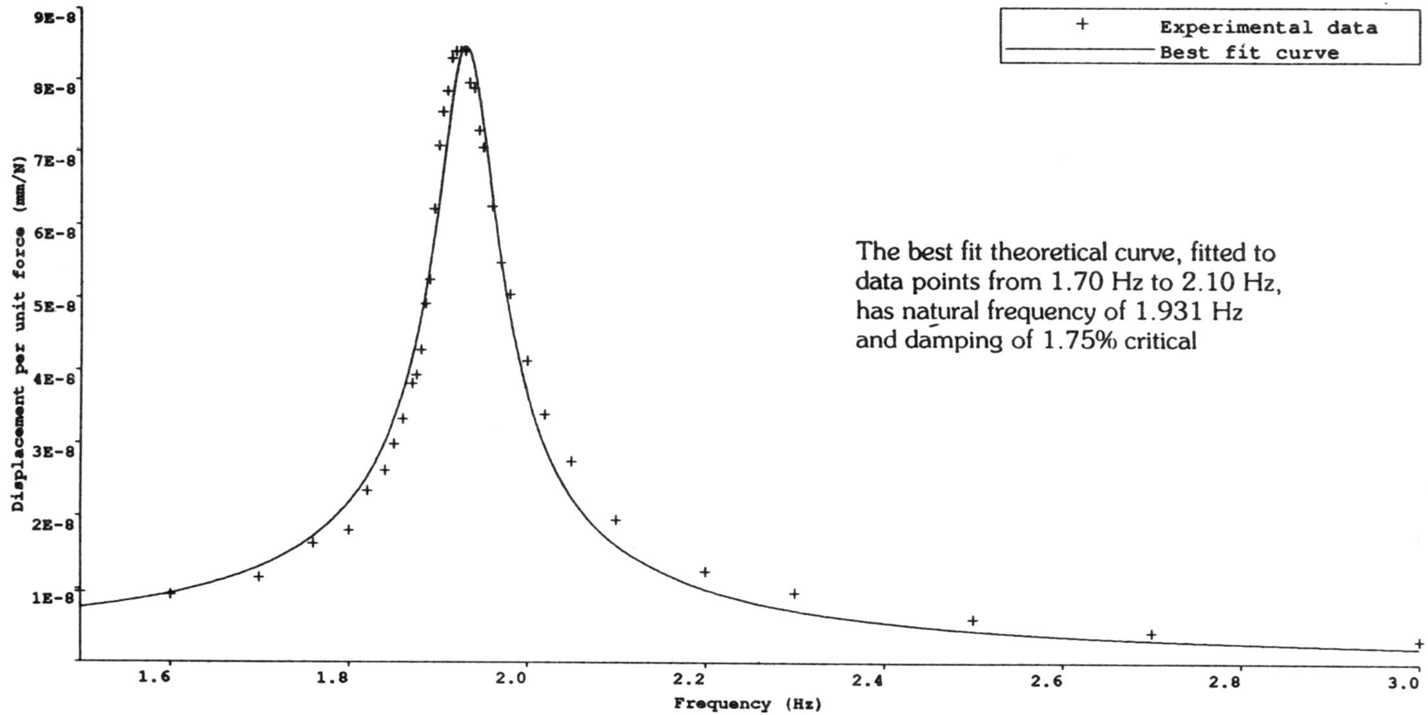

Figure 1 Response spectrum for NS1 mode at Habershon Court

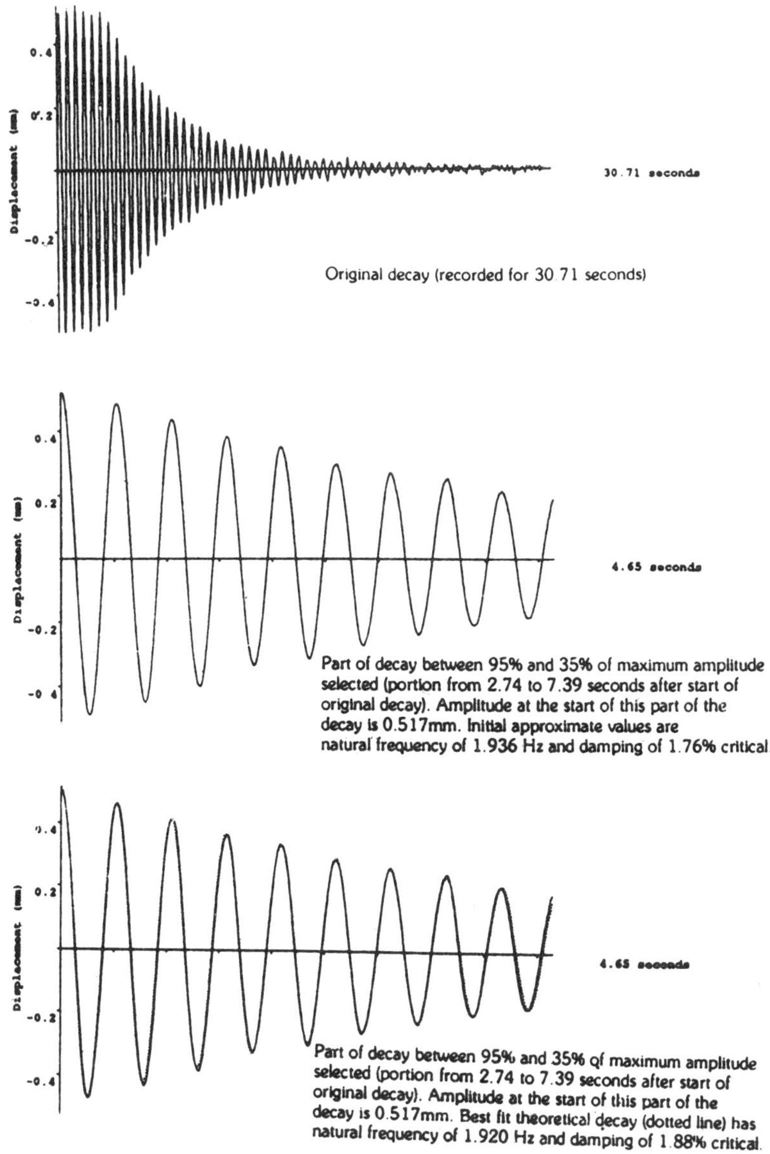

Figure 2 Decay of oscillation for NS1 mode at Habershon Court

deriving the best-fit theoretical decay. Frequency and damping are considered to have constant values throughout the theoretical decay. The lower part of Fig. 2 shows the selected part of the decay with the best-fit theoretical curve superimposed. It can be seen that the fit is excellent except at the end where some misalignment occurs. As higher weighting is applied to the points with larger amplitudes, any mismatch is likely to occur at the end of the decay, and this mismatch is generally because of the non-linear response of the structure which has been mentioned previously.

In this case, the damping values obtained by these two entirely separate methods (i.e. from the response spectrum and from the decay) are 1.75% and 1.88%, which for damping measurements can be considered to agree quite well. For tests using exciters whose stability or control is not adequate for defining the response spectrum adequately, damping values taken from decays for fundamental modes of vibration will still be acceptable, as they are less prone to experimental errors. The two methods for deriving damping are complementary and, where possible, both should be used. Any differences between the results obtained by the two methods which cannot be explained easily as modal interference etc. will show any weaknesses in experimental technique or the mathematical model used, whereas similar results from the two methods will add to the confidence that can be placed in the results.

Other Methods of Forced Vibration Testing

Other methods of forced vibration testing have been used to obtain measurements on tall buildings and similar structures. Daniels et al (1993) used both steady-state and instrumented hammer tests on a two storey building. They state that both methods have the same accuracy but that the hammer tests are considerably quicker and cheaper. However, the frequency steps used in their steady-state tests (1 Hz) are too coarse for a structure with natural frequencies of 7 to 8 Hz, as was the case in the tests reported. Hammer tests, in which both the load and the response are monitored, can be used for deriving structural characteristics, and for small structures have proved very useful. However, with larger structures it may be difficult to introduce sufficient energy over the frequency range of interest, thus creating difficulties with response measurements and thereby reducing the accuracy of any derived transfer functions

Williams and Tsang (1989) used a linear motion hydraulic exciter which could apply wide band random forces to a tall building. They acknowledge however, that there are problems getting sufficient frequency resolution with such exciters on structures with low natural frequencies, and that steady-state sinusoidal tests can generally provide better results.

SPECTRAL ANALYSIS OF AMBIENT RECORDINGS

The forced vibration tests described in the previous section, introduced a controlled force to excite the structure and thus provide a deterministic test method. But the wind also causes structural vibrations, and the analysis of these vibrations can be used to determine structural characteristics. Analysing the wind response of a structure does not require the specialist equipment to shake the structure used in forced vibration tests, and monitoring wind response will also show how the structure moves under given wind conditions, which is usually what concerns the designer. However, the analysis of a structures response to wind loading may prove to be quite a complex task, and in the next sections the problems which may be encountered and the ways of dealing with these problems are discussed.

Analysis using relatively short records

In order to examine the response of structures to wind loading, research workers have often used the spectral analysis of relatively short lengths of data, but its limitations are still not always realised. In essence, a recording of the response of the structure to the wind is made for a period of several hours. This recording is then split into a number of individual records and an FFT calculated for each one. The resulting spectra are then averaged together to give an ensemble averaged spectrum from which values of natural frequency, damping and response can be obtained for each mode. This method is usually employed when the response of the structure (e.g. r.m.s. acceleration) to the average wind speed over the recording is required in addition to the dynamic characteristics.

There are three important criteria to consider when using this method:

1. The stationarity of the data
2. Variance errors
3. Bias error

In order to use spectral analysis methods correctly, it is necessary to have stationary data for analysis. Stationary data have statistical properties which do not vary with time. If the wind direction or speed varies over the length of the recording then the response in any particular mode will vary in amplitude and will not be stationary. Averaging spectra together which were obtained under different wind conditions will not give the same spectrum that would be obtained from stationary data recorded for the average of these wind conditions, and hence it can be misleading. Therefore a test for stationarity should be performed on the whole recording. If, as is usually the case, the data are not stationary, then extreme caution should be used in extracting any parameters other than frequency. If damping values are obtained, then they will inevitably overestimate the real values.

Variance or random error results from the fact that any analysis must be performed on a finite number of sample records. On an intuitive level, the more times a random process is sampled, and the results averaged, the greater the accuracy; providing the same process is sampled each time (i.e. it is stationary). Variance error is equal to the reciprocal of the square root of the number of records averaged together. Therefore, for example, variance error is ±10% for 100 records, but only reduces to ±5% when 400 records are used. Fig. 3 shows the power spectrum for a single record of 1024 seconds duration obtained on a tall building. Fig. 4 shows the power spectrum obtained by averaging together 100 records from the same building. These 100 records were not obtained consecutively but were chosen by the selective ensemble averaging technique which is explained in the next section.

Bias errors occur when there is a rapid change of amplitude with respect to frequency and there are too few spectral lines to resolve the variation accurately. Such a situation occurs in the region of a peak in the

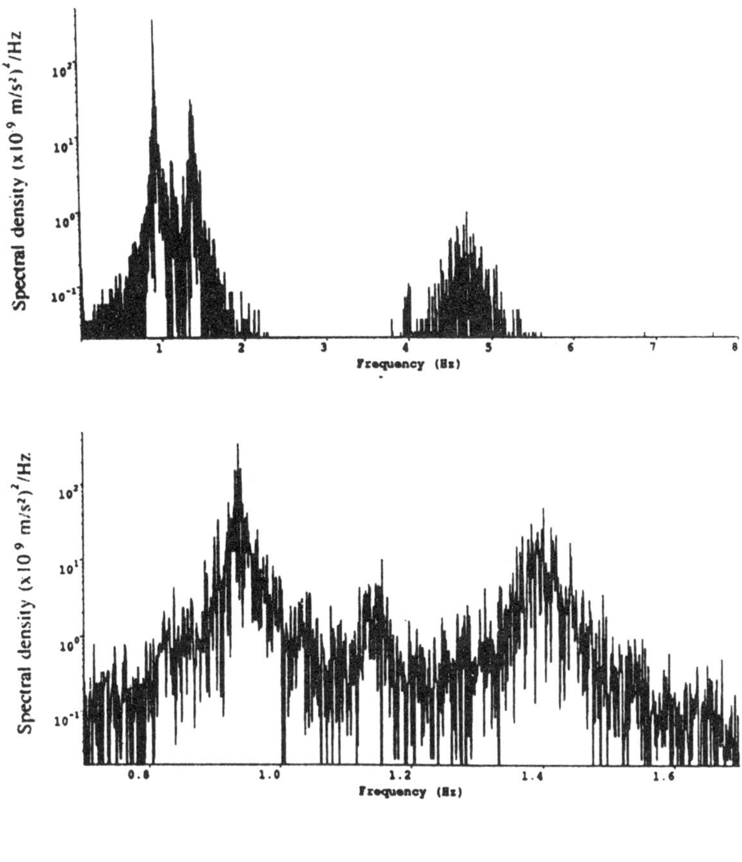

Figure 3 Power spectra for a single record of 1024 seconds duration from Hume Point

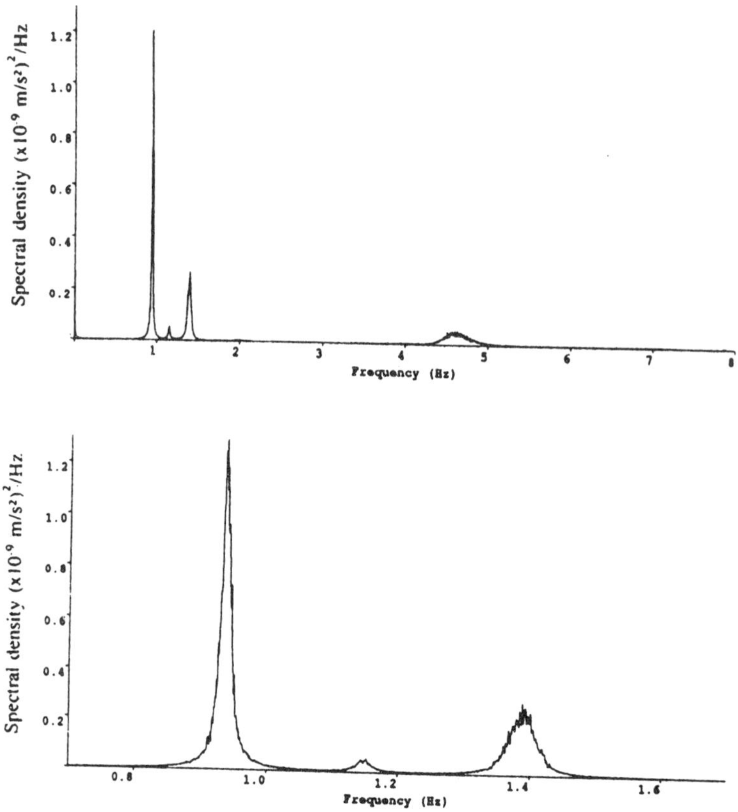

Figure 4 Ensemble averaged power spectra from Hume Point

response spectrum i.e. at a natural frequency. In order to limit bias errors to 2% there must be at least four spectral lines in the half-power bandwidth of each mode (Bendat and Piersol, 1986). There are however, two major problems in complying with this. Increasing the number of spectral lines means increasing the frequency resolution. If the number of averages is to be kept the same (in order to keep the variance error the same) then the total time of the recording must be increased. This in turn means that the data are less likely to be stationary over the increased length of time. Littler (1992) examined the extreme case of long span suspension bridges. In the case of the Humber Bridge, obtaining 100 records, each with sufficient resolution to give four spectral lines in the half power bandwidth of the fundamental vertical mode, would take approximately 96 hours. However, in the case of a moderately sized tall building such as the Sheffield Arts Tower (Littler et al, 1988) the same criteria would require a recording of at least 14 hours. The second problem is that bias error cannot be calculated until the natural frequency and damping of the mode in question are known. This is a particular problem where forced vibration tests have not been carried out on similar structures. It should be noted that bias errors always lead to damping being overestimated, but it is not possible to calculate the extent of the overestimation. Also, and most importantly, bias errors will be different for different modes in the same response spectrum. The lower the product of the natural frequency and damping for a particular mode, the larger the bias error.

The half-power bandwidth method of calculating damping from spectra is not ideal when dealing with the results of ambient excitation. This method depends on finding the peak amplitude of response for a particular mode, and then measuring the width of the spectral peak at the half-power amplitude ($1/\sqrt{2}$ of the peak amplitude). Bias errors mean that the peak response will always be underestimated, which leads to an underestimation of the half-power amplitude, thus leading to an overestimation of damping. The method of obtaining damping adopted by BRE is to obtain a response spectrum from the ambient analysis with normalised displacement as the ordinate. The best fit theoretical curve to the experimental data is calculated, in a similar fashion to that used in the forced vibration tests. The accuracy of these values can be judged by the fit between the experimental data and the fitted curve. Use of this method ensures that any damping value obtained is not entirely dependent on the value of the one spectral line closest to the peak response. It is extremely difficult to give a level of accuracy for damping values obtained by this method, and virtually impossible if the data are not stationary. If the data are stationary, then the accuracy is dependent on the inevitable errors in the spectral analysis.

Despite the problems outlined above for obtaining damping values, natural frequencies can be obtained quickly and easily by this method. Satisfactory mode shapes can also be obtained (Brownjohn et. al. 1987) even when non-stationary data are used, providing the response measurements at different positions are made at the same time. This is because the errors in the spectral analysis will be identical for every measurement position. Modal stiffness and therefore modal mass values cannot be obtained by this method as the force level is not known.

Selective Ensemble Averaging

Selective ensemble averaging tries to overcome the problem of obtaining stationary data of sufficient accuracy by forcing an ensemble to contain stationary data. This is done by making many recordings, or an extremely long continuous recording, and dividing the data into individual blocks, or records, of the length required for analysis, and storing the individual records along with the average wind speed and direction for the length of the record. Records obtained under similar wind conditions are then selecting and averaged. Fig. 5 shows an example of the data obtained by BRE from Hume Point (Littler and Ellis 1992) where this technique was used. Besides the fact that this is likely to be a long and expensive process there are two problems to consider. First in the selecting the ensembles to average, how narrow should the range of wind speeds and direction be in order to obtain accurate results. And second, how can the accuracy of the results be proven.

These problems are interrelated and for the Hume Point experiment the method was proven by producing four ensembles each containing 50 mutually exclusive records but collected under the same wind conditions. The

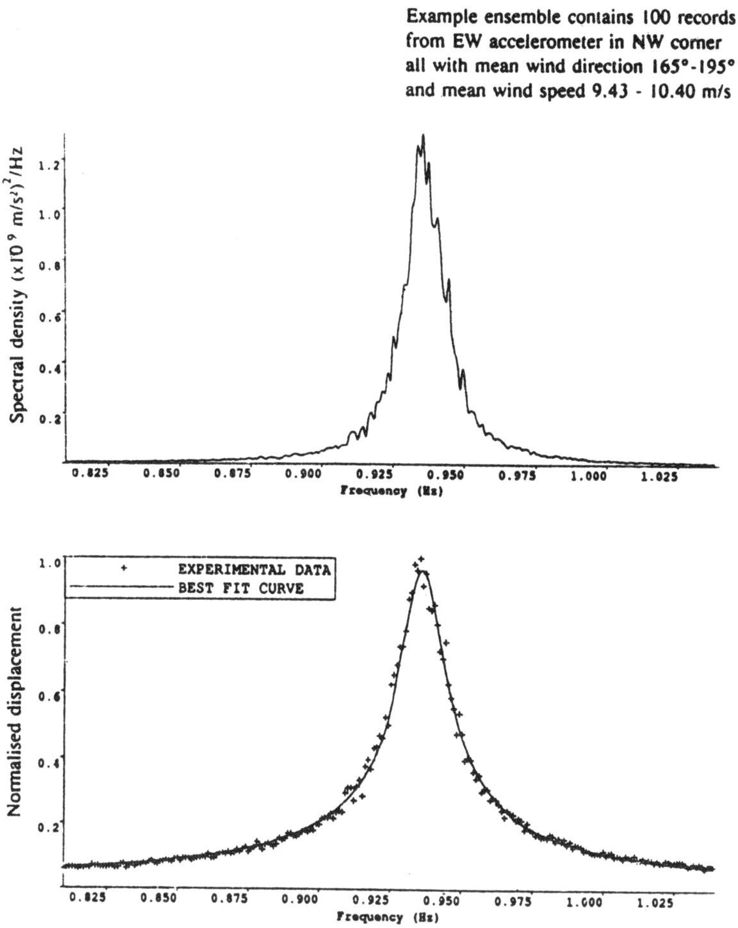

Figure 5 Ensemble averaged power spectrum from EW1 mode at Hume Point and normalised curve with best-fit curve superimposed

difference in response between the four ensembles was less than that due to variance error, so it was concluded that the limits used to define the wind conditions were sufficiently tight that the data in the whole ensemble could be considered to be stationary. With enough data, this method can give accurate figures for damping, which can be judged by the goodness of fit in Fig. 5 between the experimental data and the fitted curve. Overall, over 2,250 hours of data were recorded at Hume Point.

However, Fig. 6 from Littler (1992) shows how the technique has been used to assess the damping from the Humber Bridge. The fit is not very good here but it should be noted that this is for a mode with very low frequency and that a total of only 188 hours of data were recorded. Clearly, this technique cannot be used without an accurate way of measuring wind speed and direction at the site.

Selective ensemble averaging is primarily used for obtaining the response of a structure to a given wind to a high degree of accuracy. The results obtained using this method may be used in a comparison with those obtained from calculation or wind tunnel studies to help calibrate codes or wind tunnels. Although this method yields accurate dynamic characteristics (natural frequency, damping and mode shapes) it is unlikely to be worthwhile employing this technique if it is only these characteristics that are required from the test. The exception to this would be a structure with very low natural frequencies and damping such as a long-span suspension bridge where forced vibration tests or spectral analysis of short ambient recordings is unlikely to be successful.

Other methods of analysing wind induced response

A number of analysis methods have been developed which use much shorter lengths of data than are required by conventional spectral analysis. One of these methods is the autoregressive (AR) model which uses parameters derived by the maximum entropy method (MEM) and this method has been examined using measured data from Hume Point by Cao et al (1995). Two other related models are the moving average (MA) and the autoregressive moving average (ARMA) which both use parameters derived by the maximum likelihood method (Bendat and Piersol, 1986). With all of these types of

Wind speed 2 to 5 m/s, 0.99% damping

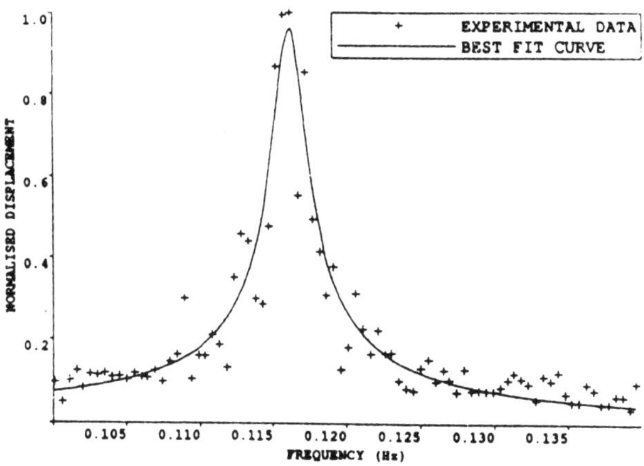

Wind speed 10 to 17 m/s, 4.76% damping

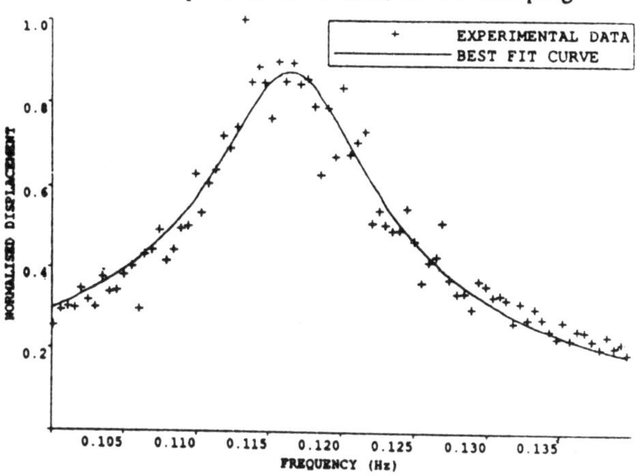

Figure 6 Ensemble averaged response spectra for the fundamental vertical mode of the Humber Bridge for different wind speeds with best-fit curve superimposed

analysis the use of stationary data is very important. Whilst all of these methods use much shorter lengths of data than standard spectral analysis, the accuracy obtained is dependent upon the model used. If an inappropriate model is selected, then the results will be much worse than those obtained by conventional analysis. At present, there are no available methods of assessing the likely accuracy of the spectra derived using these methods, and while they may be used for estimating modal frequencies, their use for estimating damping or modal response must be questioned. Nevertheless, the methods have considerable potential, although they may require further refinement and evaluation before they can be used with confidence to evaluate structural characteristics.

CONCLUSIONS

Two major ways of measuring the dynamic characteristics of real structures have been assessed. Steady-state forced vibration testing produces the most accurate assessments, but this will only be the case if both the excitation and the monitoring are carried out with due regard to the possible errors in the testing and analysis procedure. This method also has the advantage that damping can be assessed by two entirely separate methods and the answers compared. The level of discrepancy between the answers from the two methods gives a good indication of the overall accuracy of the testing procedure.

The spectral analysis of the response of structures to wind loading can also provide useful data and is especially useful for determining the natural frequencies of a structure. If the data used for the analysis are stationary and both the variance and bias errors involved in the analysis are considered, then information on damping and overall response can be obtained. The selective ensemble averaging technique avoids the problem of obtaining a sufficiently long continuous record of stationary data, but requires data to be collected for much longer lengths of time. However, it can prove useful for obtaining good quality data for code calibration.

ACKNOWLEDGEMENTS

The authors would like to thank the various members of the Dynamic Performance Section at BRE who assisted in the testing and analysis of the structures described. The authors would also like to thank the owners of the structures quoted as examples for permitting BRE to test them.

REFERENCES

Bendat, J S and Piersol, A G. (1986) "Random data analysis and measurement procedures" (2nd edition) John Wiley & Sons, New York.

Brownjohn, J M W, Dumaoglu, A A, Severn, R T and Taylor C A. (1987) "Ambient vibration measurements of the Humber Suspension Bridge and comparison with calculated characteristics" Proc. Inst. Civ. Engrs. vol. 83, pt 2, Sept, pp 561-600.

Cao, H, Ellis, B R and Littler, J D. (1995) "The use of the maximum entropy method for the spectral analysis of wind-induced data recorded on buildings" Proc. 9th Int. Conf. on Wind Engineering, New Delhi, India.

Daniels, B J, van Staalduinen, P, de Winter, P and Waarts, P. (1993) "The in-service lateral stiffness of steel-framed buildings", International Colloquium on Structural Serviceability of Buildings, Göteberg.

Ellis, B R. (1980) "An assessment of the accuracy of predicting the fundamental natural frequencies of buildings and the implications concerning the dynamic analysis of structures", Proc. Instn Civ. Engrs, Part 2, Vol 69, Sept, pp 763-776.

Littler, J D, Ellis, B R, Jeary, A P and Lee, B E (1988) "The measurement of the dynamic response of Sheffield Arts Tower to wind loading", pp 259-265, 4th International Conference on Tall Buildings, Hong Kong and Shanghai.

Littler, J D and Ellis, B R. (1992) "Full-scale measurements to determine the response of Hume Point to Wind Loading", Journal of Wind Engineering and Industrial Aerodynamics, 41-44, pp 1085-1096.

Littler, J D. (1992) "Ambient vibration tests on long span bridges", Journal of Wind Engineering and Industrial Aerodynamics, Vol. 42, pp 1359-1370.

Littler, J D and Murphy P D. (1992), "A comparison between the full-scale measured response of Hume Point and that calculated by some predictive methods", Inaugural Conference of the Wind Engineering Society, Cambridge.

Williams, C and Tsang, W F. (1989) "Determination of structural parameters from full-scale vibration tests". pp 43-59, Civil Engineering Dynamics, University of Bristol.

EVALUATION OF WIND LOADS ON LOW BUILDINGS—A BRIEF HISTORICAL REVIEW

THEODORE STATHOPOULOS

Centre for Building Studies, Concordia University, Montreal, Quebec, H3G 1M8 Canada

SUMMARY

The paper reviews some of the early research into the wind loading of low buildings starting with the earliest wind tunnel tests carried out in the last century. It refers to the difficulties of physical simulation and to the dramatic effects of turbulence and its characteristics to the loading mechanism of low building envelopes. Differences between local and area-averaged loads are highlighted along with the influence of building geometries, roof configurations and architectural features on the design wind pressures. Internal pressures and their significance are also discussed. Typical key results from several research projects are presented and discussed with emphasis on their influence on the development of wind code and standard provisions.

The impact of the progress in computational wind engineering on the evaluation of wind loads on low buildings is analyzed. In particular, the advantages and limitations of numerical simulation are discussed and comparisons between numerical results and experimental data are made. The importance of full-scale measurements in the past and at present is stressed. A glimpse in the future of the evaluation of wind loads on low buildings is also attempted.

INTRODUCTION

Wind loads on low buildings have not received the attention they deserve when the large investment in such structures is considered. Unfortunately, they have an inconvenient way of reminding us of this neglect when a hurricane or tornado strikes.

Much of the data from which present wind standards and codes of practice have been derived were obtained from model tests carried out before all the implications and the importance of correctly simulating the atmospheric boundary layer were appreciated. Only mean pressures were measured and the dominating influence of fluctuating pressure was missed. Pressure coefficients have been used with a variety of reference wind pressures associated with wind speeds averaged over several minutes or fastest mile wind speeds or gust speeds averaged over a few seconds due to the lack of understanding of the correct relationship of the pressures to the oncoming wind flow.

More recent studies have provided more useful indications of the important features of wind loading on low buildings. For example, it has been shown that the dominant loads are fluctuating and not necessarily organized either spatially over the structure or in time; effective loads are reduced with increasing tributary area but in different magnitudes depending on the location of the area vis-a-vis the wind flow; the surrounding terrain roughness and the adjacent buildings have an important influence on the distribution of pressures and the amplitude of the fluctuating components; and building shapes, in particular roof pitch and configuration, as well as architectural details (eaves, parapets etc.) are the primary factors influencing pressures. The ongoing development of computational wind engineering is also promising for the evaluation of wind loads on low buildings. Numerical solutions have the potential to overcome several restrictions of physical simulation such as the Reynolds number limitations in accurate modelling flows over curved surfaces, the inability to fully model swirling flow impacts on structures (tornadoes) and some types of gust fronts (downbursts). However, the ability of computational approaches to evaluate wind loads on low buildings has yet to be demonstrated for cases other than mean pressures for normal wind directions and simple building geometries.

HISTORICAL DEVELOPMENT

Much of the early research into wind loading was concerned with low buildings. Amongst the earliest wind tunnel tests performed were the experiments by Irminger in the 1890's on small models of gabled houses. Between the wars most western countries undertook systematic studies of the aerodynamic pressure coefficients on different building shapes in wind tunnels. Some of these studies were extremely thorough such as those by Irminger and Nokkentved (1930,1936) and those by Ackeret (1936) given in the Swiss Normen and presented in the 1965 edition of the National Building Code of Canada and other standards.

Unfortunately, the available wind tunnels and testing techniques at the time were those developed for aeronautical purposes. In these the flow was as smooth and uniform as possible; therefore it differed from the natural wind which is not only turbulent but also contains significant velocity variations with height, particularly along the height of low buildings. The effects of such lack of similarity of the flow conditions in model and full scale were highlighted by a few full scale experiments such as those by Bailey (1933) on a railway car shed and those by Arnstein and Klemperer (1936) on the Akron Airship dock. Both have indicated that the predictions of wind tunnel model tests, if they are carried out in uniform flow, can differ substantially from the full-scale realization. Typical results from Bailey's car shed are shown in Fig. 1 . The data show the average and standard deviation of the pressure coefficients Cp for 24 occasions in which the wind was westerly and normal to the ridge of the roof. For this wind direction the wind approached the car shed over approximately 500 m of open field. The diagram also shows the result of a wind tunnel test carried out in uniform flow without turbulence. Clearly in full scale the suctions recorded are generally significantly greater while the positive pressures are somewhat less than in the wind tunnel model. On the other hand, in case of Akron Airship dock, while positive pressures are similar, the peak negative suctions are significantly less in full scale than at model scale.

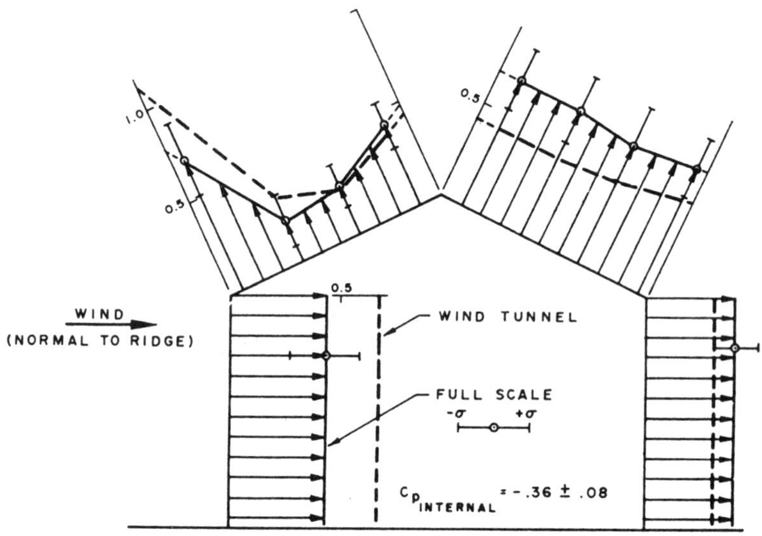

Fig. 1 Wind pressures on car shed, after Bailey (1933); comparison of full scale and model results.

Although several researchers recognized the need for adequate simulation of the turbulence of the natural wind and its velocity variation with height, the requirement was not confirmed until Jensen (1958) proposed the "Model Law for Phenomena in Natural Wind", i.e. scaling of the "roughness lengths" of the ground surface in the wind tunnel should be the same as that for the model itself. To achieve this he roughened the floor of the wind tunnel using a surface of suitably rough texture - corrugated cardboard, sand paper, wood slats - depending on the full-scale surface being considered. The results of full-scale and model experiments carried out to demonstrate the effect of this scaling are shown in Fig. 2, in which the roughness length z_0 characterizes the roughness of the surface.

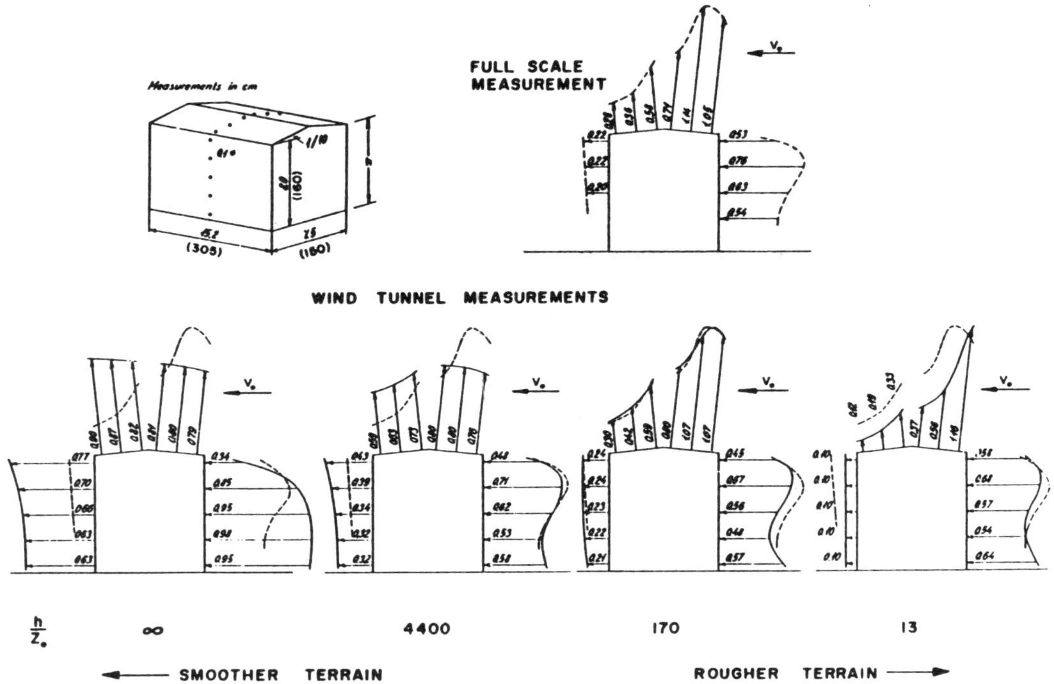

Fig. 2 Comparison of full scale and wind tunnel measurements of wind pressure; the influence of ground roughness and boundary layer profile, after Jensen (1958).

It is particularly noticeable that the uniform flow case ($h/z_0 \rightarrow \infty$) produces radically different pressure distributions from the full-scale experiments but that the latter are in agreement with the model tests for the equivalent Jensen number h/z_0. An extension of Jensen's model law is that the scales of turbulence, i.e. the characteristic eddy sizes, and the effective boundary layer thickness should also be scaled. Following Jensen's pioneering work, the wind tunnel tests generally respected these modelling laws.

It is worth mentioning that, although wind pressures on buildings reported before 1958 may not be representative for the reasons explained, some of the early ideas hold well up to date. As early as 1884 Sir Benjamin Baker stated that the mean wind pressure on a large area must be less than that on a small area because "*threads of the currents moving at the highest velocity will strike an obstruction successively rather than simultaneously*".
Having that in mind Stanton (1924) invented the wind pressure recorder which could be considered as a mechanical mean pressure transducer as shown in Fig. 3.

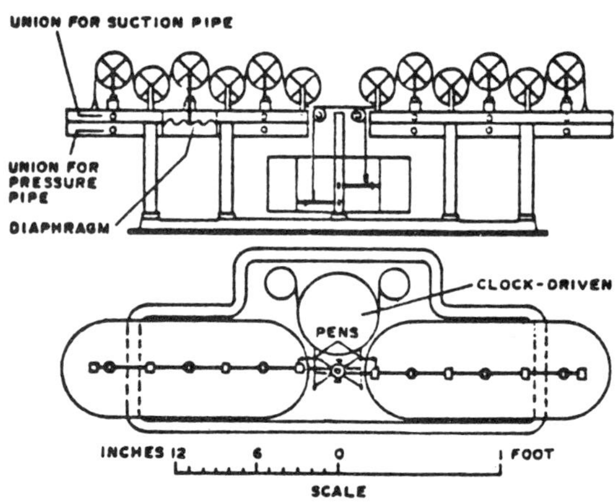

Fig. 3 Wind pressure recorder, after Stanton (1924).

The apparatus consists of a series of chambers, each containing a flexible corrugated diaphragm, the displacement of which, due to the difference of air pressure on its sides, is followed by a guide-rod carrying a light pulley. A thin platinum wire is fixed at one end and passes over the system of the pulleys shown, the free end carrying a pen in contact with a drum actuated by clock work. The motion of the pen integrates the separate motion of the diaphragms, so that a record of the average pressure differences in the chambers is obtained. It was Stanton's idea that led to the utilization of manifolds and the development of the well-known pneumatic averaging technique (Surry and Stathopoulos, 1978) used widely in the evaluation of area-averaged pressures on the building envelope.

A dramatic illustration of the tributary area effect is presented in Fig. 4 where individual instantaneous peak suction coefficients are plotted against wind direction for the four corner pressure taps indicated.

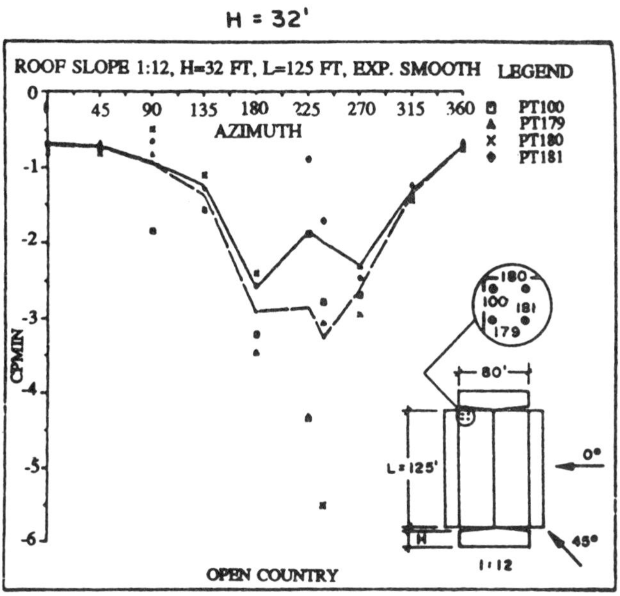

Fig. 4 The effect of averaging on peak suction coefficients on a corner of gabled roof (Symbols denote point suctions, solid lines denote true effective suctions [instantaneous spatial averages], and dashed lines denote averages of worst peaks).

The true effective averaged suction coefficients (solid line) is also given together with the arithmetic mean of the non-simultaneously registered peaks (dashed line).

The data indicate that the lack of correlation over the area and the high pressure gradients occurring over these regions (particularly for the critical azimuths) lead to significant reductions in the effective pressure coefficients applicable to corner areas.

WIND LOADING FEATURES FOR LOW BUILDINGS

Regarding wind loading of low buildings the following overall considerations are appropriate:

- Atmospheric simulation requirements are generally similar to those for tall buildings.
- There is a large variety of low building geometries.
- Special tests are usually unjustifiable economically and hence there is more reliance on generic tests.
- Aspects special to low buildings are due to their highly variable environment and the nature of their loading.

It was in the 1970's that more significant emphasis was placed on the evaluation of wind loads on low buildings. An extensive experimental study was carried out at the University of Western Ontario (Davenport et al., 1977,1978) and examined the effect of geometrical parameters (length, height, roof slope), scale and upstream roughness on a vast variety of flat and gabled-roof buildings. Data included the time-varying pressures acting over various tributary areas and time-varying moments and forces at critical structural locations, as shown diagrammatically in Fig. 5. Results of the study were expressed in codified format and were included in the Supplement to the National Building Code of Canada since 1980 and the American ANSI/ASCE-7 Standard since 1982. Typical results of the study shown in Fig. 6 indicate the effect of building height on the most critical values of pressure coefficients (peaks, mean and rms) measured at different points of the building envelope. The study generally found that the effect of roof slope and height are significant but the dependence of the load coefficients on height can be reduced considerably by referencing them to the velocity pressure at eave or mid-roof height; over the range of lengths studied, for which the length exceeded the width, there was comparatively little change in the measured loads; the dynamic component of all loading effects was dominant over the mean component. This was particularly true of local pressures, less so for distributed load effects. Finally, marked changes in the terrain roughness affected the pressure coefficients but trends were not always consistent; however, the dynamic component increased consistently with rougher terrain. In the 1980's and early 1990's research efforts were made to assess the variability of wind loads on low buildings for different surroundings (Ho, 1992) as well as to evaluate the wind loads on building roofs of different configurations.

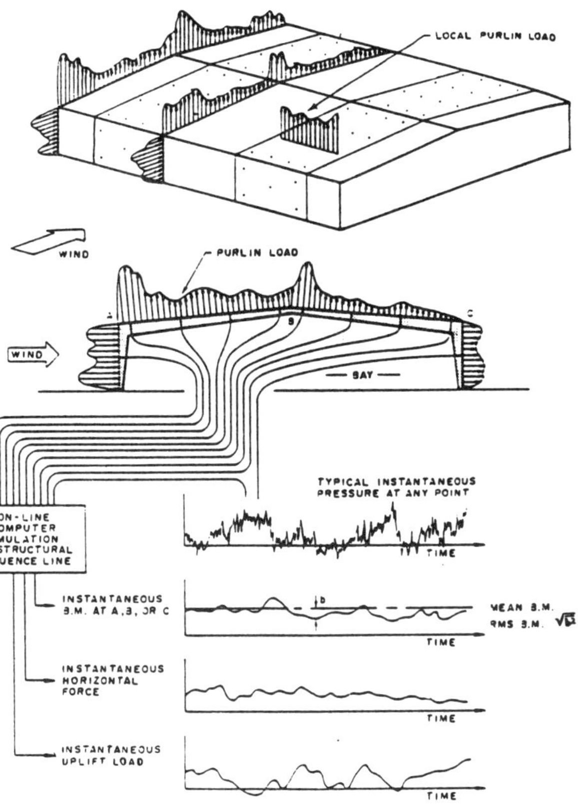

Fig. 5 Measurement of unsteady wind loads on a low building.

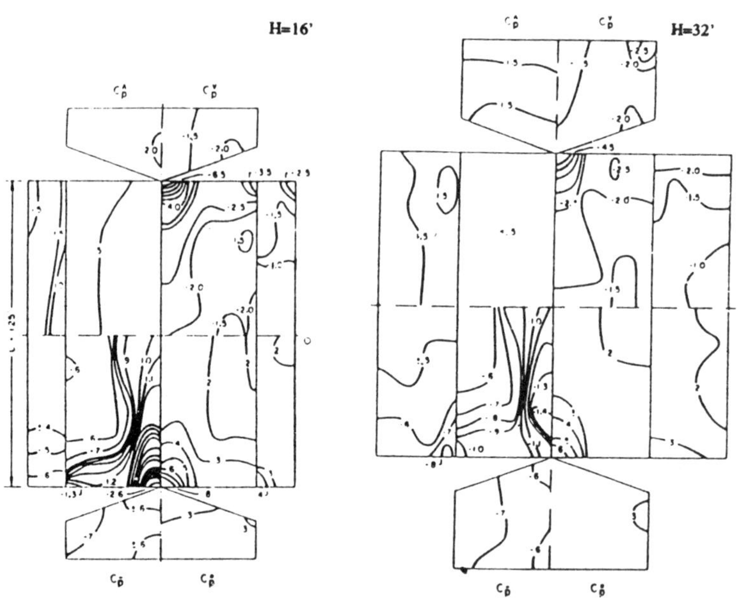

Fig. 6 Worst maximum, minimum, mean and rms pressure coefficients (for all azimuths) - 4:12 roof slope and open country exposure.

For instance, Fig. 7 taken from Meecham (1992) shows a reduction of pressure coefficients on hipped roofs in comparison to those applying on gabled roofs. This justifies the choice of the hipped roof type in several coastal areas frequently hit by violent storms. Figure 8 shows typical results of peak suction coefficients measured on a trough edge tapping of a multi-gabled (folded) roof type. The plotted data are average values of five 16-s samples; this provides a stable estimate of the instantaneous peak suction coefficient. In general, peak suctions measured on the double-span or multi-span models tend to be larger than those obtained with the single-span model. Wind pressure coefficients measured on stepped roofs are presented in Fig. 9 in contour form. A significant amount of information is included in these diagrams which are useful for design purposes since the data represent the most critical values of peak pressure and suction coefficients measured for each building configuration regardless of wind direction. Further to the substantial positive pressures occurring on the low roof of the two-level configuration, much higher suctions have also been measured on the roof

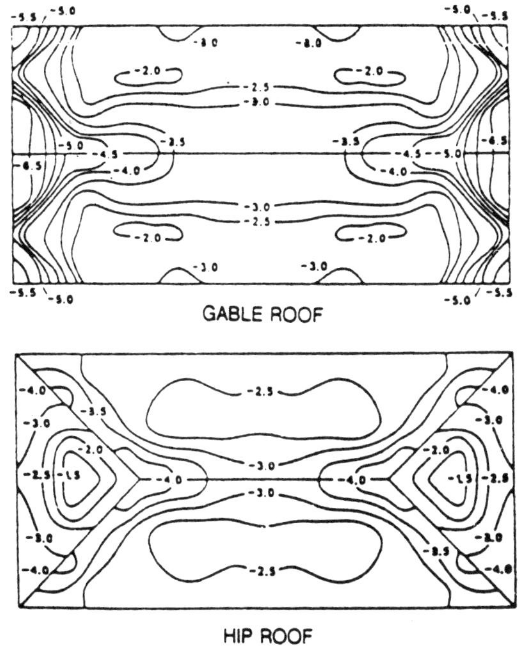

Fig. 7 Worst peak negative pressure coefficients - all azimuths, after Meecham (1992).

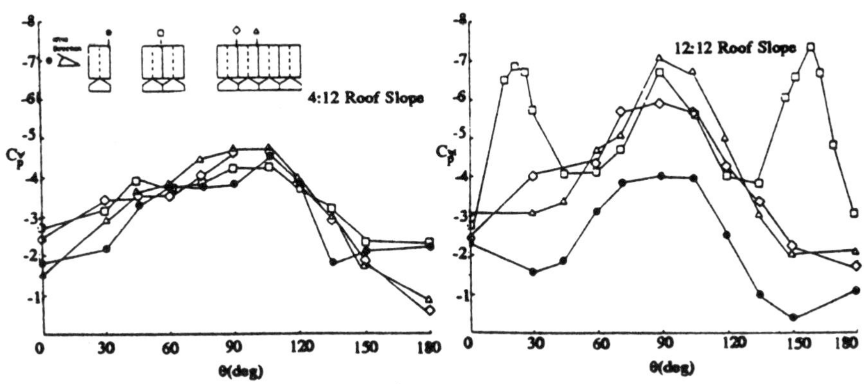

Fig. 8 Variation of peak suction coefficients with wind direction for roof corner points, after Stathopoulos and Saathoff (1991).

corner of the high roof section in comparison with those recorded on the respective corner of the low-roof section. However, these high suctions are comparable in magnitude with those obtained on the corner of the simple flat-roof configuration.

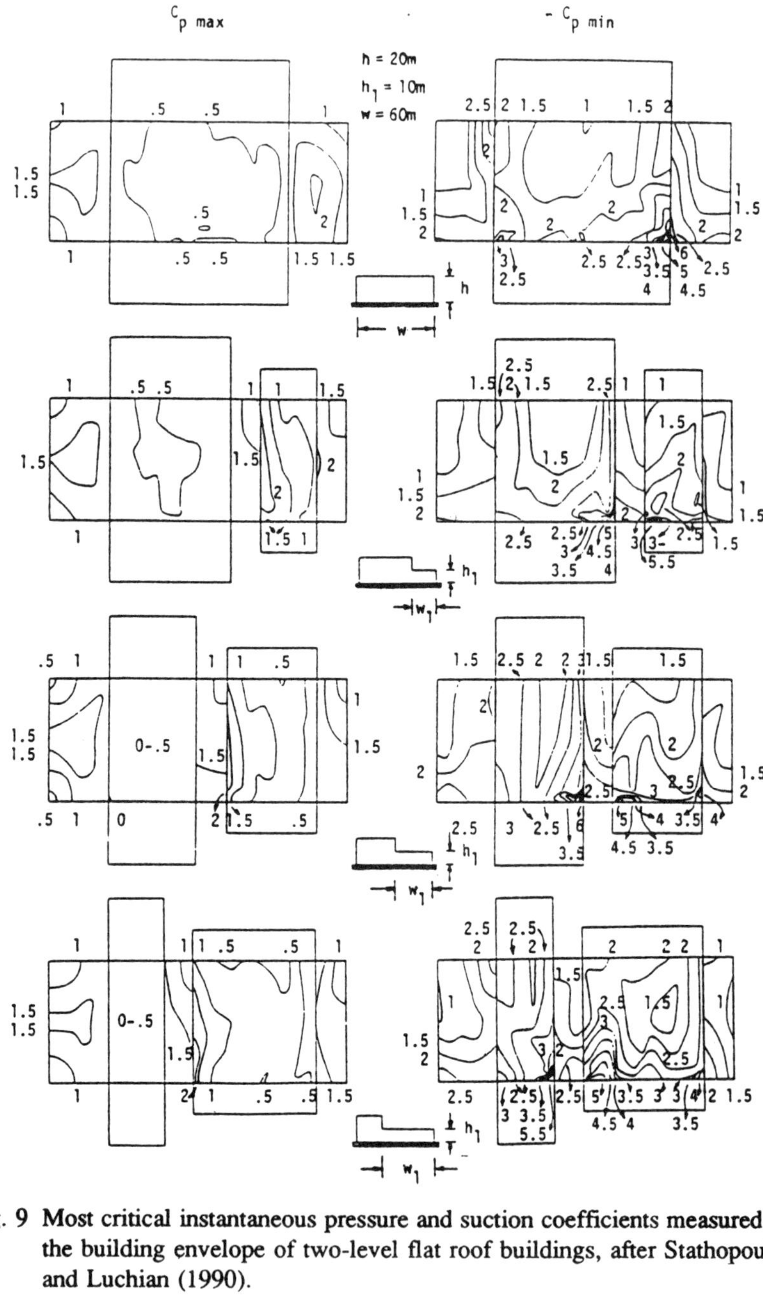

Fig. 9 Most critical instantaneous pressure and suction coefficients measured on the building envelope of two-level flat roof buildings, after Stathopoulos and Luchian (1990).

Pressures on sawtooth roofs have also been studied extensively by Holmes (1983) for low roof slopes and Stathopoulos and Saathoff (1992). Figure 10 shows the proposed codified diagram for roof angles between 10° and 30°. Design pressure coefficients have been provided for corner, edge and interior roof zones, as has been customary in the design wind standards for other building geometries. The same definition of the roof edge and corner width (z) used with other roof geometries has been applied. However, the high corner region is twice as large as the low corner region to reflect the greater area of large suctions measured in the high corner of a sawtooth roof. Simplicity and economy have led to the specification of one curve per zone for positive and negative pressures, with the exception of the roof corner, whose specifications for span A are much higher than for other spans of the sawtooth roof. Note that the vertical sections of the roof may be designed by using edge/corner specifications of the lower roof edge zone adjacent to them. As with the other geometries, the walls of the sawtooth-roof building may be designed by using coefficients established previously for walls of simple flat-roofed buildings.

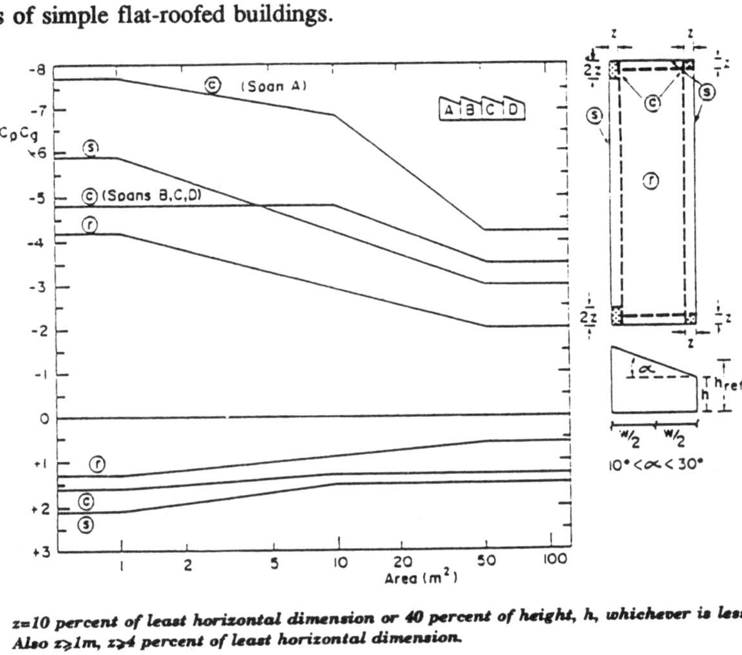

z=10 percent of least horizontal dimension or 40 percent of height, h, whichever is less. Also z≥1m, z≥4 percent of least horizontal dimension.

Fig. 10 Proposed design pressure coefficients for sawtooth roofs, after Stathopoulos and Saathoff (1992).

The question of actual magnitude of pressure coefficients appropriate for the design of cladding elements on flat roof edges and corners has been researched extensively by Kind (1986) and Stathopoulos (1987). Figure 11 shows the dramatic effect of the influence of exact pressure tap location on a roof corner. For example, for azimuths 30° to 45° pressure coefficients measured at the closest to the roof corner point have been found 2 to 3 times higher than those measured at 'usual' corner points. These results were obtained with a special flat roof model equipped with an assembly of 224 tappings distributed on one corner and along its concurrent edges. Fortunately, these high suctions affect only a very small portion of the roof corner because of the high gradient of their reduction with distance from the edge.

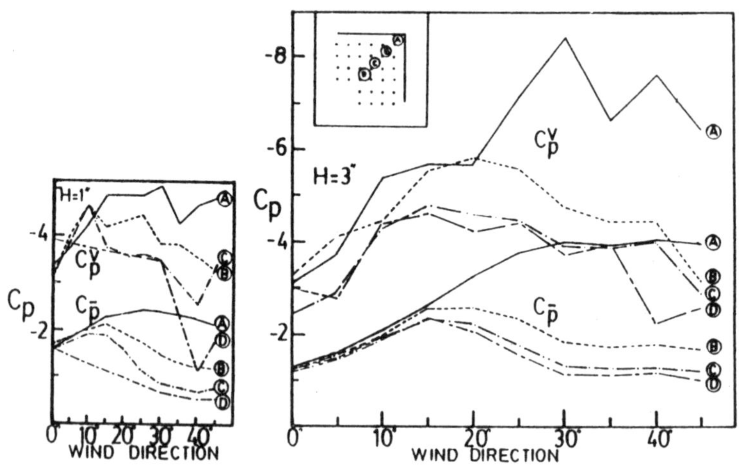

Fig. 11 Mean and negative peak pressure coefficients on roof corner points, after Stathopoulos (1987).

Furthermore, the effectiveness of these high suctions may be limited in a real building because of the edge profile of the roof which may establish a local Reynolds number modifying the separation of the flow and its characteristics. On the other extreme, Surry and Stopar (1989) have measured wind-induced suctions on the interior regions of roofs of very large low buildings and have recommended lowering the stipulations of the Supplement to the National Building Code of Canada as far as wind pressure coefficients for design purposes is concerned. Figure 12 compares the variation of peak area load coefficients measured with the current Canadian and Australian Code requirements.

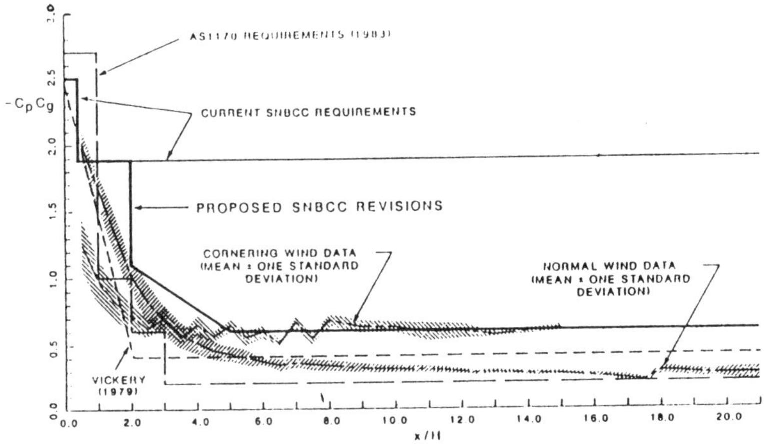

Fig. 12 Comparison of the variation of peak 12 m x 12 m area load coefficients vs normalized distance from the nearest edge with Canadian (SNBCC) and Australian (AS1170) code requirements, after Surry and Stopar (1989).

Parapets and Eaves

The effect of architectural features such as parapets, eaves, rounding the roof edges, surface roughness characteristics etc. have been studied extensively during the last decade or so. For instance, the influence of parapets on the wind loads of low buildings has been examined by Lythe and Surry (1983), Kind (1988), Baskaran and Stathopoulos (1988).

Various attempts have been made to reduce the excessive roof corner suctions caused by some parapet configurations. One such attempt was to discontinue parapets at the corner areas of the roof. Figure 13 shows roof corner pressure traces recorded for an oblique wind direction (θ=45°). The high suctions induced by a parapet 0.5 m high are clearly indicated along with their reduction caused by cutting the parapet at the ends of the edges for a length equal to the parapet height.

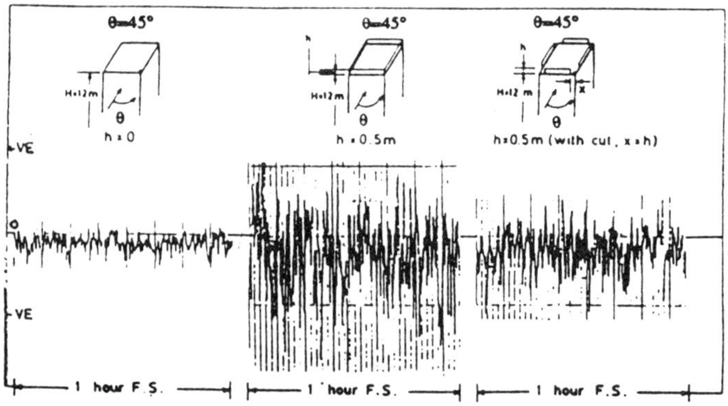

Fig. 13 Pressure traces showing the effect of parapet and cut-parapet on a flat roof corner, after Baskaran and Stathopoulos (1988).

In spite of the significant reduction of suctions caused by this configuration, suctions are still higher than those measured in the no-parapet case. This has also been shown for a number of parapet heights. Data show that regardless of parapet height, roof corner suctions are approximately equal for the cut-parapet configuration and somewhat higher than the no-parapet case. Consequently, roofs with low parapets are benefited mostly by this cut configuration. The effect of eaves on the wind loads of low building roofs has been examined by Robertson (1991) and more recently by Stathopoulos and Luchian (1994) for gabled roof buildings of different roof slopes. Figure 14 shows contours of the most critical peak pressure coefficients measured, regardless of azimuth and eave height, on the upper and lower overhang surfaces for the two different roof slopes tested. Clearly, the 4:12 sloped overhang is more severely loaded, particularly as far as suction on the upper surface near the gable is concerned. Positive pressures on the upper surface are higher for the 12:12 slope as expected from wind flow considerations. However, lower surfaces are subjected to higher pressures and suctions for the 4:12 slope. Pressures measured on lower eave surfaces have been compared extensively with values measured on the walls of the models. Although there is a general agreement, it has also been found that the highest positive peaks measured on the lower eave surface are larger than the respective values measured on the wall surfaces.

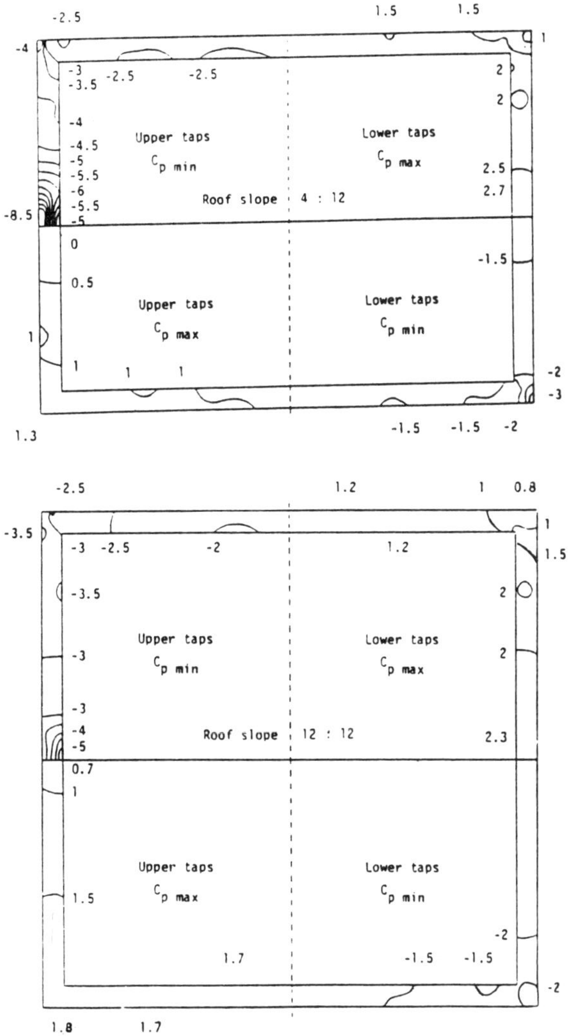

Fig. 14 Most critical local peak pressure coefficients measured on upper and lower eave surfaces, after Stathopoulos and Luchian (1994).

It is interesting to note that, although wind loads on eave sections have been measured directly, it has not been possible to measure loads on parapet sections themselves due to scaling difficulties. At present, such measurement appears feasible only in full scale.

Internal Pressures

There has been a fair amount of controversy regarding the appropriate magnitude of internal pressures for design purposes. Additional uncertainty is generated due to the influence of other factors such as envelope porosity and building openings. Further to the extensive wind tunnel studies by Kandola (1978), Holmes (1979), Kramer and Gerhardt (1983), the effect of sudden change - say due to the breakage of a window - has been studied by Stathopoulos and Luchian (1989). Figure 15 shows a typical example of the measured internal pressure in the model building during the second, in which the sudden opening occurs. The short time achieved for the realization of the sudden opening is apparent, as is the magnitude of overshooting, which is lower than the level of the subsequent instantaneous peak values. This has consistently been the case regardless of the method used for creating the sudden opening.

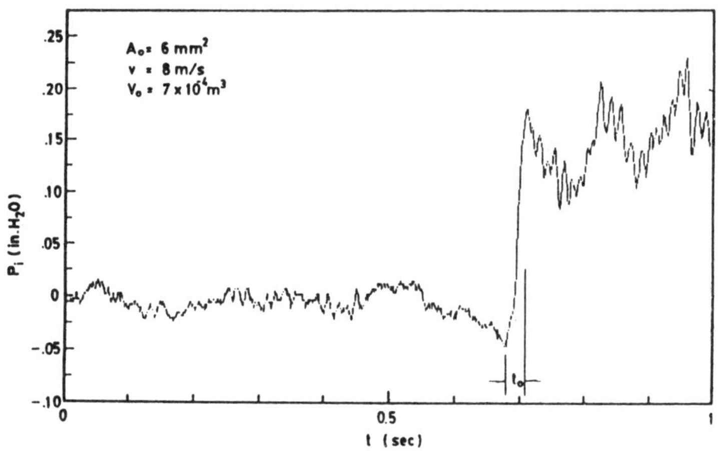

Fig. 15 Internal pressure fluctuations at the instant of sudden change, after Stathopoulos and Luchian (1989).

CODES AND STANDARDS / FULL-SCALE MEASUREMENTS

The material already presented and reviewed originates from studies in boundary layer wind tunnels and most of these findings will be included in the subsequent issues of wind codes and standards. In particular, wind loads on eaves of gabled-roof buildings as well as on roofs of various configurations (monoslope, sawtooth, multi-gabled, stepped) will be included in the 1995 issues of the Supplement to the National Building Code of Canada and the American Standard ASCE-7. Other national and international wind standards will probably follow, as in the past. Nevertheless, there still exist limitations in the extent to which physical modelling in the wind tunnel is satisfactory and provides adequate results.

Full-scale measurement results form naturally the critical test of acceptance of wind tunnel data. In this respect, Hansen and Sorensen (1986), Richardson and Surry (1992), Okada and Young (1992), Milford et al (1992) have compared full-scale measurements with wind-tunnel results and found that mean and rms pressure coefficients are in general satisfactory. However, peak pressure coefficients obtained in wind tunnel tests were smaller than the full-scale measurements depending on the measurement location. This might be due to the difference in intensity of turbulence at full-scale or to the difference in frequency characteristics of pressure measurement systems in the wind-tunnel. Although the state-of-the-art is still inconclusive in this respect, it appears that more full-scale tests will be required to shed light in a number of uncertainties still associated with the simulation of interaction between wind and low-building models in the boundary layer wind tunnel. Therefore, facilities such as the Texas Tech full-scale laboratory are extremely useful for further investigation of wind effects on low buildings.

COMPUTATIONAL EVALUATION OF WIND PRESSURES

In spite of the dramatic progress in the area of computational wind engineering (CWE), there are still several obstacles for the numerical evaluation of wind pressures on buildings. Difficulties for the numerical prediction of wind effects on low buildings include, but are not limited to numerical accuracy, boundary conditions and refined turbulence models. The most important of these areas is turbulence modelling and Murakami (1992) has exposed the shortcomings of various turbulence models and discussed their relative abilities.

Most of the CWE applications in the area of wind effects on buildings have utilized the finite difference approach and more specifically the Control Volume Method with a two-equation (k-ε) turbulence transport model in its standard form or with modifications. Results have been presented in terms of mean flow conditions with associated mean pressures on buildings or mean velocities in the building environment. Simple rectangular buildings in both plan view and cross section have been used in most cases, see e.g. Paterson (1986), for wind perpendicular to the building face; in fact most of the CWE applications refer to this particular wind direction since this is the simplest flow condition by taking advantage of symmetry. Difficulties arise for different wind directions.

Stathopoulos and Zhou (1993) have applied the k-ε turbulence model for the numerical evaluation of mean pressure coefficients acting on an L-shaped - stepped roof - building as a particular case of a computational code applying to any building configuration consisting of two different rectangular buildings. Computed and measured data for different wind directions have been compared for several roof points. Figure 16 compares the profile of wind-induced pressure coefficients along the centerline and the edgeline of the roof with the experimental data from Stathopoulos and Luchian (1990). The agreement between numerical results and experimental data is very good for the centerline, with some underestimation of suction generally shown by the computed data along the edgeline. Figure 17 shows similar results for 60° azimuth. Regarding the centerline, the general agreement is good, with some underestimation of negative pressures by the numerical simulation mainly in the low-roof region. As far as the edgeline is concerned, the pressures are not predicted properly. In particular, the suctions on low-roof sections are significantly underestimated by the numerical simulation.

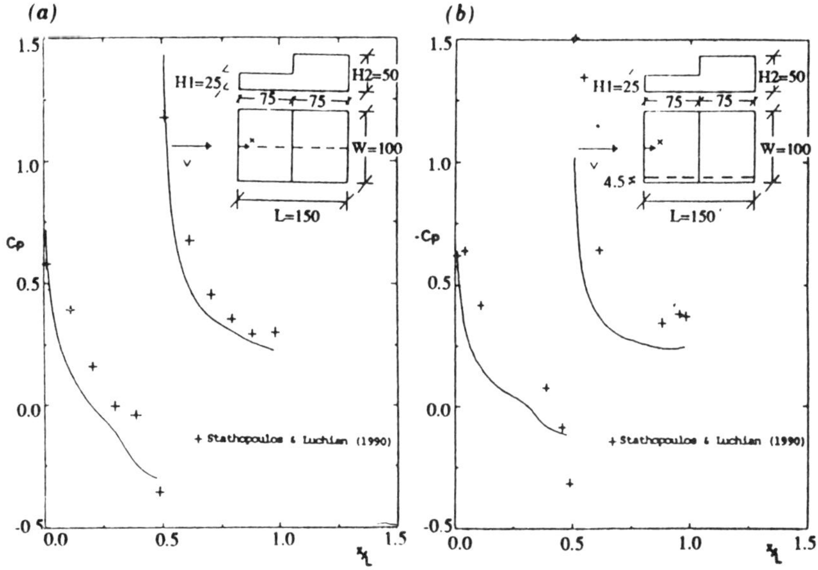

Fig. 16 Pressure coefficients along centerline and edgeline of roof (0° azimuth), after Stathopoulos and Zhou (1993).

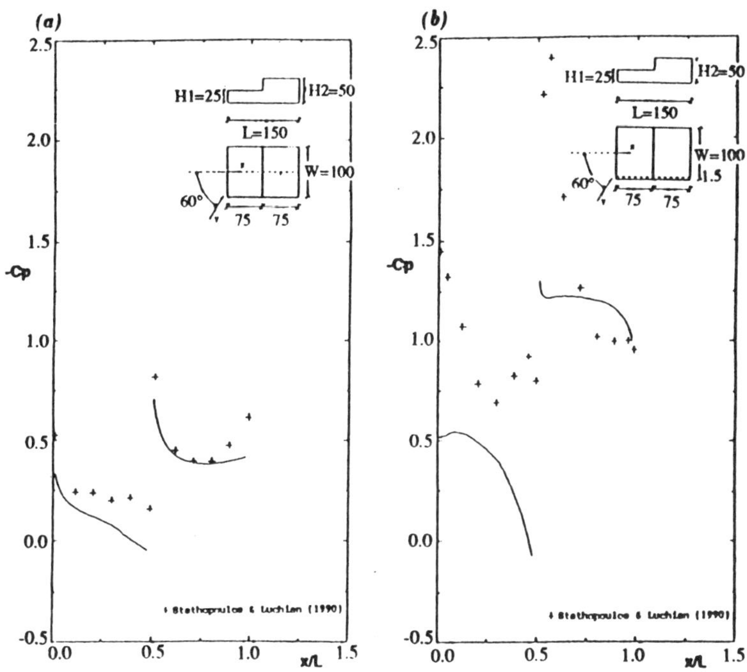

Fig. 17 Pressure coefficients along centerline and edgeline of roof (0° azimuth), after Stathopoulos and Zhou (1993).

Likewise, on the higher roof, the computed data show a smooth change, while the experimental data show a high gradient. This suggests that the k-ε model and the two-dimensional boundary conditions may not be representative for this high vorticity region and such predictions break down completely.

Other numerical simulation models and approaches are currently under development and, if successful, they may provide a significant thrust to the current state-of-the-art. However, this may require some additional time.

CONCLUSIONS

The paper has reviewed the developments in the evaluation of wind loads on low buildings after providing a historical perspective in the field. The importance of adequate physical simulation in boundary layer wind tunnel experiments has been shown along with the main wind loading attributes of low buildings. The research work leading to the wind code and standard loading provisions has been reviewed briefly. The importance of full-scale measurements has been highlighted as well as the state-of-the-art of the computational evaluation of wind loads on low buildings.

It is expected that additional field results will be very useful in order to further develop the physical and numerical simulation of wind effects on low buildings.

ACKNOWLEDGEMENTS

The assistance of my graduate student **Mr. Suresh Kumar** in the preparation of this manuscript has been greatly appreciated.

REFERENCES

Ackeret, J. (1936), "Der Winddruck auf Schornsteine mit Kreisquerschnitt", Schweizer Bauzeitung, Vol.108, No.3.

Arnstein, K. and Klemperer, W. (1936), "Wind Pressures on the Akron Airship Dock", Journal of Aeronautical Sciences, Vol.3, Jan.1936, pp 88-90.

Bailey, A. (1933), "Wind Pressures on Buildings", Selected Engineering Paper No.139, The Institution of Civil Engineers, London.

Baker, B. (1884), "The Forth Bridge", Engineering, Aug. 29,1884, pp 213.

Baskaran, A. and Stathopoulos, T. (1988), "Roof Corner Wind Loads and Parapet Configurations", Journal of Wind Engineering and Industrial Aerodynamics, Vol.29, pp 79-88, Elsevier.

Davenport, A.G., Surry, D. and Stathopoulos, T. (1977), "Wind Loads on Low Buildings: Final Report of Phases I and II- Parts 1 and 2", BLWT Report SS8-1977, The University of Western Ontario, London, Ontario, Canada.

Davenport, A.G., Surry, D. and Stathopoulos, T. (1978), "Wind Loads on Low Buildings: Final Report of Phase III- Parts 1 and 2", BLWT Report SS4 - 1978, The University of Western Ontario, London, Ontario, Canada.

Hansen, S.O. and Sorensen, E.G. (1986), "The AYLESBURY Experiment: Comparison of Model and Full-Scale Tests ", Journal of Wind Engineering and Industrial Aerodynamics", Vol.22, pp 1-22, Elsevier.

Ho, T.C.E. (1992), "Variability of Low Building Wind Loads", Ph.D thesis, The University of Western Ontario, London, Ontario, Canada.

Holmes, J.D. (1979), "Mean and Fluctuating Internal Pressures Induced by Wind", Proceedings of the 5th International Conference on Wind Engineering, Fort Collins, Colorado, USA.

Holmes, J.D. (1983), "Wind Loading of Saw-tooth Roof Buildings, Part 1: Point Pressures", Internal Report No 83/17, Division of Building Research, CSIRO, Melbourne, Australia.

Irminger, J.O.V. and Nokkentved, C. (1930, 1936), "Wind Pressure on Buildings", 1st and 2nd series, Danmarks Naturvidenskabelige Samfund, (1st) 1930 and (2nd) 1936.

Jensen, M. (1958), "The Model Law for Phenomena in Natural Wind", Ingenioren, International Edition, Vol.2, No.4, pp 121-128.

Kandola, B.S. (1978), "Wind Effects on Buildings with Varying Leakage Characteristics - Wind Tunnel Investigation", Journal of Industrial Aerodynamics, Vol.3, pp 267, Elsevier.

Kind, R.J. (1986), "Worst Suctions near Edges of Flat Roof Tops on Low-rise Buildings", Journal of Wind Engineering and Industrial Aerodynamics, Vol.25, pp 31-47, Elsevier.

Kind, R.J. (1988), "Worst Suctions near Edges of Flat Roof Tops with Parapets", Journal of Wind Engineering and Industrial Aerodynamics, Vol.31, pp 251-264, Elsevier.

Kramer, C. and Gerhardt, H.J. (1983), "Wind Loads on Permeable Roofing Systems", Journal of Wind Engineering and Industrial Aerodynamics, Vol.13, pp 347-358, Elsevier.

Lythe, G. and Surry, D. (1983), "Wind Loading of Flat Roofs with and without Parapets", Journal of Wind Engineering and Industrial Aerodynamics, Vol.11, pp 75-94, Elsevier.

Meecham, D. (1992), "The Improved Performance of Hip Roofs in Extreme Winds - A Case Study", Journal of Wind Engineering and Industrial Aerodynamics, Vol.41-44, pp 1717-1726, Elsevier.

Milford, R.V., Goliger, A.M. and Waldeck, J.L. (1992), "JAN SMUTS Experiment: Comparison of Full-Scale and Wind-Tunnel Results", Journal of Wind Engineering and Industrial Aerodynamics, Vol.41-44, pp 1705-1716, Elsevier.

Murakami, S. (1992), "Comparison of Various Turbulence Models Applied to a Bluff Body", First International Symposium on Computational Wind Engineering, Tokyo.

Okada, H. and Young, C.H. (1992), "Comparison of Wind Tunnel and Full-Scale Pressure Measurement Tests on the TEXAS TECH Building", Journal of Wind Engineering and Industrial Aerodynamics, Vol.41-44, pp 1601-1612, Elsevier.

Paterson, D.A. (1986), "Computation of Wind Flows over Three-dimensional Buildings", Ph.D Thesis, University of Queensland, Australia.

Richardson, G.M. and Surry, D. (1992), "The SILSOE Building: A Comparison of Pressure Coefficients and Spectra at Model and Full-Scale", Journal of Wind Engineering and Industrial Aerodynamics, Vol.41-44, pp 1653-1664, Elsevier.

Robertson, A.P. (1991), "Effect of Eaves Detail on Wind Pressures over an Industrial Building", Journal of Wind Engineering and Industrial Aerodynamics, Vol.38, pp 325-333, Elsevier.

Stanton, T.E. (1924), "Report on the Measurement of the Pressure of the Wind on Structures", Proceedings of the Institution of Civil Engineers, Paper No.4513.

Stathopoulos, T. (1987), "Wind Pressures on Flat Roof Edges and Corners", Proceedings of the 7th International Conference on Wind Engineering, Aachen, FRG.

Stathopoulos, T. and Luchian, H.D. (1989), "Transient Wind-Induced Internal Pressures", ASCE, Journal of Engineering Mechanics, Vol.115, No.7, pp 1501-1514.

Stathopoulos, T. and Luchian, H.D. (1990), "Wind Pressures on Buildings with Stepped Roofs", Canadian Journal of Civil Engineering, Vol.17, No.4, pp 569-577.

Stathopoulos, T. and Luchian, H.D. (1994), "Wind-Induced Forces on Eaves of Low Buildings", Journal of Wind Engineering and Industrial Aerodynamics, Vol.52, pp 249-261, Elsevier.

Stathopoulos, T. and Saathoff, P.J. (1991), "Wind Pressures on Roof of Various Geometries", Journal of Wind Engineering and Industrial Aerodynamics, Vol.38, pp 273-284, Elsevier.

Stathopoulos, T. and Saathoff, P.J. (1992), "Codification of Wind Pressure Coefficients for Sawtooth Roofs", Journal of Wind Engineering and Industrial Aerodynamics, Vol.41-44, pp 1727-1738, Elsevier.

Stathopoulos, T. and Zhou, Y. (1993), "Computation of Wind Pressures on L-Shaped Buildings", ASCE, Journal of Engineering Mechanics, Vol.119, No.8, pp 1526-1541.

Surry, D. and Stathopoulos, T. (1978), "An Experimental Approach to the Economical Measurement of Spatially Averaged Wind Loads", Journal of Industrial Aerodynamics, Vol.2, No.4, pp 385-397, Elsevier.

Surry, D. and Stopar, E.M.F. (1989), "Wind Loading of Large Low Buildings", Canadian Journal of Civil Engineering, Vol.16, No.4, pp 526-541

Vickery, B.J. (1979), "Interim 350 to AS1170: Developments in the Prediction of Wind Loads", The Profession of a Civil Engineer: Editors D.Campbell - Allen and E.H.Davies, Sydney University Press.

AERODYNAMICS OF TALL BUILDINGS

K.C.S. KWOK

Associate Professor of Civil Engineering,
Director, Wind Engineering Services, School of Civil and Mining Engineering, The University of Sydney, Australia

SUMMARY

Recent advances in the understanding of interference excitations of tall buildings and the potential beneficial effects of aerodynamic modifications of building shape are reviewed in this paper. Interference effects on the alongwind, crosswind and torsional responses of tall buildings from either an upstream or a downstream building are outlined. The beneficial effects of aerodynamic modifications of building shape in reducing the wind induced load and response of tall buildings are discussed.

INTRODUCTION

The new generation of tall buildings is taller and more slender, has futuristic shapes and sophisticated structural systems. The reduced wind velocities these buildings are required to operate in are steadily increasing and approaching, even surpassing, the threshold above which aeroelastic effects become an important strength and serviceability design consideration.

The excitation mechanisms causing alongwind, crosswind and torsional responses and the characteristics of these responses have been studied and reported extensively. Those excitation mechanisms studied include turbulence buffeting, vortex shedding, and those associated with aeroelastic effects such as galloping, amplitude dependent crosswind and torsional lock-in excitations, and the separation/reattachment mechanis,m for torsion. The presence of eccentricity and coupling, both structural and aerodynamic, have also been found to have a significant effect on the excitation and response processes.

Interference excitations caused by the presence of adjacent tall buildings, either upstream or downstream, have been shown to readily increase the alongwind, crosswind and torsional responses by 30% to 40%

at low to mid range of reduced wind velocities, and by much more at the high range of reduced velocities and at critical building separations and positions. Potential increases in response of such a magnitude are an important consideration in the design of modern tall buildings.

The role of mass, stiffness, damping and aerodynamic shape in the reduction of wind induced vibrations of tall buildings is dependent on the type of loading and is rather complex except that an increase in the damping capacity is always beneficial. Wind induced response generally is not very sensitive to changes in either the mass or stiffness although an increase in the mass or stiffness of a building can be advantageous in reducing acceleration response. Therefore, not surprisingly, a considerable amount of resources and effort has been devoted in recent years to the research of passive and active vibration control. In contrast, the potential benefits of aerodynamic modifications of building shape have so far received little attention. Relatively minor modifications to the aerodynamic shape of a building while retaining the basic plan shape, such as the introduction of chamfered corners or vertical and horizontal through building openings, have been shown to significantly reduce wind induced responses, particularly crosswind response.

This paper reviews recent advances in the understanding of interference excitations and the associated increase in response of tall buildings, and the potential beneficial effects of aerodynamic modifications of building shape in reducing the wind induced response of tall buildings.

INTERFERENCE EXCITATION OF TALL BUILDINGS

In a typical city environment where many tall buildings and structures are grouped closely together, the proximity of one building to another can influence the flow field around and the wind forces acting on the buildings. Such interference effects are particularly marked for a group of identical structures such as a line of chimneys. The interactions of neighbouring tall buildings, commonly referred to as interference or proximity effects, can cause a significant increase in wind loads and exacerbate the problem of occupant perception of motions due to enhanced dynamic vibrations. The alongwind, crosswind and torsional responses of both upstream and downstream buildings in a group may be affected for wind from all directions. It is evident the owner of a new building may be held responsible for any adverse wind effect on existing buildings, adding a new dimension in litigation and professional liability.

The effects of interference from adjacent buildings, existing or proposed, are now routinely studied during wind tunnel model testing of tall buildings and structures. A wealth of useful information has been collected through these test but rarely is this information widely available because of the proprietary nature of those tests. Nevertheless, significant advances in the understanding of the various interference excitation mechanisms and their potentials in causing significant increases in the response of tall buildings have been achieved.

Alongwind and crosswind responses of tall buildings under interference excitations

The effects of interference on the alongwind and crosswind responses of tall buildings have been investigated by Melbourne and Sharp (1976), Saunders and Melbourne (1980), Ruscheweyh (1980), Saunders (1980), Sykes (1983), Bailey and Kwok (1984, 1985), Blessmann (1985), Kareem (1987), Kwok (1989), Taniike (1991) and Yahyai et al (1992) using aeroelastic modelling techniques. These tests were undertaken in turbulent boundary layer wind tunnels in simulated wind flows ranging from that over flat open country terrain to highly turbulent urban terrain. Most of the tall building models studied, conveniently referred to as the principal buildings, were of square or rectangular cross-sections with height to breadth ratio ranging from 2 to 9. The tests covered a full range of reduced wind velocities from 2 to 14. In addition to having an identical interfering building upstream, the effects of a downstream interfering building were also investigated by Bailey and Kwok (1984, 1985), Kwok (1989) and Yahyai et al (1992). Saunders and Melbourne (1980), Saunders (1980), and Kareem (1987) also tested configurations of a pair of upstream interfering buildings. Other variables studied include interfering building height (Melbourne and Sharp, 1976), shape (Bailey and Kwok, 1984, 1985), and size (Saunders and Melbourne, 1980; Saunders, 1980; Taniike, 1991).

The increased alongwind and crosswind responses of a square cross-section tall building with a height to breadth ratio of 9, the principal building, under interference excitation and proximity effects of a neighbouring building of the same height in open country, suburban and urban terrains were investigated in a series of wind tunnel model tests conducted by Bailey and Kwok (1984, 1985) and Kwok (1989). The interference excitation mechanisms were also studied in detail. The interference effects are usualiy expressed in the form of contours of buffeting factor (BF) defined as follows:

$$BF = \frac{\text{Overturning moment (neighbouring structure present)}}{\text{Overturing moment (isolated structure)}}$$

The buffeting factor contours shown in Fig. 1 for the principal building under interference excitation by an identical building in an open country terrain and at a reduced wind velocity of 6 show large increases in alongwind and crosswind responses of up to around 80% for an upstream interference situation and more than 400% in the alongwind response when the interfering building was in a critical downwind position. A reduced velocity of 6 would be representative of a 5 to 10 year return period design wind speed for serviceability and occupant comfort consideration for typical tall buildings. The higher values of both the alongwind and crosswind buffeting factors span out diagonally from the principal building suggesting the excitation forces intensify near the edge of the wake of the upstream building. Similar distribution of the buffeting factors is evident in the results obtained by Yahyai et al (1992).

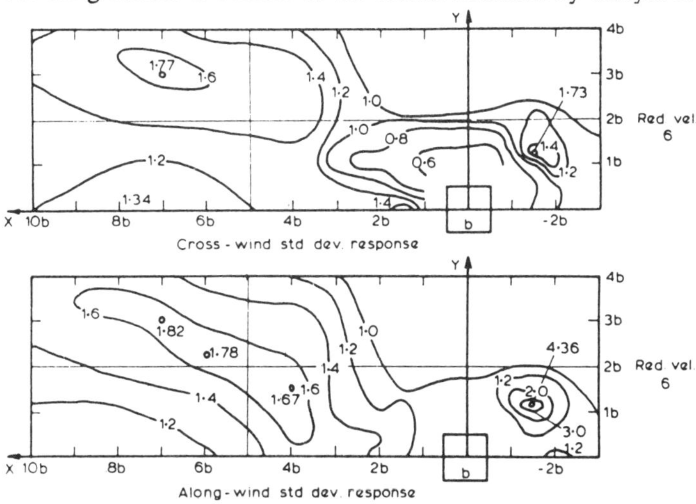

Fig. 1 Buffeting factors caused by an identical interfering building

The interference excitation mechanisms can be studied by examining the force and wake spectra of the principal building. The alongwind and crosswind force spectra of the principal building at two critical interference positions near the edge of the wake of the upstream building presented in Fig. 2 show substantial increases in excitation forces which are consistent with the observed increases in responses. Excitation forces associated with vortex shedding from the upstream building are clearly evident in the alongwind force spectrum.

A downstream interfering building generally has very little effect on the response of the principal building upstream except for a small region shown in Fig. 1. Similar results were obtained by Yahyai et al (1992). When the spacing between the buildings is at a critical width which produces a channel wide enough for the wind to converge and stream between the two buildings, the wake characteristics of the principal building are significantly altered. The spectrum of velocity

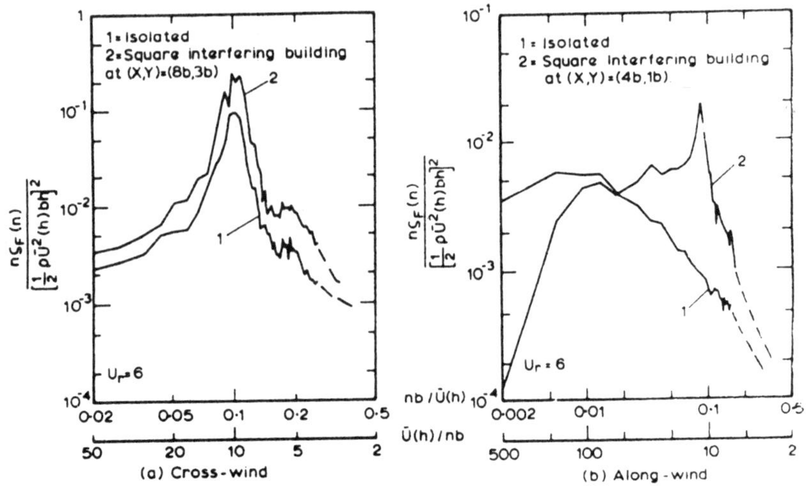

Fig. 2 Force spectra of the principal building

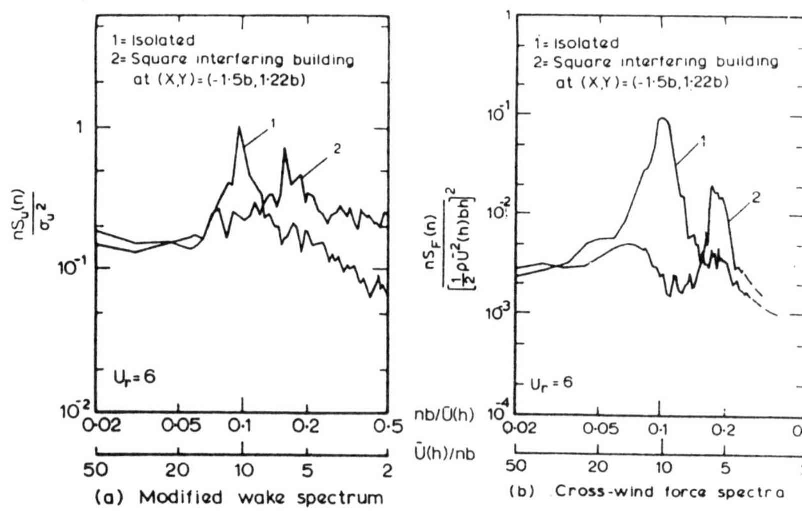

Fig. 3 Wake and force spectra casued by a downstream building

fluctuations in the modified near wake presented in Fig. 3(a) reveals a dominant reduced frequency of 0.17 which has been shown to induce resonant type response for the principal building upstream at a reduced wind velocity of around 6. Both the alongwind and crosswind responses were affected and the principal building underwent large elongated elliptical movements, as shown in Fig. 4, along the diagonal line between the two buildings. The effect on the crosswind force spectrum can be seen in Fig. 3(b) which shows a high concentration of excitation energy at the resonant reduced frequency of 0.17.

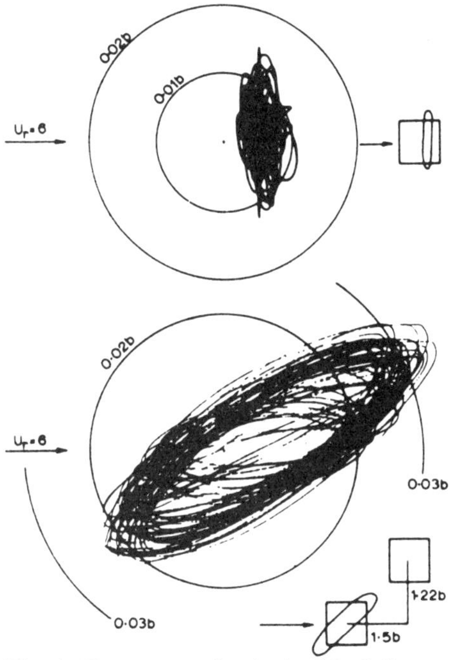

Fig. 4 Response of principal building caused by a downstream building

The interference effects on the principal building are dependent on the plan shape of the upstream interfering building. The effects of a circular cross-section building upstream of the square cross-section principal building were investigated by Bailey and Kwok (1984, 1985). Due to the difference in Strouhal Number, 0.15 for the circular and 0.1 for the square cross-section buildings, resonant buffeting occurred at a critical reduced wind velocity of 6.8 at which the vortex shedding frequency of the upstream building coincided with the natural frequency of the principal building. The corresponding increase in excitation forces is clearly evident in the alongwind and crosswind force spectra shown in Fig. 5. As a result, very large buffeting factors were recorded for both the alongwind and crosswind responses of the principal building at this critical reduced wind velocity, but at different critical interference positions, as shown in Fig. 6. Similar observations were made by Taniike (1991) using a small square cross-section upstream building which has a breadth 40% of that of the principal building.

The critical reduced wind velocity at which resonant buffeting occurs depends on the vortex shedding characteristics of the upstream interfering building and may occur at lower and more frequent wind velocities. The sensitivity of interference effects to reduced wind velocity is shown in Figs. 7 and 8 for the alongwind and crosswind responses respectively of the square cross-section principal building for a few critical

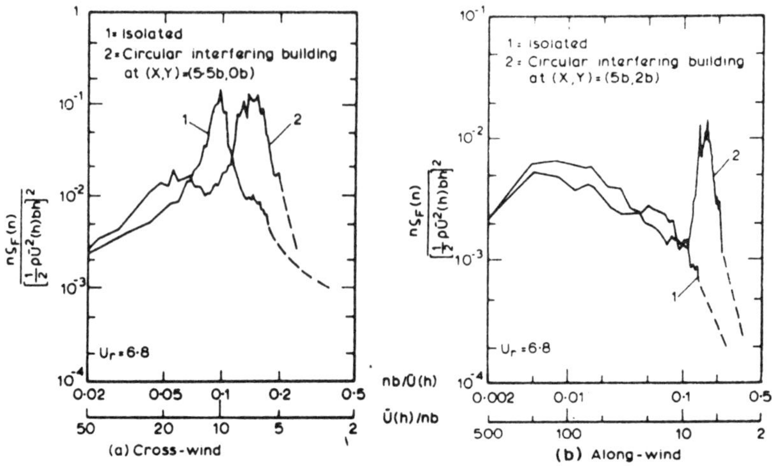

Fig. 5 Force spectra of the principal building

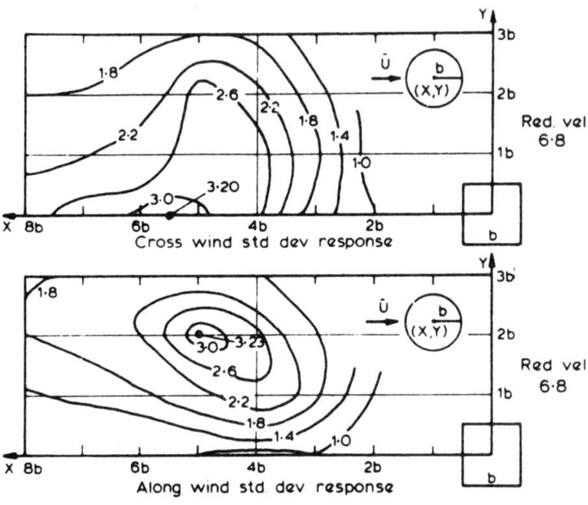

Fig. 6 Buffeting factors caused by a circular upstream building

interference positions. Large values of the alongwind buffeting factor occur at a critical reduced velocity of around 11 for a square cross-section upstream building identical to the principal building. Similarly, large values of both the alongwind and crosswind buffeting factors occur at a critical reduced velocity of 6.8 for a circular cross-section upstream building. However, the crosswind buffeting factor for a square cross-section upstream building decreases as reduced velocity increases due to the disruption of the vortex shedding mechanism of the principal building caused by the increase in turbulence created by the upstream building.

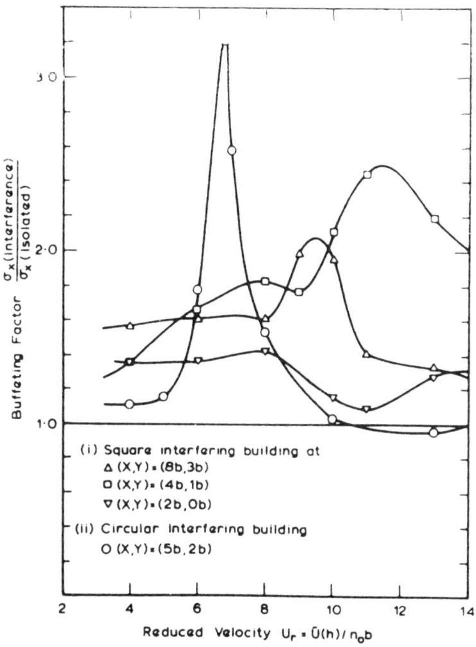

Fig. 7 Alongwind response of the principal building under interference

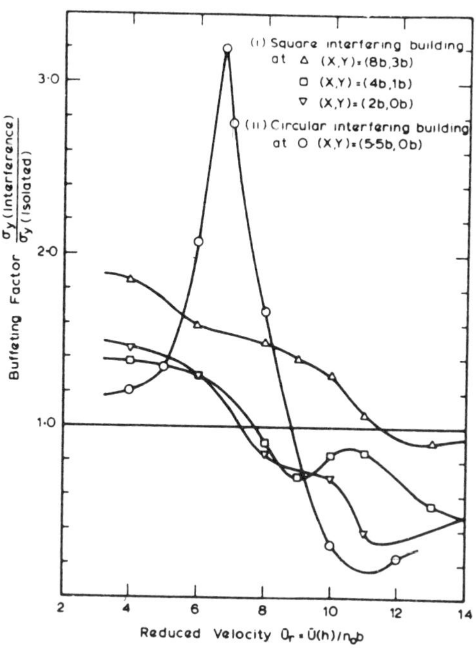

Fig. 8 Crosswind response of the principal building under interference

The effects of interference are reduced with increase in turbulence in the flow, according to Melbourne and Sharp (1976), Bailey and Kwok (1985), and Kwok (1989). Taniike (1991) suggested that interference effect caused by an upstream building is negligible when the turbulence intensity reaches 17% to 18% at two-third building height. However, Melbourne and Sharp (1976) and Kwok (1989) reported increases in response of up to 25% and 40% for a square cross-section tall building with a height to breadth ratio of 6 and 9 respectively in an urban terrain of which the turbulence intensities are significantly higher than the value suggested by Taniike (1991).

The effects of a single upstream building have been reported by Melbourne and Sharp (1976) to be evident up to 16 building breadths downstream. However, interference effects have been shown by Melbourne and Sharp (1976) and Saunders and Melbourne (1980) to be significantly reduced when the height of the upstream building was less than two-third the height of the principal building.

For a pair of upstream buildings set 5 breadth apart and at equidistance upstream of the principal building, Saunders and Melbourne (1980) reported a further amplification of the interference effects compared with that of a single upstream building, and the interference effects were noticeable even at an equivalent prototype distance of 1 km or about 30 building breadths.

In addition to those aeroelastic model tests, pressure measurement and force balance techniques have also been employed to study the effects of interference on tall buildings. Blessmann and Riera (1985) studied interference effects using a pressure-tapped model of a square cross-section building with a height to breadth ratio of 6, and Pathak et al (1989) two buildings with ratios of 2 and 8. Mean force coefficients, including torsion, were determined by integrating the mean pressure distributions. Thoroddsen et al (1985) used a high frequency force balance to determine the mean and dynamic wind loads on five tall building models of different shapes in the presence of an upstream building. Taniike (1992) employed both force balance techniques and pressure-tapped models in the study of a square cross-section building with a height to breadth ratio of 4.5 under the influence of an identical or different size square cross-section buildings. The results obtained from these studies generally agree with those obtained from aeroelastic tests.

English (1985, 1990, 1993) also studied the effects of interference using force balance technique and concentrated on the shielding effect instead of the amplification effect of square and rectangular cross-section buildings. The shield factor for the mean alongwind response was found to be sensitive to the building height to breadth ratio, wind flow turbulence characteristics, and in particular separation distance. By incorporating available data from tests performed in well-simulated atmospheric turbulent boundary layer flows, the following equation was proposed for the mean alongwind response shielding factor:

$$SF = -0.05 + 0.65x + 0.29x^2 - 0.24x^3$$

in which x = log(RSD)
RSD = reduced separation distance = d(h+w)/hw
d = separation between buildings
h = height of building
w = width of building

An approach similar to that adopted by English (1993) may be employed to derive empirical equations for the amplification effects of interference of which there is much useful data on buffeting factors.

Torsion response of tall buildings under interference excitations

Results of mean torsion coefficient obtained by Blessmann and Riera (1985) and Pathak et al (1989), and mean and dynamic torsion coefficients by Thoroddsen et al (1985) for buildings under interference suggest the amplification factor can be as high as 3.

Zhang et al (1994) employed aeroelastic technique to study the torsional response of a square cross-section building with a height to breadth ratio of 5 under interference by four types of building, a square and a circular cross-section building of the same breadth (diameter) and of 60% the breadth (diameter) of the principal building, in an open country terrain. Critical upstream and downstream interference positions and the corresponding buffeting factors were identified for a range of reduced wind velocities. The buffeting factors for the dynamic torsional response of the principal building at a reduced velocity of 6 are shown in Fig. 9 for an identical interfering building, and in Fig. 10 for the smaller square cross-section interfering building. The torsional interference excitation mechanisms are similar to those for the alongwind and crosswind responses and are dominated by resonant buffeting which occurs when the frequency of vortex shedding from the upstream building coincides with the natural frequency of the principal building. The dependence of torsional interference excitation on reduced wind velocity can be clearly seen in Fig. 11 which highlights the critical reduced velocities of 7 for the circular cross-section interfering building with the same diameter as the principal building, and 6 and 4.2 for the smaller square and smaller circular cross-section interfering buildings respectively. Torsional buffeting factors are as high as 2.2 for the circular cross-section interfering building with the same diameter as the principal building.

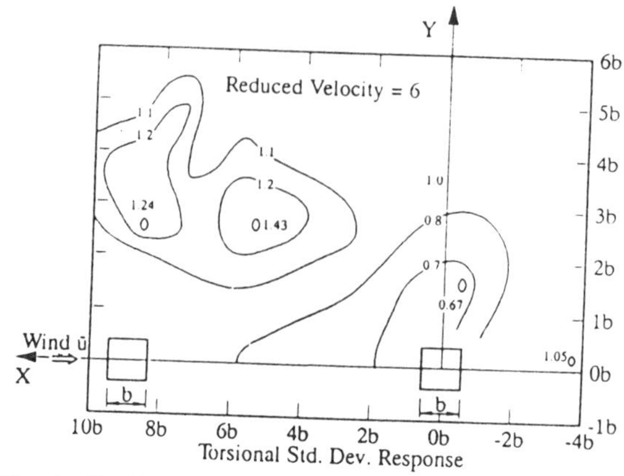

Fig. 9 Buffeting factors caused by an identical interfering building

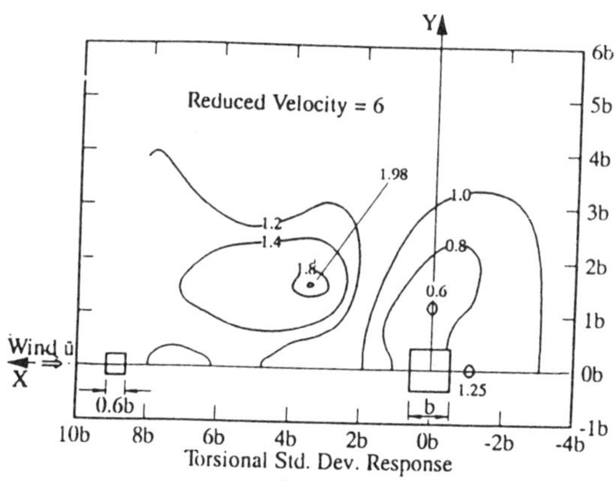

Fig. 10 Buffeting factors caused by a smaller interfering building

Zhang et al (1992) also studied the torsional interference effects on a building with a 10% eccentricity between the mass and elastic centre of the building. The torsional interference excitation mechanisms and the resultant torsional response characteristics of the eccentric principal building were found to be essentially the same as those for the normal centric building but the buffeting factors due to resonant buffeting were substantially increased, as shown in Fig. 12, to a value as high as 4.6.

Design considerations

The significance of interference effects in causing large increases in alongwind, crosswind and torsional responses of tall building has been clearly demonstrated for either an upstream or a downstream interfering building. Upstream interference can be felt at a considerable distance away and has been found to be most prominent when the principal

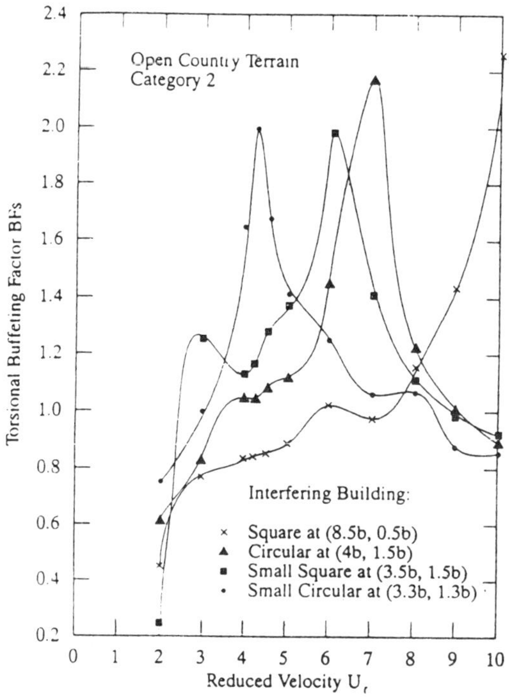

Fig. 11 Torsional response of the principal building under interference

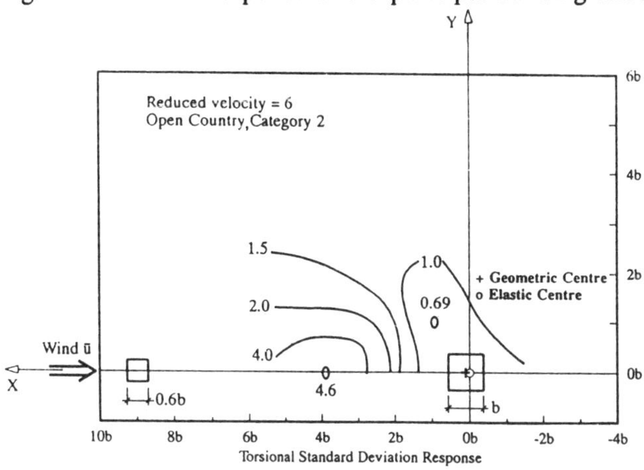

Fig. 12 Buffeting factors of an eccentric building caused by a smaller interfering building

building is located near the edge of the wake shed by the upstream interfering building. The vortex shedding characteristics of the plan shape of an upstream interfering building can cause resonant buffeting, at which the vortex shedding frequency of the upstream building coincides with the natural frequency of the principal building, at a lower and more frequent wind velocity. Downstream interference is confined to a relatively small area in close proximity which produces a critical separation between the two adjacent buildings.

The magnitude of increase in wind load and response due to interference is strongly dependent on the turbulence created by the approach terrain. Tall buildings situated at the edge of the water front or at the edge of a city centre are the most vulnerable. For tall buildings located well within a city centre with its highly turbulent wind flow characteristics, the effects of interference are significantly reduced. However, if there are a few tall buildings or other tall structures which protrude well above the surrounding building canopy, the potential for interference effects would not be diminished and should be assessed in accordance with the wind flow condition above the canopy.

The range of reduced wind velocities at which interference effects are significant encompasses design wind speed which is typical for serviceability consideration, particularly in relation to occupant comfort, to that well in excess of ultimate strength design. In terms of design, a potential increase in wind load of between 10% to 30% may be expected to be within the normal safety factor in design. However, an increase of between 30% to 50% would be of considerable concern. While an increase of such magnitude may not be catastrophic, the serviceability requirement of the building, in particular the occupant comfort criteria, will almost certainly be tested. An increase of greater than 50%, let alone the more than doubling of wind load which has been observed in some of those wind tunnel experiments, will be well outside the safety margin used in a normal design. As a result, a comprehensive program of wind tunnel model test is now routinely conducted to aid the design of modern tall buildings and structures to investigate the effects of interference from adjacent buildings, both existing and proposed.

AERODYNAMIC MODIFICATIONS OF BUILDING SHAPES

One of the best known example of aerodynamic devices employed to suppress resonant vibrations caused by vortex shedding is the addition of a helical strake to chimney stacks. This and other forms of aerodynamic treatment to structures with a circular cross-section are the subjects of a detailed review by Zdrakovich (1981). Although many types of aerodynamic devices have been found to be effective in the suppression of wind induced response of cylindrical and prismatic structures, the alterations to the shape of the structure and the substantial nature of the alterations often make them impractical and aesthetically undesirable in conventional modern tall buildings.

Up until the early 1980's, there have been comparatively few investigations into the aerodynamic treatment of prismatic structures such as conventional tall buildings (e.g. Davenport, 1971; Cermak, 1972; Naudascher et al, 1981). The feasibility of using actively controlled aerodynamic appendages has also been explored. Soong and Skinner (1981) conducted wind tunnel experiments on a simple aeroelastic model of a tall building with an aerodynamic appendage in the form of a flat plate which was actively controlled to protrude above the roof of the building model. Abdel-Rohman (1983) employed numerical techniques to examine the effectiveness of an actively controlled hinged roof type aerodynamic appendage on top of a tall building under wind excitation. Although both types of actively controlled aerodynamic devices have been shown to be effective in the suppression of wind induced vibrations, the size and aesthetics of those devices, and the engineering and cost requirements associated with their deployment require more detailed investigation before they can be widely accepted as control devices.

More recent studies have concentrated on the effects of relatively minor aerodynamic modifications of building shape in reducing wind induced response of tall buildings.

Effects of aerodynamic modifications of building shapes on wind induced response of tall buildings

Wind tunnel experiments were conducted by Kwok and Bailey (1987), Kwok et al (1988), and Kwok (1988) to investigate the effects of aerodynamic devices, building edge configuration, and through building opening on the wind induced vibrations of tall buildings. Aeroelastic models of a square cross-section building with a height to breadth ratio of 9 and the CAARC Standard Tall Building which has a rectangular plan shape, as shown in Fig. 13, were tested in open country and suburban terrains. The building model configurations tested included fins and vented fins at the corners, slotted corners, chamfered corners, and horizontal slots. The modifications to the building corners ranged from around 9% to 16% of the building breadth. It is evident that aerodynamic modifications which increase the projected area or the effective width of a building, such as fins at corners, would in general not be beneficial.

Horizontal slots, slotted corners and chamfered corners were found to be effective in causing significant reductions in both the dynamic alongwind and crosswind responses of the rectangular cross-section CAARC Standard Tall Building, as shown in Figs. 14 and 15. With chamfered corners, there was up to a 40% reduction in the alongwind response within a reduced wind velocity range of 3 to 20 compared with that for the plain building.

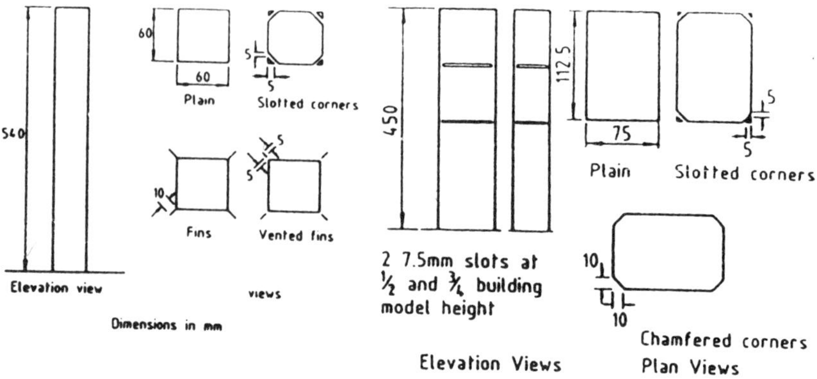

Fig. 13 Aerodynamic modifications of building shapes

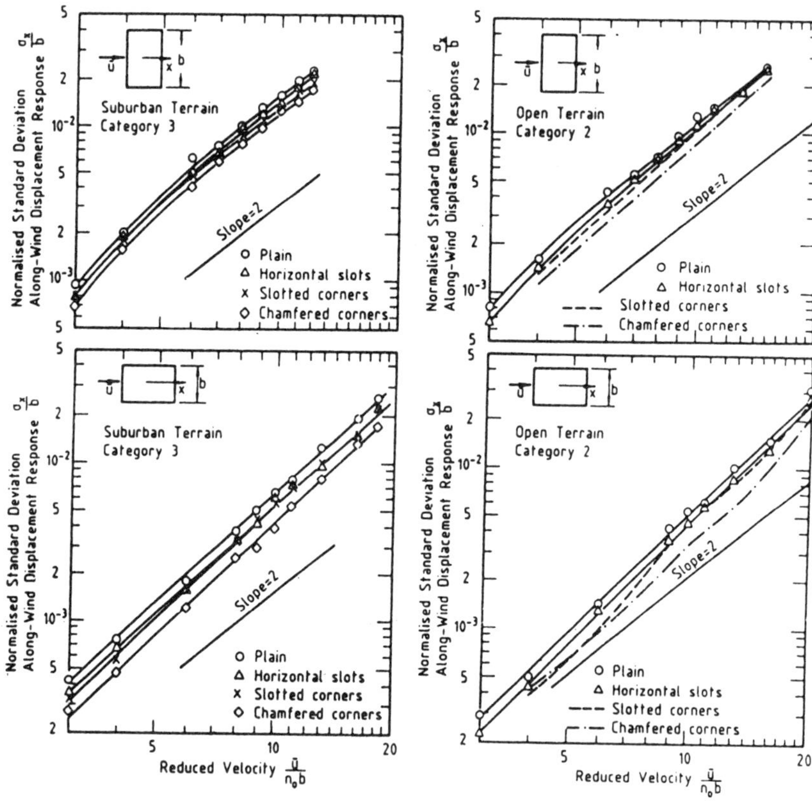

Fig. 14 Alongwind response of buildings with aerodynamic modifications

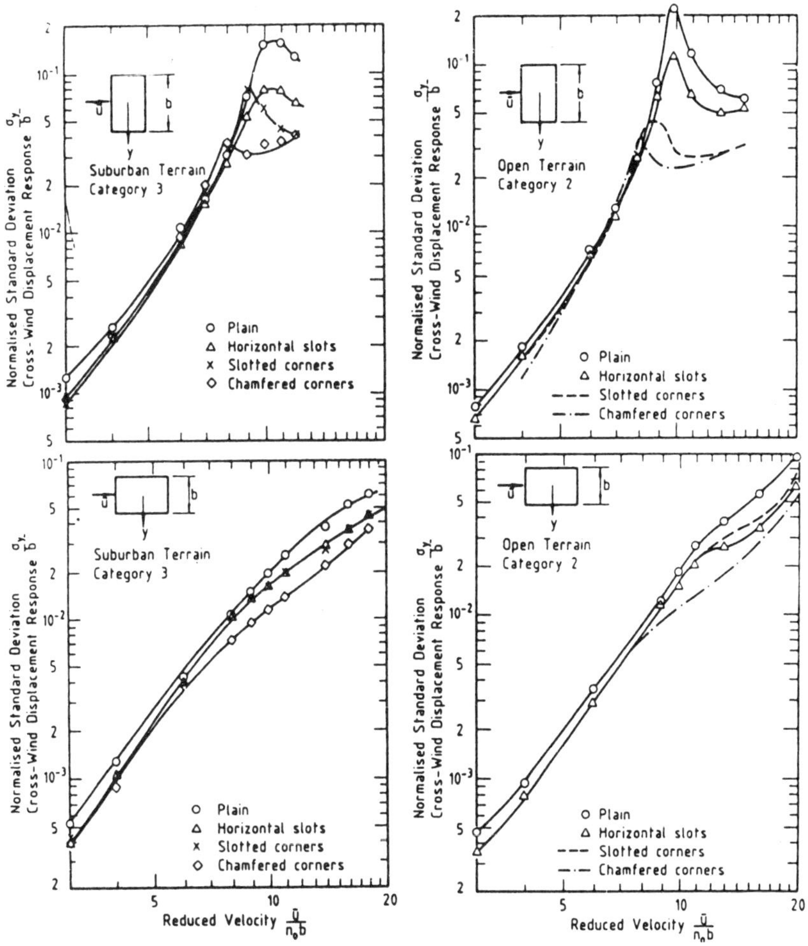

Fig. 15 Crosswind response of buildings with aerodynamic modifications

With wind normal to the wide face of the building, the dynamic crosswind response of the modified buildings were found to be up to 30% smaller than that for the plain building at the low range of reduced wind velocities. The critical reduced wind velocity changed from a value of approximately 10 for the plain building and the building with horizontal slots, to about 8 for the building with chamfered corners, and about 9 for the one with slotted corners. The crosswind response characteristics of those modified buildings changed accordingly, with resonant crosswind response occurring at the modified critical reduced velocity. The resonant crosswind response at the modified critical reduced velocities of 8 for the building with chamfered corners and 9 for the building fitted with slotted corners were found to be larger than those at the corresponding reduced velocities for the plain building.

With the incident wind normal to the narrow face of the building, the crosswind response was found to be essentially wake excited. The long afterbody of the building apparently has a disruptive effect on the vortex shedding process and there is no significant response peak associated with a dominant critical velocity effect. Building modifications such as horizontal slots, slotted corners and chamfered corners were found to result in a 30% or more reduction in the crosswind response.

The design wind velocities for modern tall building are at the low range of reduced wind velocities for serviceability requirement and at the mid to high range of reduced velocities for strength considerations. Modifications such as horizontal slots, slotted corners and chamfered corners can offer significant reductions in the magnitude of both the alongwind and crosswind responses compared with a plain building.

The effects of building plan shape on the crosswind response of tall buildings were also investigated by Melbourne and Cheung (1988) and Melbourne (1989). It was concluded from wind tunnel aeroelastic model studies and sample calculations for practical applications that for tall buildings with nominally symmetrical or square plan shapes, modest rounding or chamfering of corners up to 10% does not significantly reduce the crosswind response at low values of reduced wind velocities relevant to the serviceability acceleration criterion for occupant comfort. However, very significant reductions in the ultimate limit state design loads can be achieved at higher reduced velocities. For more substantial rounding or chamfering of corners such that the building plan shape approaches that of a rough circular shape, for example octagonal or hexagonal shape, the crosswind force spectrum was found to reduce at both the lower and higher ranges of reduced velocities. A similar characteristic was also observed for tall buildings with reduced upper level plan areas, for examples tapering, by cutting corners or dropping off corners progressively as height increases. However, for building plan shapes which are elliptical or elongated octagonal with a major to minor axis ratio of about 3:2, the critical reduced wind velocities were found to be significantly lower. The resultant high crosswind response and acceleration level makes it more difficult to meet occupancy comfort criteria. Furthermore, those plan shapes were found to exhibit significant torsional response about the vertical axis.

Dutton and Isyumov (1990) studied the effects of vertical openings on the crosswind response of a square cross-section tall building with a height to breadth ratio of 9 to 1. The crosswind base bending moment spectra obtained from force balance tests and the corresponding computed crosswind tip deflections of the tall building are shown in Figs. 16 and 17 respectively. Alongwind through building openings and in particular combined alongwind and crosswind openings were found to cause a pronounced reduction of the vortex shedding induced forces and hence the crosswind dynamic deflection of the building. The critical reduced wind velocity shifted to a slightly higher value which implied that resonant vibrations of the building would be postponed to a higher wind speed with a longer return period. However, there was a second spectral peak and hence a corresponding increase in crosswind response at a reduced velocity of around 6 which may make it more difficult to meet occupancy comfort criteria. The level of disruption to the vortex shedding process varied with the width of the opening and large reductions were observed for openings as small as 4% of the building width. Results obtained from a pressure model study suggested the mean base pressure coefficient is a good indicator of the overall aerodynamic, including the vortex shedding process, of a tall building.

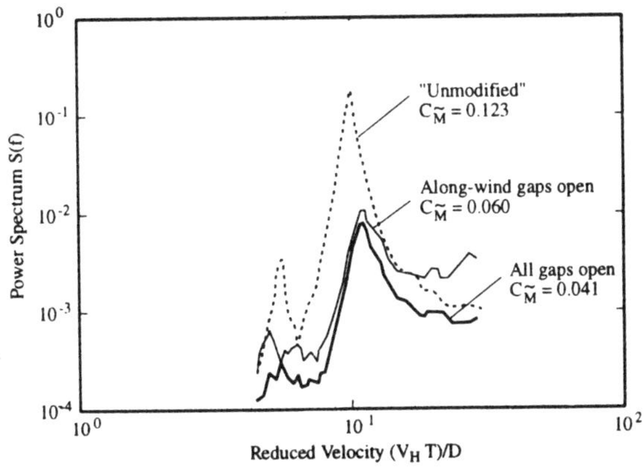

Fig. 16 Crosswind base bending moment spectra of buildings with aerodynamic modifications (from Dutton and Isyumov, 1990)

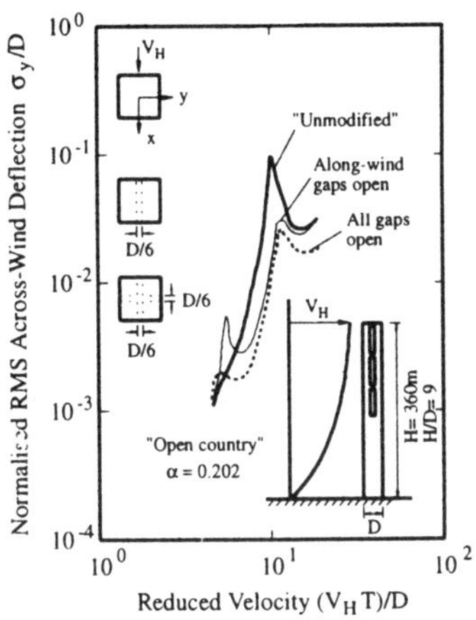

Fig. 17 Crosswind response of buildings with aerodynamic modifications (from Dutton and Isyumov, 1990)

The potential beneficial effects of a through building opening in a 390 m high office tower project were studied in a wind tunnel using both force balance and aeroelastic model techniques and reported by Isyumov et al (1992). The results show that a more sculptured building top, in this case a triangular pyramid shape for the top 15% of the building, can moderate both the alongwind and crosswind responses. A venting or "bleeding" of the building wake provided by a through building opening near the top resulted in additional reductions in these responses.

The effects of building plan shape on the aerodynamic forces and dynamic response of a 600 m high 150 storey super tall building were investigated by Hayashida and Iwasa (1990) and Hayashida et al (1992) using force balance, aeroelastic and pressure model techniques. The building plan shapes tested include a plain square shape and square shapes with notched or rounded corners, three-point star shapes with or without notched corners, triangular shapes with or without notched corners, and a circular shape with vertical ribs, each having a prototype floor area of 6,400 m^2. The crosswind force spectra obtained by force balance technique for two typical wind directions are shown in Fig. 18. The critical reduced wind velocity at which resonant vibration occurs, and which is defined by the Strouhal Number, was found to vary according to the shape and modification to the corners. The two three-point star shapes show no distinctive spectral peak associated with the vortex shedding process. Modifications to the corners were found to cause marked reductions in crosswind force around the vortex shedding frequency and at the low reduced frequencies, particularly for the square shape. The computed crosswind displacement response for the different plan shapes based on the force balance test results shown in Fig. 18 are presented in Fig. 19. For a prototype design wind speed of 64.6 m/s, the maximum crosswind displacement response was ranked in descending order of magnitude as follows: 1) square shape; 2) square shape with notched corners; 3) square shape with rounded corners; 4) circular shape; 5) triangular shape with notched corners; 6) three-point star shape; 7) triangular shape; and 8) three-point star shape with notched corners. The ranking at a lower design wind speed for occupant comfort consideration is different but the plain square shape still has the highest value of crosswind displacement and hence acceleration response. The results obtained from aeroelastic tests were found to be significantly different to those derived from force balance tests. While the plain square shape was found to generally have a larger alongwind and crosswind displacement responses than the other shapes, the circular shape clearly exhibited aeroelastic lock-in type response at a critical reduced velocity of about 8 around which the crosswind response was considerably larger than the other shapes, including the plain square shape. This highlights that aeroelastic effects can be significant for super tall buildings and force balance technique can severely underestimate the dynamic response.

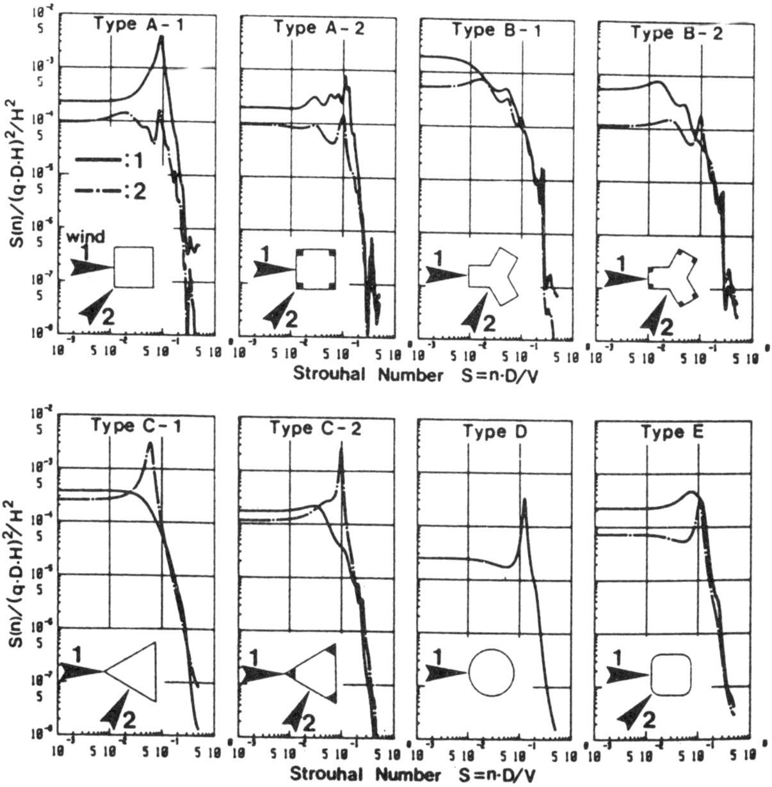

Fig. 18 Crosswind force spectra of buildings with different plan shapes (from Hayashida and Iwasa, 1990)

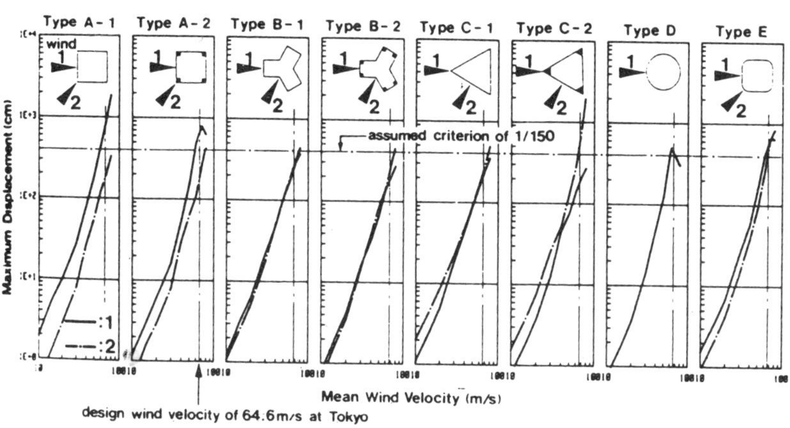

Fig. 19 Maximum crosswind displacement response of buildings with different plan shapes (from Hayashida and Iwasa, 1990)

Miyashita et al (1993) employed force balance technique to study the characteristics of both the wind forces acting on and the response of a square cross-section building shown in Fig. 20 with modified corners or openings. The building studied has a height to breadth ratio of 6. Modifications to the building included chamfered or notched corners 10% of the breadth, and through building openings 25% of the breadth of the building. Coefficients and spectra of fluctuating wind forces and their correlations were measured, and the displacement response for an ultimate limit state design wind speed of 62 m/s were determined for different angles of wind incidence. With through building openings, particularly when they were through both the X and Y building axes, substantial reductions in response were found to occur at small angles of wind incidence. However, modifications to the building corners, particularly notched corners, were found to be not particularly effective and can cause an increase in response at low angles of wind incidence.

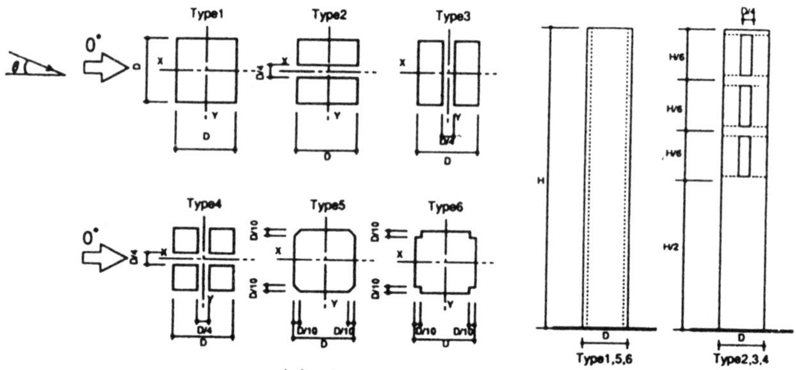

(a) Plane (b) Elevation

Fig. 20 Aerodynamic modifications of building shapes (from Miyashita et al, 1993)

Design considerations

Aerodynamic modifications of building shape, including slotted and chamfered corners, horizontal and vertical through building openings, sculptured building top, tapering, and dropping off corners have been shown to significantly reduce the wind induced loads and response of tall buildings. Modifications to the building corners such as slotted or chamfered corners need to be applied to the corner region greater than about 10% of the building breadth to be beneficial. Modifications which increase the projected area or the effective breadth of the building would in general not be beneficial.

Reductions of up to 40% in the alongwind response can be achieved by aerodynamic modifications for a wide range of reduced wind velocities which encompasses serviceability and ultimate limit state design wind speeds for most modern tall buildings. Crosswind response at serviceability design wind speed can be reduced by around 30% while very substantial reductions can be achieved at the ultimate limit state design wind speed.

The vortex shedding characteristics of a building are often altered by aerodynamic modifications. If the Strouhal Number is lowered by the modification, the critical reduced wind velocity at which the vortex shedding frequency coincides with the natural frequency of the building will be raised to a higher value with a longer return period. On the other hand, if the modification causes an increase in the Strouhal Number, resonant response will occur at a lower reduced velocity with a shorter return period which may increase the ultimate limit state design load or make it more difficult to meet serviceability requirements such as occupant comfort criteria. Occasionally secondary spectral peak may appear and the corresponding increase in crosswind response may adversely affect the performance of the building.

The wind load and wind induced response characteristics of a tall building are dependent on the basic plan shape of the building. The potential benefits of a plan shape which has a lower Strouhal Number has already been mentioned and is a parameter which can offer significant benefit when correctly selected. In the design of modern tall buildings, particular for those with an unusual plan shape and for super tall buildings, a comprehensive program of wind tunnel model test is always beneficial and provides useful information which can be used to assess the potential benefits of altering the plan shape or employing other forms of aerodynamic modifications to the building.

CONCLUDING REMARKS

Recent advances in the understanding of interference excitations and the associated increases in the response of tall buildings, and the potential beneficial effects of aerodynamic modifications of building shape in reducing the wind induced response of tall buildings have been reviewed in this paper.

Interference excitations from either an upstream or a downstream building can cause significant increases in the alongwind, crosswind and torsional responses of tall buildings. Effects of interference from adjacent buildings, both existing and proposed, should be carefully assessed in the design of tall buildings.

Aerodynamic modifications of building shape, including slotted and chamfered corners, horizontal and vertical through building openings, sculptured building top, tapering, and dropping off corners have been shown to significantly reduce the wind induced loads and response of tall buildings. For modern tall buildings, particularly the new generation of super tall buildings, these aerodynamic modifications may be employed to meet serviceability requirements such as occupant comfort criteria and to reduce the ultimate limit state design wind load.

REFERENCES

Abdel-Rohman, M. (1983), "Control of Tall Buildings Response by Aerodynamic Appendages", Journal of Wind Engineering and Industrial Aerodynamics, 13, pp. 253-260.

Bailey, P.A. and Kwok, K.C.S. (1984), "Dynamic interference and Proximity Effects between Tall Buildings", Proceedings of 3rd International Conference on Tall Buildings, Hong Kong and Guangzhou, December, pp. 305-311.

Bailey, P.A. and Kwok, K.C.S. (1985), "Interference Excitation of Twin Tall Buildings", Journal of Wind Engineering and Industrial Aerodynamics, 21, pp. 323-338.

Blessmann, J. (1985), "Buffeting Effects on Neighbouring Tall Buildings", Journal of Wind Engineering and Industrial Aerodynamics, 18, pp. 105-110.

Blessmann, J. and Riera, J.D. (1980), "Interaction Effects in Neighbouring Tall Buildings", Proceedings of the Fifth International Conference on Wind Engineering, Fort Collins, USA, July, Vol. 1, pp. 381-395, Pergamon.

Cermak, J.E. (1972), "Wind Effects on Tall Buildings - Areas for Research", Proceedings of International Conference on Planning and Design of Tall Buildings, Lehigh University, August, pp.435-440.

Davenport, A.G. (1971), "The Response of Six Building Shapes to Turbulent Wind", Phil. Trans. Roy. Soc. Lond. A. 269, pp. 385-394.

Dutton, R. and Isyumov, N. (1990), "Reduction of Tall Building Motion by Aerodynamic Treatments", Journal of Wind Engineering and Industrial Aerodynamics, 36, pp. 739-747.

English, E.C. (1985), "Shielding Factors from Wind-Tunnel Studies of Mid-rise and High-rise Structures", Proceedings of 5th U.S. National Conference on Wind Engineering, Lubboch, USA, November, pp. 4A-49 - 4A-56.

English, E.C. (1990), "Shielding Factors from Wind-Tunnel Studies of Prismatic Structures", Journal of Wind Engineering and Industrial Aerodynamics, 36, pp. 611-619.

English, E.C. (1993), "Shielding Factors for Paired Rectangular Prisms: An Analysis of Along-wind Mean Response Data from Several Sources", Proceedings of 7th U.S. National Conference on Wind Engineering, Los Angeles, May, pp. 193-201.

Hayashida, H. and Iwasa, Y. (1990), "Aerodynamic Shape Effects of Tall Building for Vortex Induced Vibration", Journal of Wind Engineering and Industrial Aerodynamics, 33, pp. 237-242.

Hayashida, H., Mataki, Y. and Iwasa, Y. (1992), "Aerodynamic Damping Effects of Tall Building for a Vortex Induced Vibration", Journal of Wind Engineering and Industrial Aerodynamics, 41-44, pp. 1973-1983.

Isyumov, N., Fediw, A.A., Colaco, J. and Banavalkar, P.V. (1992), "Performance of a Tall Building under Wind Action", Journal of Wind Engineering and Industrial Aerodynamics, 41-44, pp. 1053-1064.

Kareem, A. (1987), "The Effect of Aerodynamic Interference on the Dynamic Response of Prismatic Structures", Journal of Wind Engineering and Industrial Aerodynamics, 25, pp. 365-372.

Kwok, K.C.S. (1988), "Effect of Building Shape on Wind-induced Response of Tall Buildings", Journal of Wind Engineering and Industrial Aerodynamics, 28, pp. 381-390.

Kwok, K.C.S. (1989), "Interference Effects on Tall Buildings", Proceedings of 2nd Asia-Pacific Symposium on Wind Engineering, Beijing, June, Vol. 1, pp. 446-453.

Kwok, K.C.S. and Bailey, P.A. (1987), "Aerodynamic Devices for Tall Buildings and Structures", Journal of Engineering Mechanics, ASCE, Vol. 113, No. 3, March, pp. 349-365.

Kwok, K.C.S., Wilhelm, P.A. and Wilkie, B.G. (1988), "Effect of Edge Configuration on Wind-induced Response of Tall Buildings", Engineering Structures, Vol. 10, April, pp. 135-140.

Melbourne, W.H. (1989), "Wind Loading and Environment of Tall Building", Proceedings of International Conference on Tall Buildings and City Development, Brisbane, October, pp.159-164.

Melbourne, W.H. and Sharp, D.B. (1976), "Effects of Upwind Buildings on the Response of Tall Buildings", Proceedings of Regional Conference on Tall Building, Hong Kong, September, pp. 174-191.

Melbourne, W.H. and Cheung, J.C.K. (1988), "Designing for Serviceable Accelerations in Tall Buildings", Proceedings of 4th International Conference on Tall Buildings, Hong Kong and Shanghai, April/May, Vol. 1, pp. 148-155.

Miyashita, K., Katagiri, J., Nakamura, O., Ohkuma, T, Tamura, Y, Itoh, M. and Mimachi, T. (1993), "Wind-induced Response of High-rise Buildings", Journal of Wind Engineering and Industrial Aerodynamics, 50, pp. 319-328.

Naudascher, E., Weske, J.R. and Fey, B. (1981), "Exploratory Study on Damping of Galloping Vibrations", Journal of Wind Engineering and Industrial Aerodynamics, 8, pp. 211-222.

Pathak, S.K., Ahuja, A.K. and Mir, S.A. (1989), "Effect of Interference between Two Prismatic Buildings on Wind Loads", Proceedings of 2nd Asia-Pacific Symposium on Wind Engineering, Beijing, June,

Vol. 1, pp. 454-459.

Ruscheweyh, H. (1980), "Dynamic Response of High Rise Buildings under Wind Action with Interference Effects from Surrounding Buildings of Similar Size", Proceedings of 5th International Conference on Wind Engineering, Fort Collins, USA, July, Vol. 2, pp. 725-734, Pergamon.

Saunders, J.W. (1980), "Further Measurements of Buffeting Effects of Twin Upstream Buildings", Proceedings of 7th Australasian Hydraulics and Fluid Mechanics Conference, Brisbane, August, pp. 83-86.

Saunders, J.W. and Melbourne, W.H. (1980), "Buffeting Effects of Upstream Buildings", Proceedings of 5th International Conference on Wind Engineering, Fort Collins, USA, July, Vol. 1, pp. 593-606, Pergamon.

Soong, T.T. and Skinner, G.T. (1981), "Experimental Study of Active Structural Control", Journal of the Engineering Mechanics Division, ASCE, Vol. 107, No. EM6, December, pp. 1057-1067.

Sykes, D.M. (1983), "Interference Effects on the Response of a Tall Building Model", Journal of Wind Engineering and Industrial Aerodynamics, 11, pp. 365-380.

Taniike, Y. (1991), "Turbulence Effect on Mutual Interference on Tall Buildings", Journal of Engineering Mechanics, ASCE, Vol. 117, No. 3, March, pp. 443-456.

Taniike, Y. (1992), "Interference Mechanism for Enhanced Wind Forces on Neighbouring Tall Buildings", Journal of Wind Engineering and Industrial Aerodynamics, 41-44, pp. 1073-1083.

Thoroddsen, S.T., Cermak, J.E. and Peterka, J.A. (1985), "Mean and Dynamic Wind Loading Caused by an Upwind Structure", Proceedings of 5th U.S. National Conference on Wind Engineering, Lubboch, USA, November, pp. 4A-73 - 4A-80.

Yahyai, M., Kumar, K., Krishna, P. and Pande, P.K. (1992), "Aerodynamic Interference in Tall Rectangular Buildings", Journal of Wind Engineering and Industrial Aerodynamics, 41-44, pp. 859-866.

Zdrakovich, M.M. (1981), "Review and Classification of Various Aerodynamic and Hydrodynamic Means for Suppressing Vortex Shedding", Journal of Wind Engineering and Industrial Aerodynamics, 7, pp. 145-189.

Zhang, W.J., Kwok, K.C.S. and Xu, Y.L. (1992), "Torsional Excitation and Interference Effects of Tall Buildings with Structural Eccentricity", Research Report No. R666, School of Civil and Mining Engineering, The University of Sydney, November.

Zhang, W.J., Kwok, K.C.S. and Xu, Y.L. (1994), "Aeroelastic Torsional Behaviour of Tall Buildings in Wakes", Journal of Wind Engineering and Industrial Aerodynamics, 51, pp. 229-248.

THE RESPONSE OF CHIMNEYS AND TOWER-LIKE STRUCTURES TO WIND LOADING

B.J. VICKERY

University of Western Ontario,
London, Ontario, Canada

SUMMARY

The paper is concerned with the prediction of both the along and across-wind response of slender tower-like structures in a turbulent wind. Particular attention is paid to the efficient calculation of the along-wind response and to a model for prediction of the across-wind response to vortex shedding.

INTRODUCTION

Towers and tower like structures such as chimneys are far more amenable to theoretical analysis than buildings. The prime difference between these two classes of structure is slenderness which enables the tower to be analysed as an essentially one-dimensional structure with aerodynamic properties that are well defined by the local cross-section. Such a simplification may not be completely adequate in the vicinity of a free end or in the vicinity of significant changes in cross-section but can be accepted in terms of the overall loading. While the one-dimensional rather than three-dimensional aerodynamics is a great simplification the slenderness may introduce more complexity to the dynamic analysis in the sense that many modes of vibration may contribute significantly to the response.

Slenderness is also commonly associated with values of $V_R = \bar{V}/f_o B$, [$\bar{V}$ = a representative mean speed, f_o = a representative natural frequency, B = a representative width] which are comparatively high and certainly larger than $1/S$ [S = Strouhal Number].

In such cases, the use of quasi-steady aerodynamics is reasonable although in the vicinity of $V_R = V_{CR} = 1/S$ there is a need to consider the response to vortex excitation; in addition, at this and higher values of V_R, aeroelastic effects (aerodynamic damping etc) are commonly of importance.

EVALUATION OF DRAG LOADS

The evaluation of the response of multi-degree-of-freedom systems to the random loads induced primarily by atmospheric turbulence is well established. Of particular note is the early work of Davenport [1962] concerning the response of line-like structures and a detailed presentation of the approach through modal analysis due to Harris [1965]. The principal results for a line-like structure are summarized in the following section.

Modal Analysis

The equations of motion of the structure in terms of the normal modes are;

$$M_i\ddot{Z}_i + C_i\dot{Z}_i + K_i Z_i = Q_i(t) \tag{2.1}$$

where Z_i is the generalized modal co-ordinate and M_i, C_i, K_i and Q_i are the modal mass, damping, stiffness and force respectively of the ith mode.

$M_i \quad = \int_o^H m(z)\phi_i^2(z)dz$

$C_i \quad = 2\beta_i M_i \omega_i$

$K_i \quad = \omega_i^2 M_i$

$\omega_i \quad$ = natural (angular) frequency of ith mode

$\beta_i \quad$ = damping as a fraction of critical of the ith mode

$Q_i(t) = \int_o^H w(z,t)\phi_i(z)dz$

$w(z,t)$ = force per unit height at height z and time t

$H \quad$ = height of structure.

The above formulation implies that the equations of motion can in fact be uncoupled. In reality there will be coupling through the damping of the system. Such coupling is generally weak and commonly ignored but it can be of significance for very closely spaced modes ($\omega_i \approx \omega_j$) or in systems with unusually strong damping terms. The damping coefficient C_i has been expressed in terms of the damping as a fraction of critical. For structures with no specific damping mechanisms the structural damping is normally estimated on the basis of existing data obtained from broadly similar structures but the aerodynamic damping at large values of V_R is readily computed. For motion in the along wind direction the drag force per unit height can be approximated as;

$$w(z,t) = C_D(z)D(z)\tfrac{1}{2}\rho u_R^2(t)$$

$C_D(z)$ = local drag coefficient

$D(z)$ = local width

ρ = air density

$u_R(t)$ = relative velocity in drag direction

$= \bar{u}(z) + u'(z,t) - \Sigma \dot{Z}_i \phi_i(z)$

Ignoring second order terms the value of C_i from aerodynamic loading is then;

$$C_i = \rho \int C_D(z)D(z)\bar{u}(z)\phi_i^2(z)dz$$

and $(\beta_i)_{aero}$ is then;

$$(\beta_i)_{aero} = \frac{\rho \int C_D(z)D(z)\bar{u}(z)\phi_i^2(z)dz}{2\omega_i \int m(z)\phi_i^2(z)dz} \tag{2.2}$$

For a uniform prismatic structure in uniform flow this yields;

$$(\beta_i)_{aero} = \frac{C_D}{4\pi}\left(\frac{\rho D^2}{m}\right)\left(\frac{\bar{u}}{f_i D}\right)$$

$$= \frac{C_D}{4\pi}\frac{V_R}{M_R} \tag{2.3}$$

V_R = reduced velocity = $\bar{u}/f_i D$

M_R = dimensionless mass ratio = $m/\rho D^2$

This simple equation (2.3) provides a rapid method of assessing the significance of aerodynamic damping. Using representative values of $\bar{u}, f_i, D$ and m the damping can be roughly evaluated and if it appears significant more detailed calculations should be made.

The modal equations may be solved to evaluate the mean values of the modal co-ordinates;

$$\bar{Z}_i = \frac{1}{T}\int_o^T Z_i(t)dt$$

the variances ($i = j$) and co-variances

$$\overline{Z_i Z_j} = \frac{1}{T}\int_o^T \left(Z_i(t) - \bar{Z}_i\right)\left(Z_j(t) - \bar{Z}_j\right)dt$$

$$= \tilde{Z}_i^2 \text{ for } i = j$$

The mean values are given by;

$$\bar{Z}_i = \bar{Q}_i / K_i \tag{2.4}$$

while the co-variances are computed from a knowledge of the cross-spectrum of $w(z,t)$ $[S_w(z_1 z_2 f)]$ and the complex frequency transfer functions $\left[H_j(if) = \frac{1}{K_j}\frac{1}{\left[1-(f/f_j)^2\right] + 2i\beta_j(f/f_j)}\right]$

The relevant equations are;

$$\overline{Z_i Z_j} = \int_o^H \int_o^H \int_{f=-\infty}^{+\infty} S_w(z_1 z_2 f)\phi_i(z_1)\phi_j(z_2)H_i(if)H_j^*(if)df\,dz_1 dz_2 \tag{2.5}$$

The mean and the variance of some generalized response R (ie a stress, shear force, deflection etc) may be evaluated as;

$$\bar{R} = \sum_{Modes} \bar{Z}_i R_i \tag{2.6}$$

$$\tilde{R}^2 = \sum_{Modes} \sum \overline{Z_i Z_j} R_i R_j \tag{2.7}$$

where R_i is the value of R corresponding to unit Z_i ie; the unit mode value of R. The value of R_i may be evaluated from a knowledge of the influence line and the mode shape from the relationship;

$$R_i = (2\pi f_i)^2 \int_0^H I_R(z)\phi_i(z)m(z)dz$$

The evaluation of the response statistics is clearly computationally intensive and involves double integration over the structure, integration over frequency and a double summation over the modes. It is for this reason that attention has been paid to simpler approaches which do not significantly compromise accuracy. The errors introduced by simplification can be quite significant (say 10% or so) without having an appreciable influence on overall reliability since there are physical uncertainties (drag coefficient, structure of atmospheric turbulence, design wind speed, structural properties, damping, etc.) which will normally control the reliability of the overall evaluation. Two such simplified approaches are discussed here, the first is an approach using influence lines to evaluate the low frequency component of the response [Davenport (1987)] and the second is the well known "Gust-factor" approach due to Davenort (1967), Vikery (1971) Simiu (1973) and others.

The Use Of Influence Lines

The influence line approach is based upon the fact that the bulk of the variance in the longitudinal component of turbulence is at frequencies well below the natural (structural) frequencies of vibration and hence the loads induced by those fluctuations can be treated as quasi-static. If the dynamic force $w(z,t)$ is treated in the quasi-static fashion then the response $R(t)$ can be expressed as;

$$R(t) = \int w(z,t)I_R(z)dz$$

where $I_R(z)$ is the influence function for R ie $I_R(z)$ is the value of R for a unit load applied at height z. The mean value of $\bar{R}$ is then;

$$\bar{R} = \int \bar{w}(z,t)I_R(z)dz \tag{2.8}$$

and the variance of the low-frequency components of the fluctuating part of the response is

$$\tilde{R}_{LF}^2 = \int^H \int^H C_w(z_1,z_2)I_R(z_1)I_R(z_2)dz_1dz_2 \tag{2.9}$$

where $C_w(z_1,z_2)$ is the covariance of the fluctuating component of the loading at levels z_1 and z_2.

The resonant response in the vicinity of the natural frequencies can be determined by modal analysis with the added simplification that integration with respect to frequency is not required for lightly damped systems. The "resonant" contributions to the variance of R is given by;

$$\tilde{R}^2_{RES} = \sum_{MODES} \overline{Z^2_{i_R}} R_i^2 \tag{2.10}$$

where;

$$\overline{Z^2_{i_R}} \approx \frac{\pi}{4\beta_i} \cdot \frac{1}{K_i^2} f_i \cdot S_{Q_i}(fi) \tag{2.11}$$

[$S_{Q_i}(fi)$ in the above relationship in the single sided $(0<f<\infty)$ spectrum of the modal force Q_i for the ith mode].

The response, R, is then characterized by the mean value $\bar{R}$ given by Equation 2.8, and the variance $\tilde{R}^2$ given by;

$$\tilde{R}^2 = \tilde{R}^2_{LF} + \tilde{R}^2_{RES} \tag{2.12}$$

where $\tilde{R}^2_{LF}$ and $\tilde{R}^2_{RES}$ are given by Equations 2.9 and 2.10 respectively.

Gust Factor Methods

For many structures the dynamic response is dominated by the lowest mode of vibration and a single-degree-of-freedom approximation is acceptable. In a true single-degree-of-freedom system the mean response, $\overline{Z_1}$, is;

$$\overline{Z_1^2} = \int\int\int S_w(z_1 z_2 f)|H(f)|^2 \phi(z_1)(z_2) dz_1 dz_2 df$$

The expected peak response in some sample of time T, $\hat{Z}_1$, can be expressed as;

$$\hat{Z}_1 = \bar{Z}_1 + g_T\sqrt{\overline{Z_1^2}}$$

where g_T is a peak factor which for a Gaussian narrow-banded response is {Davenport [7]} given by

$$g_T = \sqrt{2\log_e \nu T} + \frac{0.577}{\sqrt{2\log_e \nu T}}$$

where ν is the zero-crossing rate and is equal to the square-root of the ratio of the second and zeroth moment of the spectrum of Z. The gust factor is defined as the ratio of the peak to mean and is given by;

$$G = 1 + g_T \sqrt{\overline{Z_1^2}}/\overline{Z}_1$$

The use of the gust-factor "G" to obtain an estimate of the peak loads is not completely consistent but in most applications it is applied as a multiplier to mean loads or the static response due to these loads. In the case of the American Concrete Institute's Code for concrete chimney (1988), the gust factor G is applied to the mean base moment but not to the actual mean load distribution. It is presumed that the loads producing the dynamic base moment, $(G-1)\overline{M}$, are distributed in a fashion resembling the distribution of inertia loads. Neither approach is correct and care must be taken when applying the gust-factor approach to certain types of structure. The method can produce very misleading estimates of wind-induced response for some structures eg the guyed tower [1985]

Illustrative Example

In order to illustrate the three approaches (full modal analysis, influence line approach and the gust-factor method) a simple structure consisting of a vertical, tensioned cable in turbulent uniform flow with characteristics broadly consistent with atmospheric turbulence. For the purposes of obtaining numerical results the following properties are assumed;

Height of Cable	H	=	300m
Diameter	D	=	0.05m
Mass Ratio $(m/\rho D^2)$	m_R	=	4000
Drag coefficient	C_D	=	1.0

Turbulence Spectrum (longitudinal component)

$$\frac{f \cdot S(f)}{\sigma_u^2} = \frac{2}{\pi}\frac{fL}{\overline{u}} / \left(1 + \left(\frac{fL}{\overline{u}}\right)^2\right)$$

$\overline{u}$ = 30 m/s

$L \quad = 1500\text{m}$

Normalized Cross-Spectrum

$$R(z_1,z_2,f) = \frac{1}{1+\left(\frac{Cf\Delta}{\bar{u}}\right)^2}$$

$\Delta \quad = z_1 - z_2$

$C \quad = 10$

Frequency of jth mode

$f_j \quad = jf_o$

$f_o \quad = 0.20$ Hz

mode shape $\phi_j = \sin j\pi z/H$ and it is assumed that the structural damping is negligible.

The response studied here is the reaction, $R(t)$, at one end of the cable. The mean value can be written as;

$$\bar{R} = (C_D DH\rho\bar{u}^2/2)\times F$$

and the variance as;

$$\overline{\left(R-\bar{R}\right)^2} = (CDH\rho\bar{u}\sigma_u)^2 \cdot F_2$$

where F and F_2 are dimensionless. Using the influence line approach the value of F is of course equal to the exact value of 0.50. The modal approach yields;

$$F = \sum_{i=1}^{\infty} F_i$$

which, as shown in Fig. 1, leads asymptotically to the value of 0.50. If the first mode only is considered the value is $4/\pi^2$ or 0.405; for an accuracy of 2% it is necessary to consider twenty modes of which only the odd modes contribute.

The variance coefficient F_2 as computed by the influence line approach is shown in Fig. 2 as a function of the number of modes considered. The first term corresponds to the low frequency excitation while the additional increments correspond to the "resonant" peak in each of the modes. The inclusion of one mode only yields a variance coefficient of 0.197 and the computation converges to a value of 0.202. Similar values are shown in Fig. 3 for the computation by modal analysis. In this case the coefficient F_2 is obtained from the double summation;

$$F_2 = \sum_{j=1}^{N} \sum_{k=1}^{N} F_{2jk}$$

After twelve modes the coefficient has attained a value of 0.196 and it is clear that the rate of increase is not negligible. The matrix of the modal contributions F_{2jk} is shown in Table 1. This table clearly demonstrates that the primary contributions are from the cross-terms F_{2jk} where either j or k is unity. The contribution from these terms is virtually all from the low frequency or quasi-static portion of the spectrum is very inefficiently computed by modal analysis.

If the gust factor approach were applied the factor F would be 0.50 and the value of F_2 would be taken as being equal to $\left(\bar{F}^2 \times F_{2j}/F_j^2\right)$ for $j=1$. This yields the result that

$$F_2 = 0.5^2 \times \frac{0.143}{(4/\pi^2)^2}$$

$$= 0.218$$

Consideration Of Non-Linearities

The analysis in the previous sections is derived from the force/velocity relationship;

$$\bar{w}(z) = C_D \cdot D \cdot \tfrac{1}{2}\rho(\bar{u}(z) + u'(z))^2$$

$$= C_D D \tfrac{1}{2}\rho\left(\bar{u}^2 + \overline{u'^2}\right) + C_D D\rho\bar{u}u' + C_D D\frac{\rho}{2}\left(u'^2 - \overline{u'^2}\right)$$

in which the term $\overline{u'^2}$ was ignored in the computation of the mean and the term $C_D D\frac{\rho}{2}\left(u'^2 - \overline{u'^2}\right)$ was ignored in regard to the computation of the variance. In the illustrative example the error in the mean is $(\sigma_u/\bar{u})^2$ which is generally small and, for a structure of

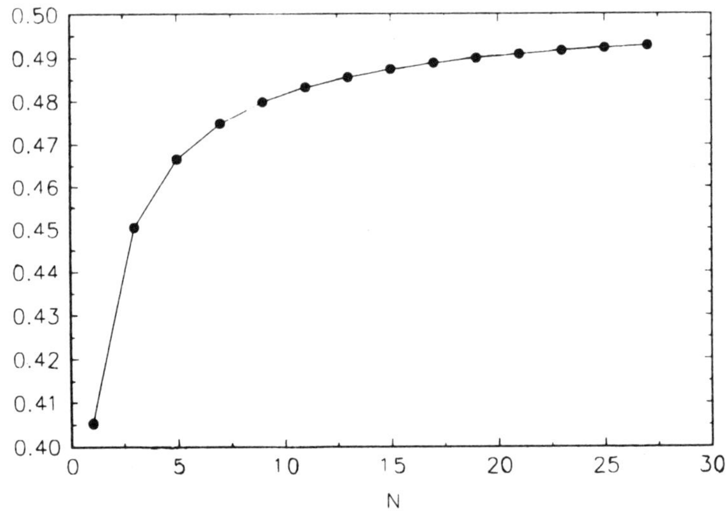

Fig. 1: **Convergence of the mean response by modal analysis**

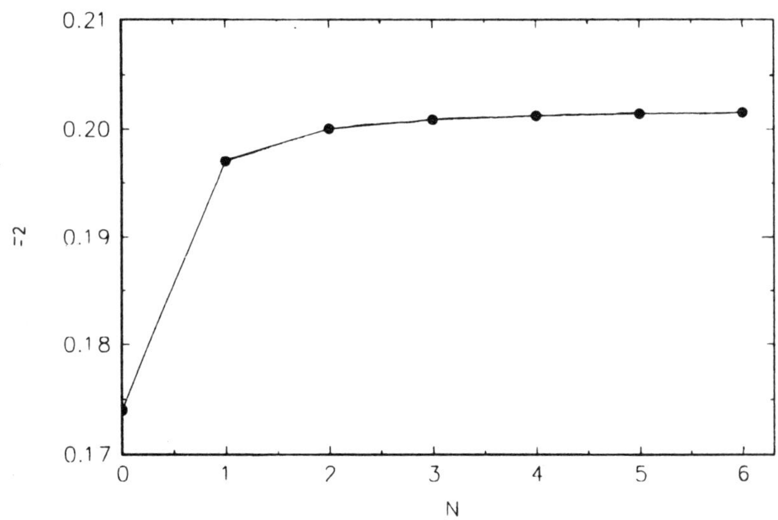

Fig. 2: **Convergence of the variance coefficient computed by the influence line approach and "diagonal" resonant terms**

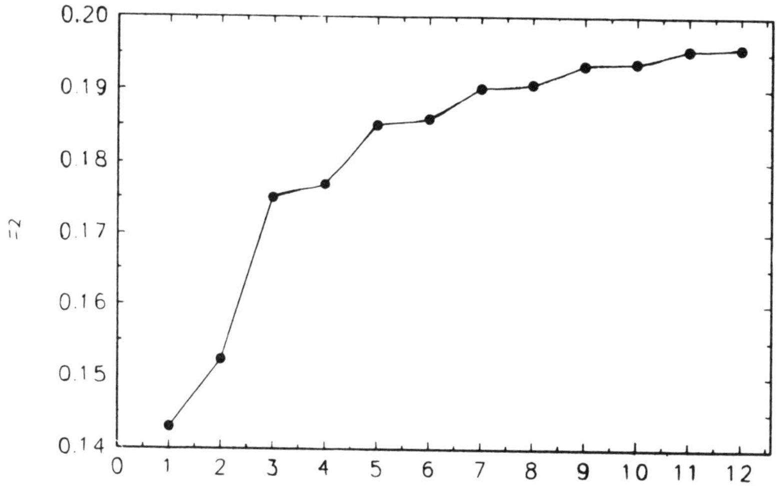

Fig. 3 **Convergence of the variance coefficient computed by modal analysis**

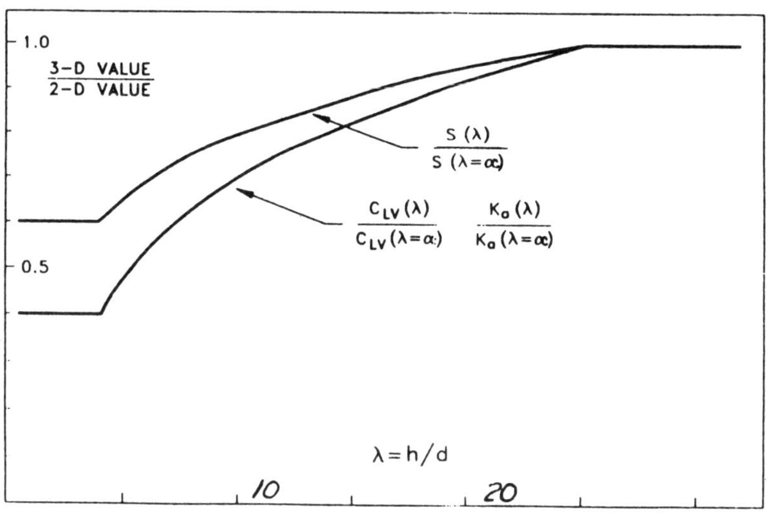

Fig. 4 **Dependence on aspect ratio of the vortex shedding parameters for the circular cylinder**

	1	2	3	4	5	6	7	8	9	10	11	12
1	.143	0	.00948	0	.00347	0	.00178	0	.00108	0	7.223×10^{-4}	0
2	0	.00938	0	6.914×10^{-4}	0	3.144×10^{-4}	0	1.785×10^{-4}	0	1.147×10^{-4}	0	7.986×10^{-5}
3	.00948	0	.00262	0	4.51×10^{-4}	0	2.321×10^{-4}	0	1.409×10^{-4}	0	9.454×10^{-5}	0
4	0	6.914×10^{-4}	0	5.036×10^{-4}	0	8.77×10^{-5}	0	5.092×10^{-5}	0	3.275×10^{-5}	0	2.281×10^{-5}
5	.00347	0	4.51×10^{-4}	0	2.972×10^{-4}	0	8.598×10^{-5}	0	5.221×10^{-5}	0	3.503×10^{-5}	0
6	0	3.144×10^{-4}	0	8.77×10^{-5}	0	1.016×10^{-4}	0	2.326×10^{-5}	0	1.518×10^{-5}	0	1.057×10^{-5}
7	.00178	0	2.321×10^{-4}	0	8.598×10^{-5}	0	7.66×10^{-5}	0	2.688×10^{-5}	0	1.803×10^{-5}	0
8	0	1.785×10^{-4}	0	5.093×10^{-5}	0	2.327×10^{-5}	0	3.222×10^{-5}	0	8.635×10^{-6}	0	6.014×10^{-6}
9	.00108	0	1.409×10^{-4}	0	5.222×10^{-5}	0	2.687×10^{-5}	0	2.787×10^{-5}	0	1.055×10^{-5}	0
10	0	1.147×10^{-4}	0	3.275×10^{-5}	0	1.516×10^{-5}	0	8.634×10^{-6}	0	1.303×10^{-5}	0	3.869×10^{-6}
11	7.223×10^{-4}	0	9.454×10^{-5}	0	3.502×10^{-5}	0	1.803×10^{-5}	0	1.055×10^{-5}	0	1.209×10^{-5}	0
12	0	7.986×10^{-5}	0	2.28×10^{-5}	0	1.057×10^{-5}	0	6.014×10^{-6}	0	3.87×10^{-6}	0	6.164×10^{-6}

Table 1: Matrix of terms contributing to the convergence of the variance coefficient shown in Fig. 3

the dimensions chosen and situated in open country might amount to 1% or 2%.

The inclusion of the non-linear term in the analysis of the fluctuating response will result in errors in both the background response and the "resonant" contributions. For a Gaussian distribution of u the error in the background contribution to the variance of the local load per unit height is;

$$\frac{\left(C_D D\frac{\rho}{2}\right)^2 \overline{\left(u'^2-\overline{u'^2}\right)^2}}{(C_D D\rho\bar{u})^2\overline{u'^2}}$$

$$=\frac{1}{4}\cdot\frac{\int u'^4 p(u)du-\sigma_u^4}{\sigma_u^2\bar{u}^2}$$

The error in standard deviation of the load at a given height is then equal to $\sigma_u^2/2\bar{u}^2$ or about 1% in the illustrative example. The distortion of the spectrum due to inclusion of the non-linear terms will have an influence on the resonant response but again this will be small. The errors associated with the linearization of the square law relationship are not large for either the mean or the variance calculations but should not be ignored when considering peak values. The resonant component of the response will be well represented by a Gaussian Distribution but the background or low frequency part is not and the peak factor "g" may be seriously in error if the fluctuating response is due primarily to low frequency loads which are relatively well correlated over the structure. In the simplest case of a fully correlated load with no resonant effects the relationship between peak load and peak velocity is;

$$\hat{w}/\bar{w}=1+g_w\sigma_w/\bar{w}$$

$$=(1+g_u\sigma_u/\bar{u})^2$$

$$g_w=g_u\{1+g_u\sigma_u/2\bar{u}\}$$

For a typical value of g_u, say 3.7, and for the illustrative example with $\sigma_u/\bar{u}$ about 0.12 we have;

$$g_w\approx 1.2g_u$$

For small structures or rougher terrains in which the typical level of turbulence would be loser to 0.2 the corresponding relationship is;

$$g_w \approx 1.4 g_u$$

The failure to recognize the influence of the square-law relationship on the distribution of the background or low-frequency response can be very significant. The formulation of the Gust factor in the Australian Code [1989]recognizes this potential error and has adopted an empirical adjustment to the peak factor for the background response. The form adopted can be expressed as;

$$G = 1 + 2\frac{\sigma_u}{\bar{u}}\sqrt{g_B^2 B + g_R^2 R}$$

where

$$g_B \quad = g_u(1+w)$$

$$w \quad = g_u\sqrt{B}\,\sigma_u/2\bar{u}$$

g_R is a peak factor related to the resonant response and R a measure of resonant response.

For a small stiff structure with little resonant contribution $B \to 1$ and $R \to 0$ which yields the result;

$$G = 1 + 2\frac{\sigma_u}{\bar{u}} g_u(1 + g_u\sigma_u/2\bar{u})$$

$$= (1 + g_u\sigma_u/\bar{u})^2 = (\hat{u}/\bar{u})^2$$

ACROSS-WIND RESPONSE

Introduction

The across-wind (or lift) response of slender tower like structures is induced by the incident turbulence, by the unsteady forces associated with the turbulent wake (ie; vortex shedding) and by interactions between structural motion and the flow (eg aerodynamic damping and stiffness). Unlike bridge structures, for example, most tower like structures have neither strong lift coefficients nor strong lift coefficient derivatives and hence loads induced directly by either component of the incident turbulence are not likely to play a critical role in design since they will in general be less than the along wind load fluctuations and almost always significantly less than the combined mean and fluctuating components in the along-wind direction. Even in combination with along-wind loading the across-wind loads due to incident turbulence rarely result in a significant increase in the peak design load for a particular element or cross-section.

The more significant sources of across-wind response are associated with vortex excitation and/or negative aerodynamic damping. Negative damping in a single degree-of-freedom vibration (eg galloping) or in a coupled mode (eg classical flutter) cannot be ignored but are likely only at speeds well beyond the critical speed and, of course, for cross-sections susceptible to such behaviour. For isolated circular cross-sections only vortex excitation need be considered but other instabilities may be of considerable significance for multiple structures in close proximity.

Vortex Induced Across-Wind Loads

The forces associated with shedding from a stationary bluff body can be characterized by four aerodynamic parameters, a lift coefficient, a central frequency, a spectral bandwidth and a measure of the spanwise correlation. All four are functions of the cross-sectional shape, the Reynolds Number, the turbulence scale and intensity and, for structures with a height to width ratio less than about twenty, the aspect ratio. For sharp-edged bodies the parameters can be evaluated by experiment but for the circle and other shapes having curved surfaces the strong dependence on both Reynolds Number and turbulence scale and intensity greatly reduces the value of any model scale measurements and reliance must be made on full scale data which are limited in both quantity and quality. The following sections are concerned with the definition of the key parameters for circular cross-sections in large scale turbulence and at Reynolds Numbers consistent with full scale structures (generally in excess of 10^7).

The Spectrum

The spectrum of the vortex-induced force per unit length is expressed as;

$$f \cdot S_w(f)/\sigma_w^2 = \frac{f/f_s}{B\sqrt{\pi}} \exp\left\{-\left(\frac{1-f/f_s}{B}\right)^2\right\} \qquad (3.1)$$

σ_w = rms lift force per unit length (w)

f_s = shedding frequency

$S_w(f)$ = spectrum of w

B = a bandwidth parameter

$f_s D/\bar{u}$ = Strouhal Number (S)

σ_w = $\tilde{C}_{LV}\frac{1}{2}\rho\bar{u}^2 d$

$\bar{u}$ = local mean speed

d = local breadth of structure

Equation 3.1 defines the spectrum of lift forces induced by shedding and the general form normally provides an adequate fit to measured spectra over the frequency range $f_s \pm Bf_s$. There is commonly a low frequency component in the spectrum due to forces induced by lateral turbulence but these lateral velocity components have little direct influence on the spectrum near f_s or on the maximum response which is induced when f_s is near a natural frequency of the structure.

The spectrum is defined by three parameters;

the rms lift coefficient	$\tilde{C}_{LV}$
the bandwidth parameter	B
the Strouhal Number	S

The parameter B depends primarily on the large scale turbulence which produces slow variations in the local velocity. Vickery & Basu [1983] argue that the B should be equal to $\sqrt{B_o + 2i^2}$ where B_o is the bandwidth in smooth flow and i the local intensity of turbulence however limited full-scale data suggest that a simpler relationship;

$$B = 0.10 + 2i \qquad (3.2)$$

is a better predictor for the intensities encountered in atmospheric shear flows. Small scale turbulence, such as produced by a nearby slender structure, will not have a significant effect on B.

The Strouhal Number "S" is dependent upon surface roughness, Reynolds Number, turbulence and the aspect ratio (height/width) of the structure. At Reynolds Numbers beyond about $2\mathrm{x}10^6$ there is very little data but S tends to decrease slightly with roughness and increase with turbulence intensity. The dependence on aspect ratio is strong and the suggested relationship is shown in Fig. 4. This relationship is;

$$S(\lambda)/S(\infty) = 1.0;\ \lambda > 25$$

$$= 0.6;\ \lambda < 4$$

$$= 0.6 + 0.22 \log_e\left(\frac{\lambda}{4}\right);\ 4 < \lambda < 25$$

λ = $h/\bar{d}$

$\bar{d}$ = average diameter of upper third of structure

h = height of structure

$S(\infty)$ = Strouhal Number for two dimensional flow

The value suggested for $S(\infty)$ is 0.23 to 0.24

ie $$S(\lambda) = 0.14 + 0.05 \log_e \frac{\lambda}{4}$$

for $4 \leq \lambda \leq 25$ and constant above $\lambda = 25$ (3.3)

The value of $\bar{C}_{LV}$ has been shown from full-scale measurements {Vickery & Basu [1984], Sanada et al [1992],Weldeck[1992]} to be strongly dependent on the scale and intensity of turbulence. Although the full-scale data indicates this strong dependence on intensity, there is no reason to expect the lift forces coefficient to be changed by large scale turbulence which, in effect, is responsible only for a slowly varying local wind speed. It can be argued that only scales of the same order or less than the diameter could have the observed effects. This suggests that a modified intensity i^* which excludes the very large scales should be employed, such an intensity can be roughly defined as;

$$i^* = i(d/L)^{1/3} \quad (3.4)$$

where L is the integral scale of turbulence.

Using i^* as the independent variable the data suggest a relationship;

$$\tilde{C}_{LV} = [0.15 + 0.55i^*] - [0.09 + 0.55i^*]e^{-(20i^*)^3} \quad (3.5)$$

This relationship is shown in Fig. 5 along with data collected from a variety of sources by Vickery & Basu [1984] and more recent full-scale data reported by Waldeck [1992].

In cases where the incident turbulence is not that associated with turbulent atmospheric shear flow (eg; when the structure is subject to wake turbulence from a nearby structure) the value of i^* should be estimated as the intensity of turbulence considering only those scales (wavelengths) less than $10d$. For a chimney which is part of a comparatively closely spaced group all with similar diameters the value of i^* will be very similar to the intensity of turbulence within the wake of the upstream structure and may well be beyond the range of i^* for which data are available and the fitted relationship [Eq'n 3.5] may well be in error.

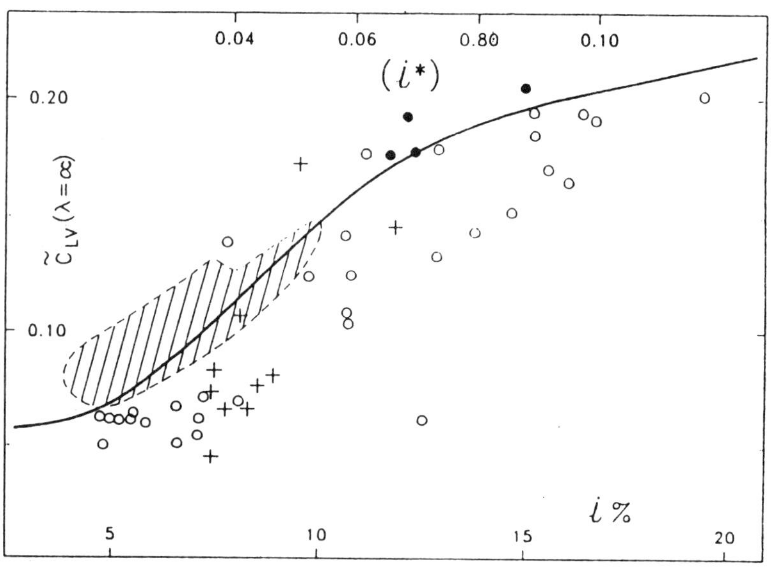

Fig. 5 Dependence of the vortex induced lift coefficient on turbulence [circular cylinder at Re > 10^7]

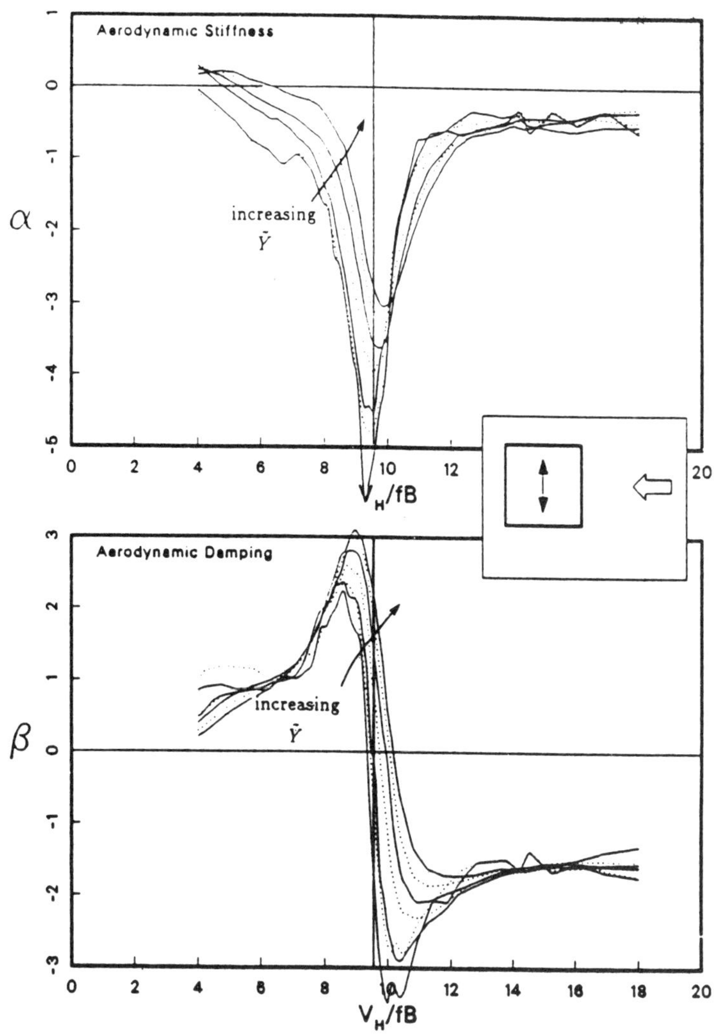

Fig. 6 **Aerodynamic mass (stiffness) and damping parameters for a square prism in turbulent shear flow**

The values of $\tilde{C}_{LV}$ defined by equations 3.4 and 3.5 are for nominally two-dimensional conditions and must be modified to account for aspect ratio (h/d). There is an overall reduction in $\tilde{C}_{LV}$ with decreasing h/d and, in addition, the value of $\tilde{C}_{LV}$ reduces rapidly to zero in the immediate vicinity of the tip. The suggested reduction with aspect ratio is as follows;

$$\tilde{C}_{LV}(h/d)/\tilde{C}_{LV}(h/d=\infty) = \tilde{C}_{LV}(\lambda)/\tilde{C}_{LV}(\infty)$$

$$\begin{aligned} &= 0.4 && , \lambda < 4 \\ &= 1.0 && , \lambda > 25 \\ &= 0.4 + 0.33 \log_e \lambda/4 && , 4 < \lambda < 25 \end{aligned} \qquad (3.6)$$

Near the tip of the lift coefficient can be further reduced in accordance with the relationship;

$$\tilde{C}_{LV}(tip)/\tilde{C}_{LV} = 1 - e^{-\Delta z/\beta d} \qquad (3.7)$$

$\tilde{C}_{LV}(tip)$ = value of $\tilde{C}_{LV}$ at a section located Δz from the tip

Δz = distance from tip; d = tip diameter; β = ½ for circular cross-section

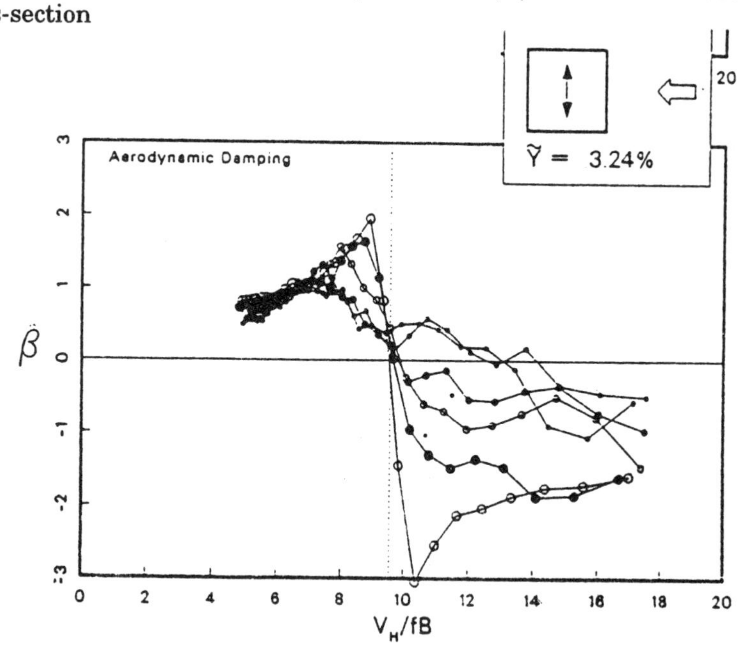

Fig. 7 **Influence of turbulence on the aerodynamic damping parameters**

Cross-Spectra

Section 3.2.1 defines the spectrum of the vortex-induced loads at a particular cross-section but the relationship between spectra and two cross-sections is also required. The relationship suggested by Vickery & Basu [1984] is

$$\frac{C_o(z_1,z_2,f)}{[S(z_1f)S(z_2f)]^{1/2}} = R(z_1,z_2,f)$$

$$= \cos\{2r/3l\}\cdot\exp\left\{-(r/l)^2\right\} \tag{3.8}$$

where;

$C_o(z_1,z_2,f)$ = co-spectrum of forces at position z_1 and z_2

$S(z,f)$ = spectrum at z

r = $(z_1 - z_2)/0.5(d_1 + d_2)$

l = a non-dimensional correlation length with a value near unity or perhaps a little greater.

In most computations the actual form adopted to define $R(z_1,z_2,f)$ is not important and it is only the correlation length, l, which is of significance.

Motion-Induced Forces

The discussion in Section 3.2 concerns the nature of the forces acting on a stationary structure.

If sinusoidal or narrow-band resonant motion is introduced then there are additional forces induced together with changes in the pre-existing forces. The additional forces are motion correlated and comprise a component in phase with the motion (an aerodynamic mass or stiffness force) and an out-of-phase component (the aerodynamic damping, which is commonly negative when the shedding frequency and the frequency of motion are nearly equal). The forces that existed on the stationary body are also influenced by motion in that span-wise correlation is improved and the spectrum is distorted. While these latter effects are significant they do not play an important role in determining the response of the structure. The changes in spectral distribution and correlation are very noticeable at amplitudes in excess of about five percent of the body width but at these amplitudes the response is determined primarily by the phase related forces, particularly the damping.

The motion correlated forces that develop when the shedding frequency (f_s) and the frequency of motion (f_v) are roughly equal have been studied by numerous workers [1971, 1969] but the most comprehensive study for tower-like structures was conducted by Stecley [1989, 1993, 1994] who examined both the local and the integrated forces on a square cross-section moving in a base-pivotted mode in turbulent shear flow and the results which follow are derived primarily from this work.

The wind induced base moment may be written as;

$$M(t) = C_M(t)\tfrac{1}{2}\rho V^2 BH^2 - \tfrac{2}{3}\omega^2 \rho B^2 H^2 \left\{\alpha y + \tfrac{\beta}{\omega}\dot{y}\right\} \tag{3.9}$$

where y and $\dot{y}$ are the tip displacement and velocity respectively and α and β are dimensionless coefficients. In the case of two-dimensional flow the "damping" parameter β is equivalent to $\dfrac{m\beta_a}{\rho B^2}$ where β_a is the aerodynamic damping as a fraction of critical and the "stiffness" parameter α is equivalent to $-C_m/2$ where C_m is the "added mass" coefficient. Although the relationship in Equation 3.9 suggests a linear aerodynamic system both α and β are amplitude dependent. Values of α and β are shown in Fig. 6 and the dependence of these on turbulence level is shown in Fig. 7. The marked reduction in the magnitude of the negative damping with increasing turbulence is of great practical significance. At turbulence intensities above 20% the negative damping in the vicinity of the critical speed is completely eliminated and it is only at higher reduced velocities that the "negative lift slope" of the square cross-section results in the negative damping associated with galloping.

The aerodynamic damping parameters for a number of simple cross-sections are shown in Fig. 8. All data were obtained from base pivotted models in shear flow with a "power-law" exponent of 0.11 and a turbulence intensity over the upper part of the model of approximately 8%. The results of Fig. 8 clearly demonstrate the adverse characteristics of the square cross-section.

CONCLUSION

Slender structures such as chimneys and towers are amenable to analysis primarily because a comparatively simple aerodynamic

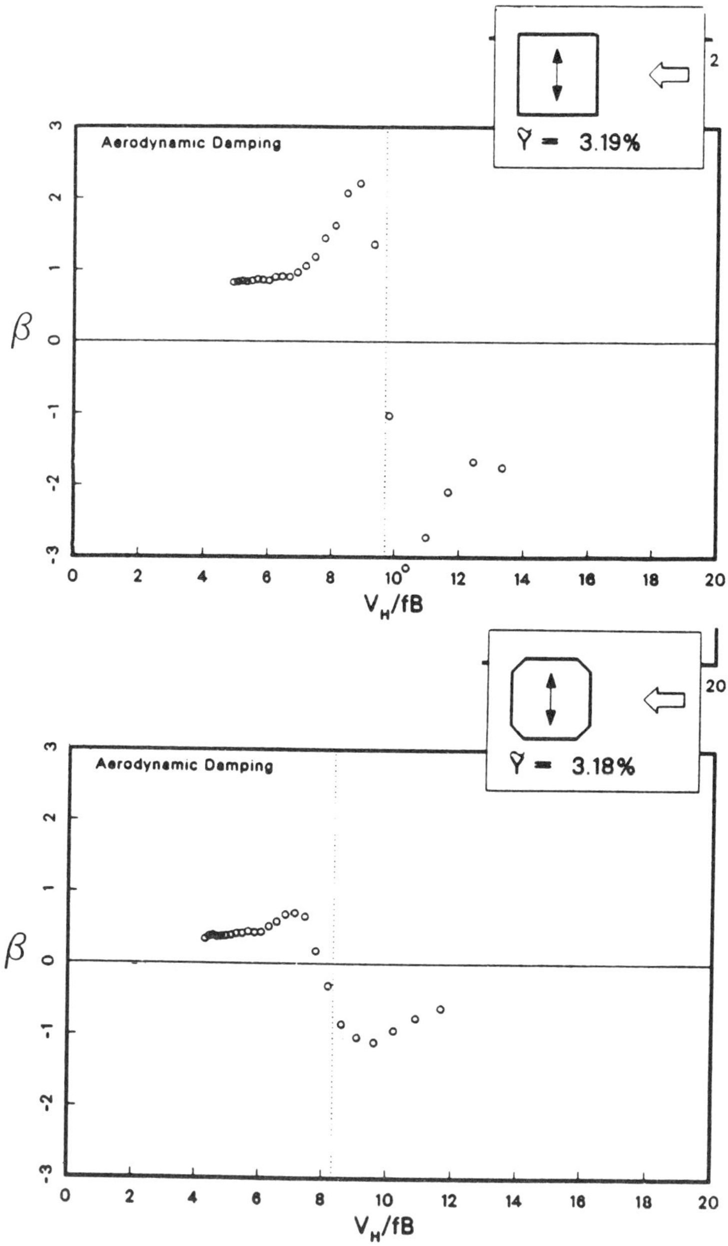

Fig. 8 **The aerodynamic damping parameter for a variety of prismatic structures in turbulent shear flow**

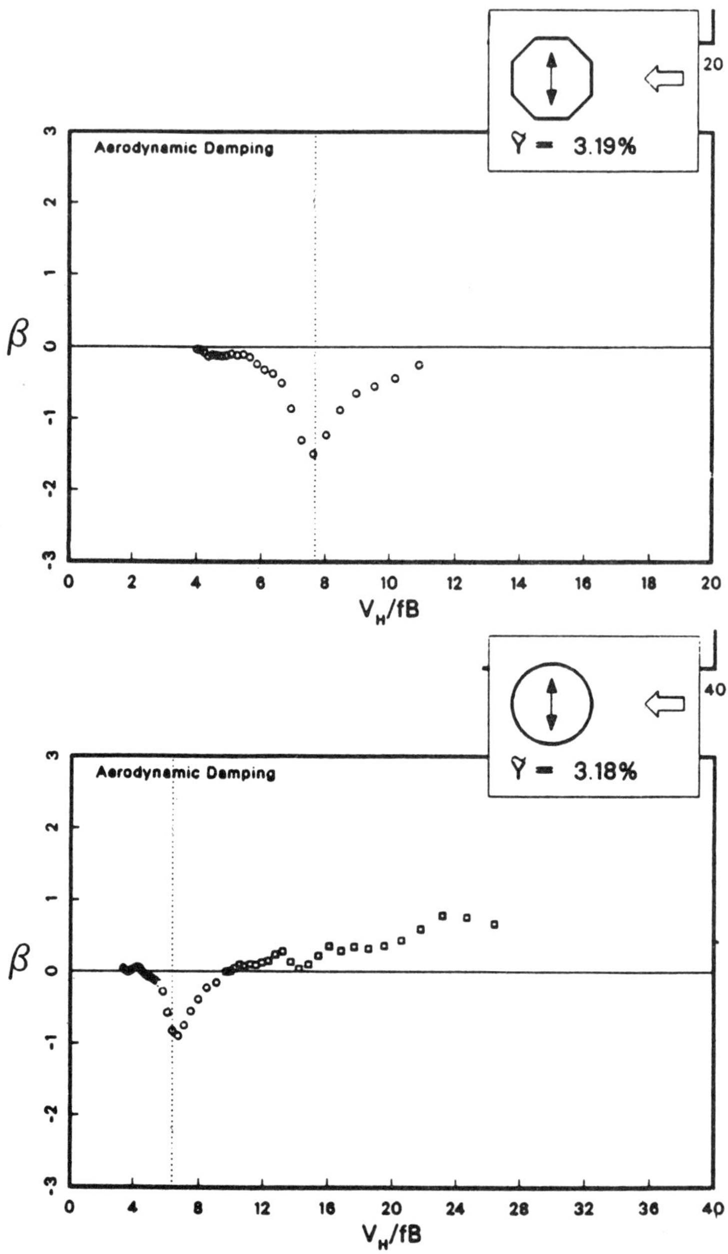

Fig. 8 **The aerodynamic damping parameter for a variety of prismatic structures in turbulent shear flow**

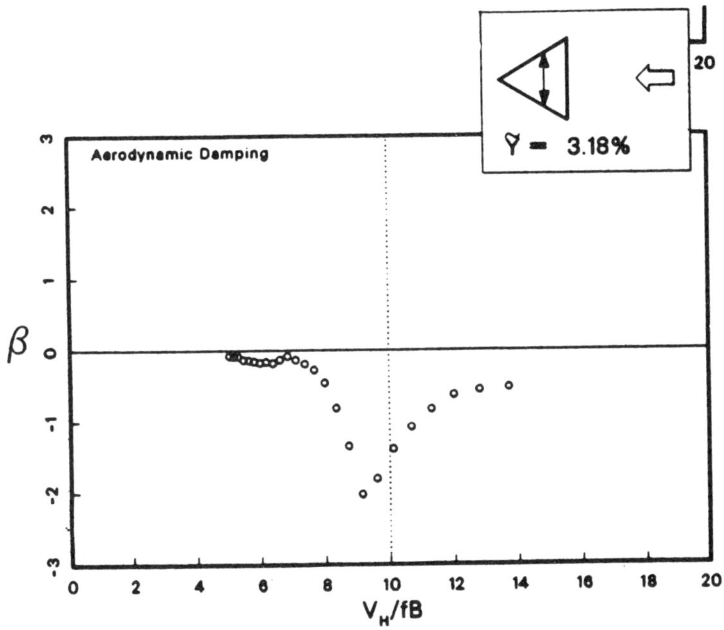

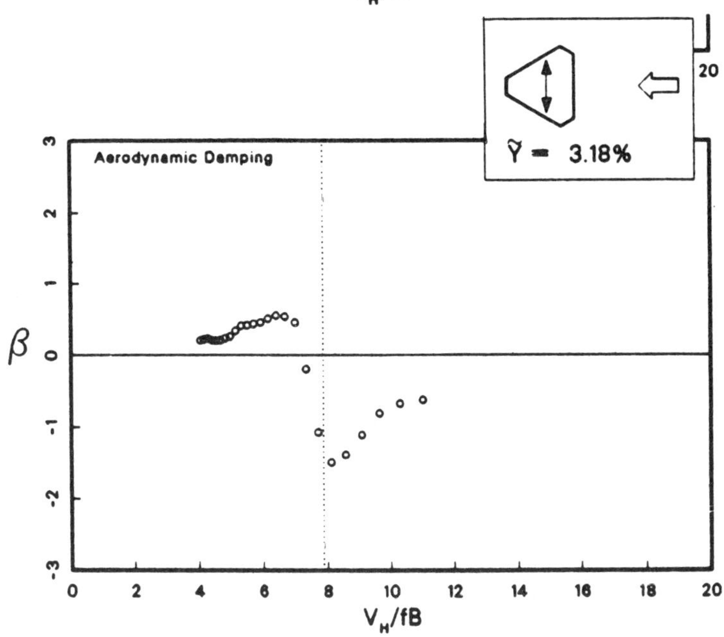

Fig. 8 The aerodynamic damping parameter for a variety of prismatic structures in turbulent shear flow

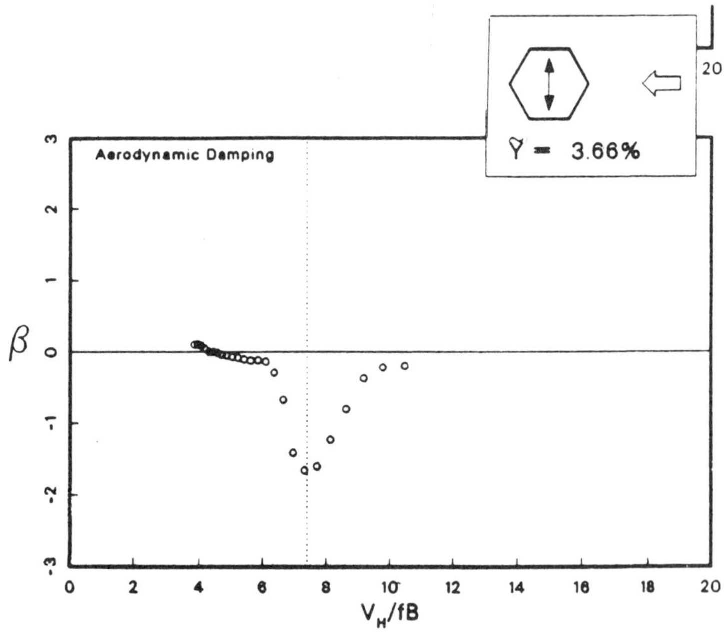

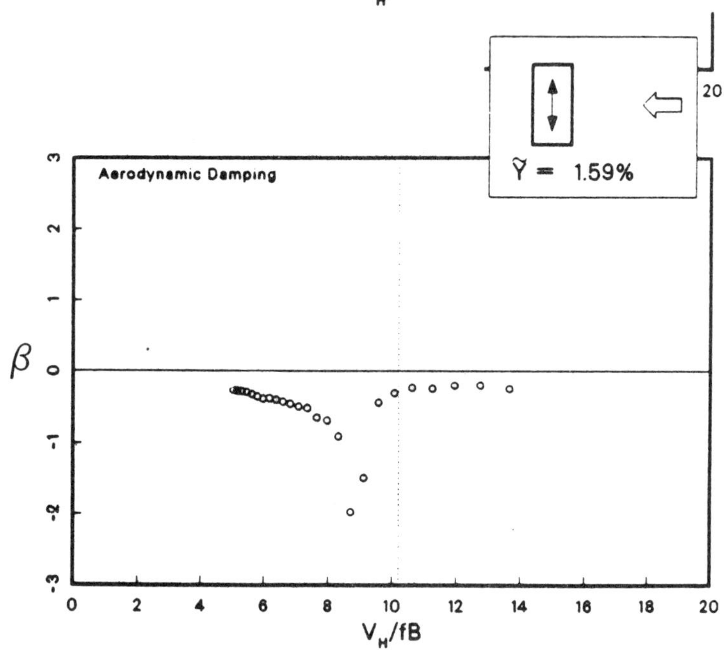

Fig. 8 **The aerodynamic damping parameter for a variety of prismatic structures in turbulent shear flow**

approximation (strip theory) yields acceptable results and that quasi-steady assumptions are often but not always acceptable. The present paper attempts to define approaches suited to the prediction of the along-wind or drag response and to provide some data to aid in the prediction of the across-wind response due commonly to vortex excitation.

In the prediction of the along-wind response it is demonstrated that the separation of the computation into a low frequency or background component and a series of modal resonant components is accurate and computationally efficient. The use of conventional modal analysis to evaluate the background response can be extremely inefficient since not only can many modes contribute, the cross-terms between modes are far from insignificant.

In regard to across wind motion the paper outlines a general approach and provides a data base for circular cylinders at high Reynolds Numbers. The question of motion-induced or aeroelastic effects is addressed with examples derived from a study of a prism of square cross-section in turbulent shear flow. The results demonstrate the very important (beneficial) effect of free stream turbulence on the motion induced forces and consequent large amplitude motions.

REFERENCES

Davenport, A.G., "The response of slender line-like structures to a gust wind", Proc. Inst. Civ. Eng., p. 389, Vol. 23, 1962.

Harris, R.I., "The random vibration of distributed parameter mechanical systems with reference to wind loading problems", Elec. Res. Assoc. U.K., Rep. No. 5110, 1965.

Davenport, A.G., "The response of slender structures to wind" Lecture notes for short course "Application of wind engineering principles to the design of structures", Lausanne, Switzerland, Feb. 1987.

Davenport, A.G., "Gust loading factors", J. Struct. Div., Proc. A.S.C.E., Vol. 93, No. ST3, June, 1967.

Vickery, B.J., "On the reliability of gust loading factors", Civil Eng'g Transaction, Inst. Eng. Aust. pp. 1-9, Vol. 13, 1971.

Simiu, E., "Gusts factors and along-wind pressure correlations", J. Struct. Div., Proc. A.S.C.E., pp. 773-783, Vol. 99, No. ST4, April, 1973.

Davenport, A.G., "Note on the distribution of the largest value of a random function with application to gust loading", Proc. Inst. Civ. Eng., pp. 187-196, Vol. 28, 1964.

American Concrete Institute, "Standard practice for the design & construction of cast-in-place reinforced concrete chimneys", ACI 307-88, ACI Detroit, USA, 1988.

Gerstoft, P. and Davenport, A.G., "A simplified method for the dynamic analysis of a guyed mast", Proc. 6th Coll. on Indust. Aero., Aachen, Germany, June,,1985.

Standards Australia, "S.A.A. Loading Code, Part 2 Wind Loads", Australian Standard AS1170.2-1989.

Vickery, B.J. and Basu, R.I., "Across-wind vibrations of structures of circular cross-section", Parts I and II, J. Wind Eng'g & Ind. Aero, pp. 49, Vol. 12, 1983.

Vickery, B.J. and Basu, R.I., "The response of reinforced concrete chimneys to vortex shedding", J. Eng'g. Str., pp. 324-333, Vol. 6, Oct. 1984.

Sanada, S., Suzuki, M. and Matsumoto, H. "Full scale measurements of wind force acting on a 200m concrete chimney and the chimney response", J. Wind Eng'g & Ind. Aero., pp. 2165-2176, Vol. 43, 1992.

Waldeck J.L., "The measured and predicted response of a 300m concrete chimney", J. Wind Eng'g & Ind. Aero., pp. 229-240, Vol. 41, 1992.

Yano, T. and Takahara, S., "Study on unsteady aerodynamic forces acting on an oscillating cylinder", pp. 737-746, Proc. 3rd Int. Conf on Wind Effects on Buildings and Str., Tokyo, 1971.

Jones, G.W., Cincotta, J.J. and Walker, R.W., "Aerodynamic forces on a stationary and oscillating cylinder at high Re", NASA Tech Rep. R-300, NASA Washington, DC, USA, 1969.

Steckley, A., "Motion induced wind forces on chimneys and buildings", Ph.D. Thesis, The University of Western Ontario, London, Ontario, June 1989.

Vickery, B.J. and Steckley, A., "Aerodynamic damping and vortex excitation on an oscillating prism in turbulent shear flow", J. Wind Eng'g & Ind. Aero., pp. 121-140, Vol. 49, 1993.

Vickery, B.J. and Steckley, A., "Motion induced forces on prismatic structures in turbulent shear flow", p. 49-54, Proc. 2nd Int. Conf. on Engineering Aero-Hydroelasticity, Pilsen, Czech Rep., June, 1994.

A REVIEW OF SOME KEY DEVELOPMENTS IN THE AEROELASTICITY OF LONG-SPAN BRIDGES

ROBERT H. SCANLAN

Professor of Civil Engineering, The Johns Hopkins University, Baltimore, Maryland 21218, USA

SUMMARY

At their present stage aeroelastic studies of long-span bridges have undergone a little over a half century of development. Certain key elements in that development have aeronautical origins of an analytical nature. The present paper comments on this legacy and notes some of the newer experimental counterparts that are particular to bridges.

INTRODUCTION

For over 100 years prior to the pivotal Tacoma Narrows bridge disaster of 1940 flexible bridges had been noticeably disturbed— and some severely damaged— by the wind. However, the late 1930's and early 1940's witnessed introduction of new consciousness and appreciation of what kinds of forces were at work in such events. The relatively new field of *aeroelasticity*— not yet known by that name— had been developing since the late 1920's in the aeronautical field. By 1935 the United States NACA had in hand the pioneering theoretical papers of Theodorsen and Garrick (1935, 1940) on airfoil flutter, and comparable theory was developed in Germany by Küssner (1936) and others. When the Tacoma Narrows incident came along it was natural to attempt to infer some connection between bridge and airfoil flutter, which idea Bleich (1948, 1949) proceeded to exploit. Unfortunately Bleich's effort was wide of the mark in that it implicitly attributed airfoil characteristics to a bridge deck cross section. This easy but quite inappropriate transposition of ideas stood as somewhat of a discouragement to sound analytical attempts at describing bridge flutter, since the latter, as compared to the airfoil variety, is in most cases a distinctly different, bluff-body phenomenon. Pugsley (See Bleich 1949), in a commentary on Bleich's work, suggested the possibility of measurement of motional, wind-induced forces. Halfman (1952), using an airfoil, demonstrated a method by which such measurement could be accomplished.

Initial analytical inadequacies were of course compensated for early-on by recourse to testing. Wind tunnel models of bridges— particularly the representative deck-section model— were early introduced from as early as 1940 by Farquharson (1949–54), Scruton (1952), Kármán and Dunn (see Farquharson 1949–54) and others. Section models were extensively exploited by Steinman (1943–50, 1954), Selberg (1961), Klöppel (see Klöppel and Weber 1963), and by Walshe (1963). Later models took on a variety of partial and full-bridge forms. For a number of years now this strongly empirical approach has dominated the field.

Alongside it, however, has grown up a body of analytical theory that has begun to reduce this empiricism, explain some of the physical mechanisms at work, and exhibit good accuracy and productive capability. This theory— coupled with appropriate basic experiment— is presently proving quite valuable in the design studies of new flexible bridges.

THEORETICAL DEVELOPMENTS: FLUTTER

The central points of departure of the theory around which its more recent greater degree of success has turned have been recognition of the following:
a) the value for design of the aerodynamic force coefficients;
b) the intrinsic relations, and differences, between *static* aerodynamic force coefficients and their *dynamic* (motion-related) counterparts (i.e. the "flutter derivatives");
c) the need and means to make separate *experimental* determinations of each;
d) the need to draw sharp and clean distinctions between, on the one hand, the *mechanical*, and on the other, the *aerodynamic* aspects of any type of test model;

e) the need to distinguish between the *dynamics* and *aerodynamics* of a given model and those of the prototype bridge structure; and
f) the nature of the turbulence in the approach flow and its influence upon the distribution of forces over model or prototype.

A section model of a typical bridge deck section has classically been used to acquire static lift, drag and moment coefficients of the prototype (Farquharson 1949). This was time-honored practice. On the other hand, the same model, allowed to move in proper degrees of freedom (vertical, torsion and sway), can lend insight into the sectional "flutter" forces. Such forces, dependent on near-wake flow phenomena, cannot be inferred from steady-state conditions although they become asymptotic to the latter at zero frequency. Test methodology to measure such forces was unused for bridges until the mid-1960's. It requires, for its exploitation, various sophisticated techniques of realizing model displacement and of implementing system identification to interpret the aerodynamic force results, which partially explains its relative slowness in coming into common use. Following Halfman's airfoil studies, the practice of measuring flutter derivatives was generally introduced to bridge applications, to the writer's knowledge, by a paper of Ukeguchi, Sakata and Nishitani (1966). This was followed by papers by Scanlan and Sabzevari (1967, 1969) and by Scanlan and Tomko (1971),and since then by many others of increased sophistication, of which those by Széchényi (1973, 1987), Larose, Davenport and King (1993), Raggett and Scanlan (1993), Sarkar, Jones and Scanlan (1992), Curami and Zasso (1993) are more recent representatives.

The notion of flutter derivative (as of the earlier-recognized static force coefficients) derives directly from the legacy of the aeroelastic theory of aeronautics, particularly the Theodorsen (1935) work. However, *measuring* bridge deck flutter derivatives began to reveal a special and quite typical trend among bridges: one derivative in particular, the torsional aerodynamic damping derivative, consistently reversed sign with increasing (dimensionless) velocity, in contrast to the consistently stabilizing action of the analogous derivative for airfoils. This outstanding effect characterized most bridge flutter as strongly driven by single-degree negative damping in torsion. This tendency has continued to typify bridge aerodynamics until recently when modern deck profiles have become more streamlined. Classic coupled, vertical/torsion flutter of the airfoil type can occur in bridges, but the coupling effects, revealed via the "cross" flutter derivatives, are relatively weak. In certain respects most bridge flutter is closer to the stall flutter of airfoils than to its classical counterpart.

Although from the beginning in aeronautics flutter derivatives were consistently employed in the time domain to detect the critical (oscillatory) flutter condition, their proper interpretation within the frequency domain was not emphasized in the bridge context until much later (Scanlan 1984). The practice of using the sectional model uniquely as an analog source of both static force coefficients and of flutter derivatives— and not as an (inevitably improper) simulator of the full prototype bridge— was a step finally acknowledging the kind of information correctly to be inferred from it.

When a section model is oscillated, the principal type of frequency-domain information to be extracted from it consists of the motion-related flutter derivatives. When, on the other hand, the frequencies of interest lie in the turbulence components of the approach flow, an alternate kind of frequency-domain information to be extracted from the model is the *aerodynamic admittance*. In the airfoil case the corresponding associated frequency-domain function is the *Sears* function (Sears 1941). This complex lift function is related in a Fourier-transform manner to the Küssner vertical-gust-wall-penetration lift function of the time domain (Küssner 1936). In the airfoil case, the time-domain lift-growth function which corresponds to an inverse Fourier transform of the Theodorsen flutter circulation function is known as the Wagner function (Wagner 1925). The corresponding (but usually very differently-shaped) function for bridge decks was shown (Scanlan and Budlong 1971) to be derivable from certain of the measured flutter derivatives.

Y.K. Lin and his co-workers (see discussion and reference listing in Scanlan 1993) have pursued exploitation of such *indicial* lift functions in the bridge context with the aim of observing, in time-domain studies, the effect of oncoming turbulence upon the flutter stability. These studies to date, concerned with the basic question of how the stochastic nature of coefficients governs the stability of response, have been exclusively of a mathematical nature and have not benefitted from measurement of the subtle physical changes in aerodynamic forces that can be occasioned by turbulence in the oncoming flow. Alternate experimental approaches (Huston 1986, Sarkar 1992, Larose, Davenport and King 1993) have directly inferred flutter derivatives under turbulent approach flow, using identification techniques. Such methods have facilitated flutter analysis in the frequency domain. Typical results have shown sharper differences between smooth-flow and turbulent-flow flutter derivatives mainly in those cases where the bridge deck forms have received little or no aerodynamic treatment.

Turbulence in the approaching wind has been demonstrated to increase the velocity at which flutter occurs and, in general, the trend into instability of model bridges, making it more gradual (Irwin and Schuyler 1978). Testing has shown repeatedly that the wind speed of entrance into full instability for full-bridge models is higher and less abrupt when testing occurs under turbulent flow, as compared to section or full-bridge models of the same form under laminar approach flow. This trend to higher instability velocities can be accounted for analytically (Scanlan and Jones 1990) by postulating a spanwise diminution of lateral coherence of the flutter or motion-induced forces, such as has analogously been measured in the case of vortex-induced forces (Toebes 1967, Ehsan and Scanlan 1990).

Often in analyses for flutter, these reduction-of-correlation effects of motion-induced forces under turbulence can conservatively be neglected, to the end that oncoming turbulence is not counted upon during design as a stability enhancement, since the estimated level may not always be assured in the field. However, no experimental cases have been reported wherein turbulence has occasioned actual flutter onset at a *lower* wind velocity than occurred under laminar approach flow. Clearly the coherence properties of force effects under smooth flow will inevitably be stronger than those under turbulent flow.

The primary preoccupation with motion-induced forces (Scanlan and Tomko 1971) and their effect upon instability has been a step toward a state of the art where the nature of the phenomena involved are now quite well understood. Once a set of flutter derivatives has been measured for a given deck geometry, accurate and conservative prototype design forecasts are routinely being made with available modern analytic methods. With the insight gained from examining the evolution of the flutter derivatives with dimensionless velocity, the geometric forms of several modern bridge decks have been strongly influenced toward wholly stable contours. However, an adjacent problem— that of bridge buffeting response— presently remains of strong concern, whether the deck form is intrinsically stable or not.

THEORETICAL DEVELOPMENTS: BUFFETING

Conventional structural studies of bridges have encountered little difficulty in assessing conservative wind pressures for use in basic static design, but for fluctuating wind such as occurs due to mechanical stirring of the earth's boundary layer under high winds its effects upon structural response have demanded more detailed analysis. First, the spectral nature of the natural atmospheric turbulence itself has required assessment. (See review in Simiu and Scanlan 1986.) Second, the manner in which the turbulent wind 'seizes the structure' has been a preoccupying concern. This phenomenon is itself basically an unsteady one (or at least a frequency-dependent one), its time-dependent variations arising from two factors: the turbulent approach flow and the 'signature' distortion of the flow by the bridge itself.

Davenport (1962) was among the first to develop an analytic approach, based on steady-state aerodynamic theory, to the response of extended-span flexible bridges to gusty wind. In certain respects this employed concepts used by Liepmann (1952) for aircraft gust response analysis. Holmes (1975) also considered the problem, enlarging on Davenport's treatment. Early formulations of the problem were based upon static force coefficients; considerations of flutter derivatives were ignored, although 'aerodynamic damping'— a feature later seen to be more explicitly incorporated via the flutter derivatives— was empirically included from steady-state information. Scanlan and Gade (1977) and Scanlan (1988) discussed the gust response problem, identifying aerodynamic damping via the flutter derivatives, though still employing a quasi-steady formulation for the gust forces, which was equivalent to assuming unit aerodynamic admittance. This is further discussed below.

An important consideration required by early works to bring calculation into line with experimental results was the modification of the analysis based upon steady force coefficients to a more realistic unsteady force format. This was empirically done by inclusion of the 'aerodynamic admittance' in the spectral response equations. For want of better information, in the power spectral form of the problem the square of the absolute value of the airfoil-related Sears function (Sears 1941) was typically employed to this end (Holmes 1975), although Holmes commented on its shortcomings in this context. Jancauskas and Melbourne (1986) later showed how near (or far) from the mark this device may come for some aerodynamic shapes. In a few cases it appears remarkably close; in others it is highly unconservative. Appropriate experimental evaluations of aerodynamic admittance have become an additional asset to bridge investigations (Kumarasena 1989).

The writer (Scanlan 1993) has delineated basic two-dimensional aspects of the combined stability and gust response problems using a frequency-domain approach, demonstrating the roles of flutter derivatives, the aerodynamic admittances, the implied relationships among them, and their links to antecedents in aeronautical aeroelasticity. The (3D) lateral coherence characteristics of the natural, turbulent wind have been quite well documented. (See Simiu and Scanlan 1986.) This information serves as a preliminary guide to the lateral coherence characteristics of buffeting forces over the span of a full bridge. Studies by Larose (1992) have revealed a somewhat greater degree of coherence among spanwise buffeting forces than estimated from atmospheric coherence alone. The lateral coherence factor is one of the most important in analytical buffeting response estimations, as it indicates a very great net reduction in generalized force as compared to that calculated assuming fully coherent effects.

All of the work on bridge buffeting until recently has considered single-mode response. Recent studies at Johns Hopkins University (Jain, Jones and Scanlan 1994) have extended both buffeting and flutter analyses to multi-mode response.

Studies associated with the Straits of Messina Bridge project (Diana et al 1991, Falco et al 1992, and Curami and Zasso 1993, Diana 1993) have strongly espoused, modified, and advanced the state of the art as sketched in the foregoing remarks.

CONCLUSION

This paper has in no way attempted a comprehensive compilation covering the full state of the art in bridge aeroelasticity. Since the 1980's, in particular, the field has expanded extremely rapidly with contributions worldwide. The writer (Scanlan 1979) provided a paper outlining fairly broadly the state of the art up through the 1970's. It would be remiss here, however, not to list some of the noteworthy contributions that have demonstrated advances in the field.

For a number of years, since the 1960's, the great thrust in long-span bridge building in Japan, including the great Honshu-Shikoku projects, has given important impetus to bridge aeroelasticity. The writer is not in possession of an exhaustive list of related publications, which would be lengthy. A few representative papers will be cited, however, to acknowledge some of the recurring names of pioneers and strong and repeated contributors to the field: Hirai (1956); Hirai, Okauchi, Miyata (1966); Konishi, Shiraishi and Matsumoto (1975); Nakamura (1975); Ito (1987); Miyata and Yamada (1990); Miyata, Matsumoto and Yasuda (1991); Tanaka, Yamamura and Tatsumi (1989); Fujino, Iwamoto, Ito and Hikami (1992); Tanaka, Yamamura and Shiraishi (1993).

Some recent doctoral theses have collected, extended, and applied the present considerable range and body of theory: Kimura (1991), Tanaka (1993), Stoyanoff (1993).
Notable studies at Chengdu (Zhou 1993) and Shanghai (Xiang, Lin and Song 1993) have recently appeared.

The volume of invited papers published by the International Symposium on Aerodynamics of Large Bridges in Copenhagen (Larsen, Ed., 1992) contains a useful overview of the present state of the art.

Bridge deck model testing and theory have developed side-by-side to the point where interpretation of model tests, on the one hand, has led to more insightful understanding of their significance, and on the other, has strengthened the independent character of the developing theory. Each successive series of empirical model test results has, when subjected to analysis, permitted the sharpening of method and the delimiting of tests to their minimum essentials. Today, analysis, based on these test essentials, constitutes a powerful body of knowledge directly useful in design studies (Scanlan et al 1993; Scanlan 1994). This paper has traced a few key points linking bridge aeroelastic theory to its aeronautical antecedents and mentions some of its leading features that today are fast multiplying around the world.

A number of points remain in need of further clarification. It is well known that the flutter derivatives are dependent upon the amplitude of motion of the structure. The effect of this in perhaps limiting , or slowing the onset of, flutter response beyond the incipient stage could benefit from further study.

The present analytical models of self-excited response can be extended to include so-called vortex-induced response, on the few occasions where this becomes an important issue.

While it is clear that flow around bluff bridge deck sections is quite complex during structural oscillation the linear flutter derivative model of deck response appears to remain reasonably accurate in this context. This is apparently due to the fact that the deck integrates the locally nonlinearly varying pressure forces and, with the preponderance of structural elastic and inertial forces over aerodynamic ones, the net response retains a strong sinusoidal character, so that the current linear flutter derivative analytic model remains fairly descriptive of observed fact. What occurs to the actual flow forces (and consequent actual indicial function representations thereof) under transient physical conditions remains to be clarified. For example, does an actual physical indicial response function continue to resemble the inverse Fourier transform of a flutter derivative? What is the physical validity of indicial response functions so obtained when used in time-domain response formulations?

Experimental turbulence simulations initiated by passive means for models most often develop integral scale lengths that are low compared to the atmospheric scale lengths they are intended to match. Means for active turbulence simulation for section models (Bienkiewicz et al 1983; Kobayashi and Hatanaka 1992; O'Connell 1994) are in need of further development. In view of orders-of-magnitude Reynolds numbers discrepancies, can the force and flutter coefficients of greatly reduced-scale models be considered reliable? The question of the aerodynamic equivalence of small-scale models is to date not wholly answered, nor for that matter, is that of consistency among the results of different examples of such models. This question remains delicate (Curami, Zasso 1993).

Today the state of the art— at its best— has arrived at a considerable degree of refinement, but this stage of development also allows the profession to view more objectively some of the gaps that remain to be closed before fully consistent results are achievable.

REFERENCES

Bienkiewicz, B., Cermak, J.E., Peterka, J.A. and Scanlan, R.H. (1983) "Active Modeling of Large-Scale Turbulence", *Proc. 6th Int'l Conf. on Wind Engrg.*, Australia, March.

Bleich, F., (1948–49) "Dynamic Instability of Truss-Stiffened Suspension Bridges under Wind Action" *Proc. ASCE*, **74** (8) Oct.1948 :1269–1314; **75** (3) Mar.1949, :413–416; **75** (6) Jun.1949, :855–865.

Curami, A and Zasso, A. (1993) "Extensive Identification of Bridge Deck Aeroelastic Coefficients: Average, Angle of Attack, Reynolds Number and Other Parameter Effects", *Proc. Asia-Pacific Symp. on Wind Effects III*, Hong Kong, December.

Davenport, A.G. (1962) "Buffeting of a Suspension Bridge by Storm Winds", *Jnl. Struct. Div.* ASCE **88** (3) June :233–252.

Diana, G. (1993) "Analytical and Wind Tunnel Simulations for the Aeroelastic design of the Messina Straits Bridge", Int'l Seminar on Utilization of Large Boundary Layer Wind Tunnels. Public Works Research Institute, Tsukuba, Dec.

Diana, G., Cheli F., Collina, A., Zasso, A. and Brownjohn, J.M. (1991)

"Suspension Bridge Parameter Identification in Full Scale Test", *Proc. 8th Int'l Conf. on Wind Engrg.* London, Ontario.

Ehsan, F. and Scanlan, R.H. (1990) "Vortex-Induced Vibrations of Flexible Bridges", *J. Engrg. Mech.*, ASCE **116** (6) June :1392–1411.

Falco, M., Curami, A. and Zasso, A. (1992) "Non-Linear Effects in Sectional Model Aeroelastic Parameter Identification", *Jnl. Wind Engrg. and Indust. Aerodyn.* **42** :1321–1332.

Farquharson, F.B., Ed. (1949–1954) **Aerodynamic Stability of Suspension Bridges**, University of Washington Engineering Experimental Station, Bulletin No.116, Parts 1–5.

Fujino, Y., Iwamoto, M., Ito, M. and Hikami, Y. (1992) **Progress in Wind Engrg.** (A.G. Davenport, Ed.) Elsevier, :1333–1346.

Halfman, R.L. (1952) "Experimental Aerodynamic Derivatives of a Sinusoidally Oscillating Airfoil in Two-Dimensional Flow", **NACA TR-1108**.

Hirai, A. (1956) "Aerodynamic Stability of Suspension Bridges under the Influence of Wind Load", *Bauingenieur*, Berlin, **31** (11) Nov. :402–408.

Hirai, A., Okauchi, I. and Miyata, T. (1966) "On the Behavior of Suspension Bridges under Wind Action", *Proc. Symp. on Suspension Bridges*, Lisbon, Nov.

Holmes, J.D. (1975) "Prediction of the Response of a Cable-Stayed Bridge to Turbulence", *Proc. Fourth Int'l Conf. on Wind Effects on Bldgs. and Struct.* (K.J. Eaton, Ed.) Cambridge Univ. Press, London :187–197.

Huston, D.R. (1986), "The Effects of Upstream Gusting on the Aeroelastic Behavior of Long, Suspended-Span Bridges", Doctoral Dissertation, Dept. of Civil Engrg., Princeton University.

Irwin, P.A. and Schuyler, G.D. (1978) "Wind Effects on a Full Aeroelastic Bridge Model", Paper No.3268, Spring Convention ASCE, Pittsburgh, PA.

Ito, M. (1987) "Measures against Wind-Induced Vibrations of Bridges", *Proc. Struct. Cong. '87, Struct. Div.* ASCE.

Jain, A., Jones N.P. and Scanlan, R.H. (1994) "Fully-Coupled Buffeting Analysis of Long-Span Bridges", *Proc. 9th Int'l Conf. on Wind Engrg.* New Delhi, India (to appear 1995).

Jancauskas, E.D. and Melbourne, W.H. (1986) "The Aerodynamic Admittance of Two-Dimensional Rectangular Section Cylinders in Smooth Flow", *Jnl. Wind Engrg. and Indust. Aerodyn.* **23** :395–408.

Kimura, K. (1991) "Flat Plate Buffeting in Yawed Wind with Possible Applications to Bridges", Doctoral dissertation, Univ. of Ottawa, Canada.

Klöppel, K. and Weber, G. (1963) "Teilmodelversuche zur Beurteilung des aerodynamischen Verhaltens von Brücken", *Der Stahlbau*, Berlin, Heft 3 and 4.

Kobayashi, H. and Hatanaka, A. (1992) "Active Generation of Wind Gust in Two-Dimensional Wind Tunnel", *Progress in Wind Engrg.* (A.G. Davenport, Ed.) Elsevier, Amsterdam, Part 2 :959–970.

Konishi, I., Shiraishi, N. and Matsumoto, M. (1975) "Aerodynamic Response Characteristics of Bridge Structures", *Proc. 4th Int'l Conf. on Wind Effects on Bldgs. and Structures*, (K.J. Eaton, Ed.) Cambridge Press :199–208.

Kumarasena, T. (1989) "Wind Response Prediction of Long-Span Bridges", Doctoral Dissertation, Dept. of Civil Engrg., Johns Hopkins Univ.

Küssner, H.G. (1936) "Zusammenfassender Bericht über den instationären Auftrieb von Flügeln", *Luftfahrt-Forschung* **13** :410–424.

Larose, G.L. (1992) "The Response of a SuspensionBridge Deck to Turbulent Wind: the Taut Strip Approach", Master's thesis, Univ. of Western Ontario, March.

Larose, G.L., Davenport, A.G. and King, J.P.C. (1993) "On the Unsteady Aerodynamic Forces on a Bridge Deck in Turbulent Boundary Layer Flow", Paper No.36, *Proceedings, 7th U.S. National Conference on Wind Engineering*, UCLA, June :373–382.

Larsen, A. (Ed.) (1992) **Aerodynamics of Large Bridges**, Balkema, Rotterdam.

Liepmann, H.W. (1952) "On the Application of Statistical Concepts to the Buffeting Problem", *Jnl. Aeron. Soc.* **19** (12) Dec. :197–201.

Miyata, T. and Yamada, H. (1990) "Coupled Flutter Estimate of a Suspension Bridge", *Jnl. Wind Engrg. and Indust. Aero.* **33**.

Miyata, T., Matsumoto, H. and Yasuda, M. (1991) "Circumstances of Wind Resistant Design Examination for Very Long Suspension Bridge", *Proc.*

8th Int'l Conf. on Wind Engrg. London, Ontario.

Nakamura, Y. (1975) "Some Research on Aeroelastic Instabilities of Bluff Structural Sections", *Proc. 4th Int'l Conf. on Wind Effects on Bldgs. and Structures*, (K.J. Eaton, Ed.) Cambridge Press :359–368.

O'Connell, S.D.M. (1994) "Experimental Study of Turbulent Flow Generated by Pulsating Grids", M.S. Thesis, Colorado State University.

Raggett, J.D. and Scanlan, R.H. (1993) "Identification Procedure for 8 Flutter Coefficients from Section Model Tests in Two-Dimensional Flow", *Proceedings, 7th U.S. Nat'l Conf. on Wind Engrg.* June :533–542.

Sarkar, P.P., Jones, N.P. and Scanlan, R.H. (1992) "System Identification for Estimation of Flutter Derivatives", **Progress in Wind Engineering** (A.G.Davanport, Ed.) Elsevier :1243–1254.

Sarkar, P.P. (1992), "New Identification Methods Applied to the Response of Flexible Bridges to Wind", Doctoral Dissertation, Dept. of Civil Engrg, Johns Hopkins Univ., Baltimore, MD.

Scanlan, R.H. and Sabzevari, A. (1967) "Suspension Bridge Flutter Revisited", presented at the May, 1967 ASCE Structural Engineering Conference, held in Seattle, WA (Preprint 468).

Scanlan, R.H. and Sabzevari, A. (1969) "Experimental Aerodynamic Coefficients in the Analytical Study of Suspension Bridge Flutter", *Jnl. of Mech. Engrg. Sci.* U.K., **11** (3) :234–242.

Scanlan, R.H. and Budlong, K.S. (1971), "Flutter and Aerodynamic Response Considerations for Bluff Objects in a Smooth Flow", *Proc. Symp.* Paper D1, Karlsruhe, Germany; **Flow-Induced Structural Vibrations,** (E. Naudascher, ed.,) Springer-Verlag, N.Y. (1974) :339–354.

Scanlan, R.H. and Tomko, J.J. (1971) "Airfoil and Bridge Deck Flutter Derivatives", *Jnl. Engrg. Mech. Div.* ASCE **97** (6) :1717–1737.

Scanlan, R.H. and Gade, R.H., (1977) "Motion of Suspended Bridge Spans under Gusty Wind", *J. Struct. Div.* ASCE **103** (9) :1867–1883.

Scanlan, R. H., (1979). "On the State of Stability Considerations for Suspended-Span Bridges under Wind" , **Practical Experiences with Flow-Induced Vibrations**, E. Naudascher and D. Rockwell, eds., Springer, Berlin, :595-618.

Scanlan, R.H., (1984) "Role of Indicial Functions in Buffeting Analysis of Bridges", *J. Struct. Engrg.* ASCE **110** (7) :1433–1446.

Scanlan, R.H. (1988) "On Flutter and Buffeting Mechanisms in Long-Span Bridges", *Probabilistic Engrg. Mech.* **3** (1) March :22–27.

Scanlan, R.H. and Jones, N.P. (1990) "Aeroelastic Analysis of Cable-Stayed Bridges", *J. Struct. Engrg.* ASCE **116** (2) :279–297.

Scanlan, R.H. (1993) "Problematics in Formulation of Wind-Force Models for Bridge Decks", *Jnl. Engrg. Mech.* ASCE 119 (7) July :1353–1375.

Scanlan, R.H., Stroh, S.L. and Raggett, J.D. (1993) "Methods of Wind Response Investigation Employed for the Kap Shui Mun Bridge", *Proc. 3rd Asia-Pacific Symp. on Wind Engrg.* Dec. :95–100.

Scanlan, R.H. (1994) "Toward Reduction of Empiricism in the Wind Design of Long-Span Bridges", *Proc. Deauville Conf. on Cable-Stayed and Suspension Bridges* Oct. (to appear).

Scruton, C. (1952) "Experimental Investigation of Aerodynamic Stability of Suspension Bridge with Special Reference to Severn Bridges", **Proc. 5th Civil Engrg. Conf.** London, **1** Part 1, (2) :189–222.

Sears, W.R. (1941) "Some Aspects of Non-Stationary Airfoil Theory and its Practical Application", *Jnl. Aero. Sci.* **8** (3) :104–108.

Selberg, A. (1961) "Oscillation and Aerodynamic Stability of Suspension Bridges", *Acta Polytechnica Scandinavica* No.308, Stockholm, :69.

Simiu, E. and Scanlan, R.H. (1986) **Wind Effects on Structures**, 2nd Ed., Wiley-Interscience, New York, N.Y.

Steinman, D.B., (1943–1950) in *Proceedings*, ASCE **69–71, 75, 76** Nov.1943 through Jan.1950 (10 papers in all).

Steinman, D.B., (1954) "Suspension Bridges, the Aerodynamic Problem and its Solution", *American Scientist* **42** (3) July.

Stoyanoff, S.T. (1993) "Wind-Induced Vibrations of Cable-Stayed Bridges", Doctoral dissertation, Kyoto University, Japan.

Széchényi, E. (1973) "Etude du comportement aéroélastique du pont de Saint-Nazaire à Saint-Brévin", NT 2/3044 RY (unpublished document), ONERA, Châtillon, France.

Széchényi, E. (1987) "Pont de Normandie: effets du vent: étude aéroélastique-essais", ONERA Rapport No.10/3588 RY 070, R 370 G Aug.

Tanaka, H., Yamamura, N. and Tatsumi, M. (1989) "Interpreting Section Model Tests to Predict 3-dimensional Flutter Behavior of Long-Span Bridges", *Proc: Canada-Japan Workshop* NRC Report 31871, Ottawa, Canada, Sept. :239–248.

Tanaka, H. (1993) "Aerodynamic Stability of Long-Span Bridges due to Unsteady Aerodynamic Forces", Doctoral dissertation (in Japanese), Kyoto University, Japan.

Tanaka, H., Yamamura, N. and Shiraishi, N. (1993) "Multi-Mode Flutter Analysis and Two- and Three-Dimensional Model Tests on Bridges with Non-Analogous Mode Shapes", *Jnl. Struct. Mech. and Earthquake Engrg.* No.471/I-24 Japan Soc. of Civil Engrg. July :35–46.

Theodorsen, T., (1935) "General Theory of Aerodynamic Instability and the Mechanism of Flutter", *NACA Rept. 496.*

Theodorsen, T. and Garrick, I.E. (1940) "Mechanism of Flutter – A Theoretical and Experimental Investigation of the Flutter Problem", **NACA TR-685.**

Toebes, G.H. (1967) "Fluidelastic Features of Flow Around Cylinders", *Proc/ Int'l Research Seminar on Wind Effects on Bldgs. and Structures* **II**, Univ. of Toronto Press :323–334.

Ukeguchi, M., Sakata, H. and Nishitani, H. (1966) "An Investigation of Aeroelastic Instability of Suspension Bridges", *Proc. Symp. on Suspension Bridges*, Lisbon, Portugal.

Wagner, H., (1925) "Über die Entstehung des dynamischen Auftriebes von Tragflügeln", *Zeit. Angew. Math. u. Mech.* **5** :17–35.

Walshe, D.E. (1963) "The Use of Models to Predice the Oscillatory Behavoir of Suspension Bridges", Paper No.17, *Proc. Int'l Conf. on Wind Effects on Buildings and Structures*, Nat'l Phys. Lab., Teddington, U.K.

Xiang, H.F., Lin, Z.X. and Song, J.Z. (1993) "Study on Aerodynamic Selection of Cross Section for Suspension Bridges", *Proc. 3rd Asia-Pacific Symp. on Wind Engrg.* **I** :101–106.

Zhou, X.H. (1993) "Three-Dimensional Nonlinear Simulation Analysis of Buffeting Response of Long-Span Suspension Bridges", Doctoral dissertation (in Chinese), Southwest Jiaotong Univ., Chengdu, P.R. China.

FULL MODEL TESTING OF LARGE CABLE-SUPPORTED BRIDGES

TOSHIO MIYATA

Professor, Dept. of Civil Engineering, Yokohama National University, Yokohama 240, Japan

SUMMARY

Focusing on how to approach to full model wind tunnel testings and how to get better findings from those tests, some of governing factors interactions and influenced behaviors of wind effects are illustrated and discussed with particular reference to a very long-span suspension bridge and partly to a very long-span cable-stayed bridge, both being under construction. Flutter instability as well as gust responses of a truss-stiffened girder are characterized by quite three-dimensional behaviors interacted with mean deformations due to wind loads. Some of analytical approaches to those three-dimensional behaviors are also presented, and compared with experimental data of a full model wind tunnel testing.

INTRODUCTION

Recent needs of going for the record to stretch the spans much farther are hardly competitive in the construction of suspension bridges as well as cable-stayed bridges. To provide longer, uninterrupted spans over terrain the structural systems of cable-supported crossings are more complicated to design and build. As a matter of fact, the stretch of longer span length leads to more flexible structural features, which may cause a couple of never-experienced, complicated behaviors, governing its safety, in static as well as dynamic responses to wind action.

Among various plans of those very long span bridges, constructions of the Akashi Kaikyo Bridge, a suspension bridge of span of 960 + 1,990 + 960m (Miyata et al 1992a), and the Tatara Bridge, a cable-stayed bridge of span of 270 + 890 + 320m (Endo et al 1991), are well under way for their completion in 1998 and 1999 respectively, when they will be the longest of each type of bridge in the world. Because of their span length never-experienced so far, numerous investigations on wind effects have been, as a matter of course, carried out from a couple of viewpoints accumulated from the past understandings. In case of the Akashi Kaikyo Bridge, a series of full aeroelastic model wind tunnel testings have just finished at a newly built large boundary layer wind tunnel with 41m width and 30m length (Miyata et al 1992b, 1992c; Miyata and Yamaguchi 1993).

Those behaviors, such as flutter instability and gust responses, observed in the full model wind tunnel testings have been realized in three-dimensional fashion under statically deflected phase caused by wind loads, particularly in the higher wind speed range associated with the limit state. Among them, the flutter instability occurred in the truss stiffening girder of the Akashi Kaikyo Bridge, showed a complicated, three-dimensional mode shape consisting of predominant torsional as well as vertical motions, slightly accompanied by horizontal motion, of which the description could be successfully done by means of a newly-developed three-dimensional flutter analysis method. Thus, looking at the fact that a specified behavior was quite three-dimensionally correlated with the deflection and vibration, taking place at the same time, rather well performance in building the full aeroelastic wind tunnel model should be strictly verified again. Here, in this contribution, some of governing factors interactions and influenced behaviors of wind effects on above-mentioned two bridges will be illustrated and discussed, focusing on how to approach to full model wind tunnel testings and to get better findings from those tests (Miyata et al 1994a, 1994b, 1994c).

FLEXIBILITY OF LONG-SPAN CABLE-SUPPORTED BRIDGES

Looking at the reduction of natural frequency and critical wind speed of flutter with the stretch of span length, it is quite easy to understand how flexible are the long-span bridges in their structural and aerodynamic characteristics. Fig. 1 shows the change of natural frequencies of the lowest torsional and vertical bending modes with increase of midspan length in a couple of existing and proposed suspension bridges, and also that of critical wind speeds of so-called coupled flutter, calculated by using the preceding natural frequencies on assumption of flat plate unsteady aerodynamic forces, which may be often used for a good index to present wind resistibility of a specified long-span bridge. It can be seen that it decreases from higher wind speed than 80m/s in conventional class of 1,000m midspan length to 60m/s in 2,000m class, and furthermore down to 40m/s in 3,000m class, such lower level as expected frequently to occur.

Primary reason of this decrease can be referred to a significant reduction of natural frequencies, particularly to that of torsional ones because of the greater polar moment of inertia caused by increasing mass of main cables to stretch the longer spans over terrain. The torsional frequency tends to approach to the vertical heaving one, extremely to the limit of two parallel suspended cables from which a non-stiffened girder simply hangs. Then, it is quite indispensable for a proposed very long-span bridge to overcome such an inevitable and flexible characteristics, in

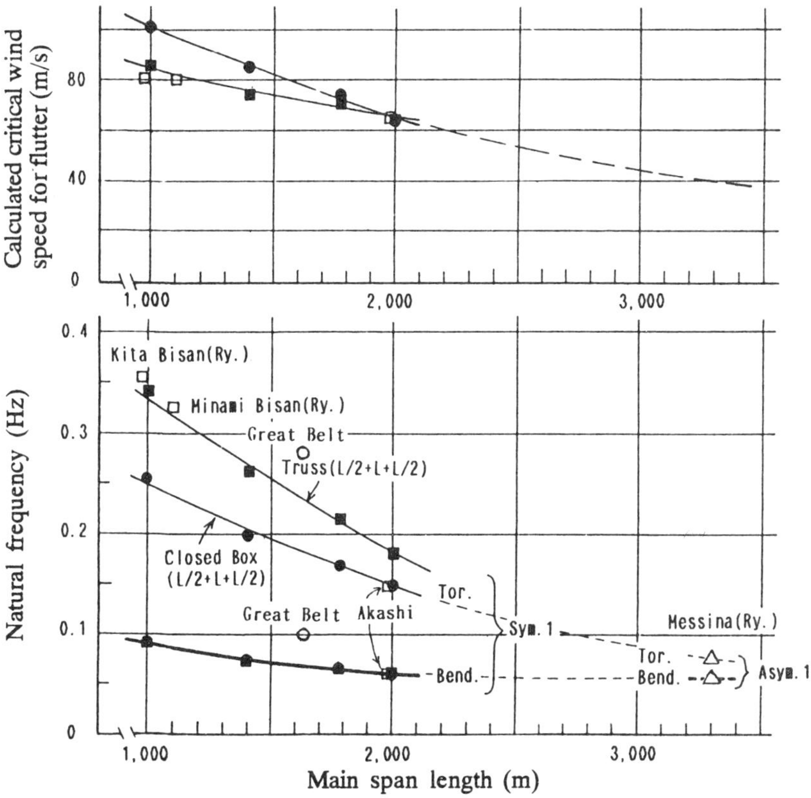

Fig. 1 Reduction of natural frequency with increase of span length in suspension bridge

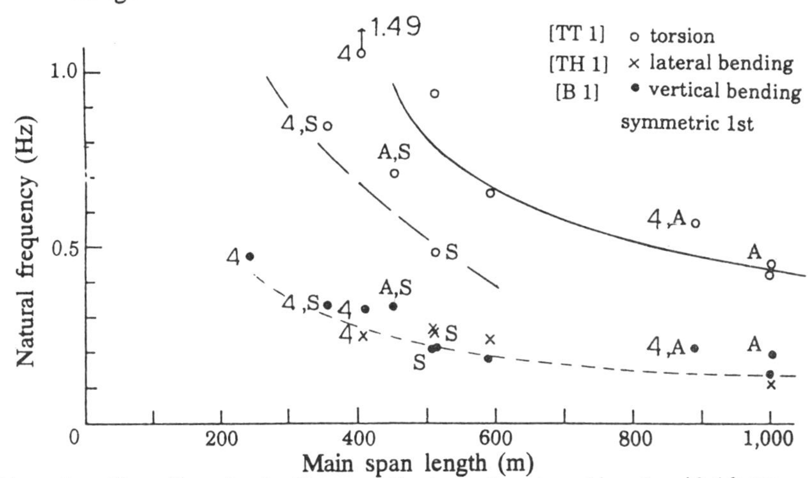

Note: A : with auxiliary piers in side spans; S : single plane stay cables; 4 : with 4 lanes; and none : with double plane stay cables and 6 lanes without auxiliary piers in side spans

Fig. 2 Natural frequencies of long-span cable-stayed bridges cooperation with developing aerodynamically more stable cross section of stiffening girder other than the flat-plate-like if not enough. In order to realize an appropriate high level of wind speed against possible flutter instability, say more than 70m/s, a couple of ideas for countermeasures have been actually developed and investigated (Miyata et al 1992a; Diana 1993; Astiz 1993; Reinhold 1993).

A nondimensional reduced wind speed U/fB, where U is wind speed, f torsional natural frequency and B width of stiffening girder, is a key factor to identify a specified aeroelastic behavior on objects in wind. Calculating that of a certain flutter instability at design wind speed to be verified for the safety, usually anticipated at same level in any case though, it increases nearly in proportion to midspan length, because the torsional natural frequency reduces inversely but the width is nearly constant. Higher reduced wind speed in case of a very long-span bridge, never-experienced so far, should require aerodynamic investigations from different viewpoints. This is a kind of evidence to show the change of targeted behavior due to that of reduced wind speed. A wind tunnel test by the use of sectional model indicated that torsional flutter, usually observed in the truss stiffening girder of conventional 1,000m midspan class suspension bridges, changed to so-called motion-coupled flutter, recognized in a shallow flat-plate-like girder, at such high wind speed range as verified for the Akashi Kaikyo Bridge of 2,000m class (Miyata et al 1992c).

Along with the reduction of natural frequencies with the stretch of span length, a significant structural coupling can be found out in natural vibration mode shapes between vertical, horizontal and torsional components of motion, particularly when deflected and twisted under the action of drag, lift and pitching moment forces by a strong wind. For instance, there happens a contribution of horizontal and vertical motions in the torsion-predominant modes, and that of torsional motion in the horizontal sway-predominant modes. Then, paying an attention to above mentioned existence of motion-coupled flutter on a very long-span suspension bridge, it is quite natural that those effects of essential coupled vibration mode shapes must be carefully taken into account in its verification by means of full model wind tunnel testings.

Stretching the spans of cable-stayed bridges longer, those natural frequencies are also reduced. Fig. 2 illustrates that a couple of lower modes are compared in the midspan range of 300 to 1,000m for normal three-span bridges. Cable-stayed bridges can have a wider choice of play with the form structurally as well as aesthetically. According to the type of their towers and cables, those bridges are classified to several ways of combination; single, delta or twin tower type, two-vertical-planes, two-inclined-planes or single-plane position of cables in space, number of cables and their arrangement (Leonhardt and Zellner 1991; Billington and Deodatis 1991). And, structurally, a couple of other selections are provided with support conditions of the decks, including existing or no auxiliary piers in sidespans, and those between the decks and the towers. Among them, it is quite clear that the reduction of torsional frequencies in case of a single-plane stay is the most remarkable with the stretch of span length. Significant structural coupling in the natural vibration mode shapes is also seen, particularly between horizontal and torsional motions (Miyata 1991).

Now, to verify and measure the aerodynamic vibrations and forces on a bridge, there are a couple of combinations of wind tunnel model of a bridge and a smooth/simulated turbulent flow (+ a topographic model). In process of practical experiments the purpose of the wind tunnel testings is to find out desirable configurations of bridge deck proposals first and then the screened proposal becomes to be brushed up. Although the combination of three-dimensional full bridge model and simulated turbulent flow is the most preferable in the modeling, the cost of this choice is very high. Besides it may be very difficult to find out the best configuration of a bridge deck. Therefore, usual two-dimensional sectional model test is a practical selection for that purpose, although interpretation of the wind tunnel test results might be complicated. Usually, this complicated interpretation must be very essential in the wind resistant design. On stretching the span length of a bridge beyond the past experiences, however, the interpretation is probably not so simple, because of possible correlation caused by some characteristics mentioned above, that full aeroelastic bridge model wind tunnel testing comes to be carried out, also by modeling the topography if necessary (Davenport and King 1993; Miyata et al 1994c).

CHOICE OF TRUSS STIFFENING GIRDER ON AKASHI KAIKYO BRIDGE

The Akashi Kaikyo Bridge, which crosses over a open sea terrain of the Akashi Kaikyo Straits with a shore-to-shore distance of approximately 4 km in an inland sea of Japan, is a very long-span truss-stiffened suspension bridge for six road lanes with a midspan of 1,990 m and two sidespans of 960 m. Construction has started in the spring of 1988 to bring to complete in 1998. The main cables erection is now under way to terminate in the spring of 1995.

As mentioned in the preceding, with the suspended deck crossings stretched to longer, more flexible structural features are so indispensable that static as well as dynamic wind effects are of primary importance for the structural design of this bridge. First of all, flutter instability has been a serious problem for possible selection of deck configurations, because a very long-span, never-experienced, made the entire structure extremely flexible, as seen in predicted lower torsional natural frequency, and prescribed design code required a rather high critical wind speed which naturally stemmed from wind environment at frequent typhoon attack region. Therefore, from the very beginning of the project, numerous investigations have been repeated on the aerodynamic problem by means of conventional two-dimensional sectional model wind tunnel testings in a smooth flow.

After passing beyond preliminary stage, along with conventional truss girders, a couple of box girders and their variations were also proposed, aiming at the improvement of aerodynamic stability as well as sophisticated design (Miyata et al 1992a). Single closed box girders failed to meet the code requirement for flutter, provided that conventional design was made for prescribed loadings and design conditions. To raise critical wind speed, it should be necessary either to increase the use of steel for greater torsional rigidity in spite of economical disadvantage, or to adopt aerodynamic devices such as open slots and a median barrier. Some double box girders with a wide open slot and a median barrier met the requirement, but they were not advantageous because estimated dead load of the girder was heavier than that of a truss structure. In addition, they also suffered from severe vortex excitation. Compound stiffness girder was an attempt to ensure aerodynamic stability by using conventional girder configuration, which consisted of two single box sections of same width and different depth. They were arranged in such a way as shallow sections at middle portions of each span to attain sufficient stability and deep sections at end portions to provide appropriate rigidity. Through these investigations, finally, the decision was made to adopt a properly designed truss structure, taking account of various factors such as structural details, erection methods over frequent navigation route, maintenance cost, etc.

Further efforts have been continued to improve the performance of the truss stiffening girder from viewpoints of economy, aerodynamic stability and design practice, for instance;

(a) effects of truss width, carrying the same floor deck,
(b) location and numbers of built-in members such as inspection ways, electric power line and water supply piping,
(c) configuration of detailed members as perforated curb,
(d) interaction effects of main cables with the girder, resulting to make the flutter more unstable with the relative distance reached to the same as truss height,
(e) effects of a kind of stabilizer, increasing means of flutter stability, with solid short plates installed at the center on and/or under road deck.

In general, the flutter instability, recognized in sectional wind tunnel tests at this stage, was of a heaving-and-torsion coupled type in the most stable case, or in the higher range of wind speed close to that for design verification, although it turned to a torsion predominant type, or so-called torsional flutter, resulting in lower critical wind speed of 40 to 50 m/s, in the most unstable case. The torsional flutter used to be observed in the truss girder of conventional span length of 1,000m class. The change to coupled type in the most stable case was really significant.

The cross section of a truss stiffening girder, finally judged the most stable and reasonable for adoption, was such a simplified, contemporary truss type with a kind of stabilizer installed at the center as shown in Fig. 3.

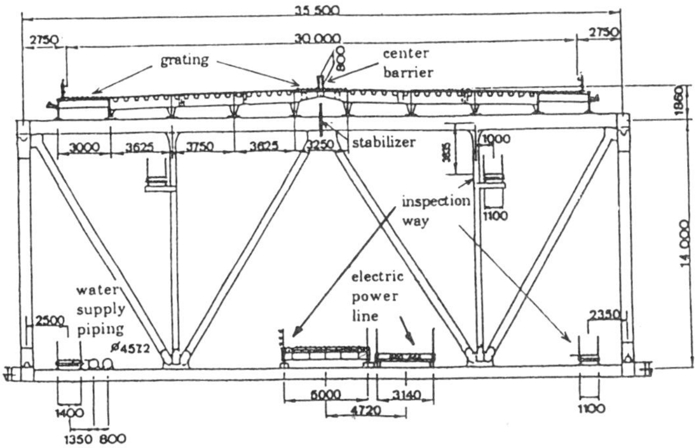

Fig. 3 Cross section of stiffening truss girder for full model wind tunnel tests

Meantime, the code requirement for wind effects has also been investigated for this specified bridge in the form of improvement and revision of the former code (1976) and manual (1980), applied to conventional class-span bridges constructed so far (Honshu-Shikoku Bridge Autority 1976, Honshu-Shikoku Bridge Autority 1980), along with considering extension of the scope which never-experienced scale of the bridge brought out. A new code was established in 1990 (Honshu-Shikoku Bridge Autority 1990), where the requirement against flutter instability was prescribed in the form of critical wind speed U_C satisfied with the relationship in the 10 min. mean;

$$U_C \geq 1.2\mu_F U_z = 78 \text{ m/s}; \quad -3^\circ < \alpha < 3^\circ \tag{1}$$

in which $U_z = (z/10)^{1/8} U_{ref} = 60$ m/s ; wind speed at mean height of the girder z = 80 m and U_{ref} = 46 m/s; a reference wind speed at 10 m height over open sea level for 150 years return period; and α, angle of attack of air flow, specified by taking into account of the effects of wind speed fluctuation over open sea, probable torsional displacement caused by wind loads and experimental error on mounting sectional wind tunnel models as well as the fact that flutter instability is quite susceptible to it. The coefficient 1.2 is a principal safety factor considering reliability of experimental testings, design and construction of the structure, social importance of the structure and a destructive feature of flutter instability, while the coefficient μ_F =1.08 is a conditional safety factor considering spatial gust effect of wind speed fluctuation under assumption of the built-up time (30 sec. for this case) of flutter compared with 10 min. mean wind speed. Here, as a result of requirement of two-dimensional sectional model test in a smooth flow, most of effects of wind speed fluctuation come to be dealt with as quasi-static effect. In case of the wind tunnel test in a simulated turbulent flow, the conditional coefficient must be $\mu_F = 1.0$ because of its original definition, and then the critical wind speed shall be larger enough than 72 m/s, only for a horizontal angle of attack if the torsional displacement of the girder is not so significant.

In the old code (1976) the wind load includes both time averaged wind effect and influence of gust response in stress of structural members. However, as far as calculated results for the truss girder proposed are concerned, after loading of those wind loads, allowable torsional amplitude to be added is not large enough to permit significant vibration, particularly in the lateral bracings of truss girder. In comparison among a couple of gust effects, usually, horizontal gust response is predominant. In case of this very long-span bridge, however, the gust effects due to vertical and torsional responses are not negligibly small. These two results made us add a separate detailed consideration on vertical and torsional gust effects on the various members to the new code (1990). It took into account these effects in two steps of the design procedure. At the first step as static wind resistant design, a simple wind load was selected in the form of a typical lateral load, very similar to the old. There may be other procedures for detailed wind load especially in the loading pattern. Predominant wind load effects, however, are still due to lateral static load and horizontal gust response. Therefore, it will be very convenient to propose the deck configuration under this load and to make various structural and aerodynamic comparisons before the detailed gust analysis. After this designing procedure, detailed gust response estimation of the full bridge system shall be carried out as one of aerodynamic verifications. This process shall be repeated until all the related design requirements in the code come to be satisfactory.

FULL AEROELASTIC MODEL WIND TUNNEL TEST PROGRAM

As mentioned, a much simpler stiffening truss girder of contemporary type was finally proposed, instead of any other massive one of conventional type, after satisfying a couple of requirements for wind effects, such as wind-induced displacements, flutter instability and gust responses to be verified, although those were carried out only by sectional model wind tunnel test in a smooth flow, but also through a conventional analytical calculation for individual motion, upon an appropriately assumed analytical model of the bridge structure. In order to raise the critical wind speed of flutter instability over the prescribed wind speed, however, a kind of stabilizer and/or center barrier had to be installed at the center part of road deck.

As for such a proposal of stiffening truss girder to be proceeded towards its final design in detail, a series of full aeroelastic model wind tunnel tests were to be planned in order to carry out final verification of the wind resistant safety, because of following reasons;

(1) Models and methods for the requirements to be verified have been never applied, even if the best at that time, to such a very long-span suspension bridge of very long period never-experienced so far. Those have been derived only from the understanding and experiences of conventional long-span bridges though.

(2) Behaviors of the bridge may be of quite three-dimensional fashion in its deformation and vibration mode shape, at higher wind speed range in particular associated with its limit state, showing such a way as, first a large lateral deflection along with windward-down twist due to wind loads, on which superposed random gust responses of primary three directions, and finally a coupled flutter instability of complicated mode shape taking place on them. Some of those three-dimensional effects are caused by followings;

    (a) Deformation of horizontal deflection and twist accompanied with vertical deflection due to wind loads are quite varying along the bridge axis, resulting in the largest at the center of midspan.

    (b) Aerodynamic interference between truss girder and main cables depends on their interval, which may also be varying by their relative location partly caused by wind loads.

    (c) There exists a complicated coupling in natural vibration mode shapes, between torsional and horizontal motions in particular, also including the contribution of vertical motion in case of laterally-deformed structure due to wind loads, according to the three-dimensional frame model analysis for the bridge.

    (d) Effect of stabilizing devices to suppress the flutter instability must be confirmed with all the spans installed or only the midspan.

    (e) Onset of coupled flutter instability in three-dimensional fashion may come to be quite different from that observed in two-dimensional sectional model.

Since the aim of the program was verification of the safety directly from the model responses, special care was taken to realize a good geometric and elastic simulation of the long-span bridge model as well as a good natural wind simulation. To realize a good simulation in a couple of key points such as geometric configuration and elastic characteristics of the bridge model as well as natural turbulent wind, a lot of examinations and comparisons were made along with the consideration for cost constraint of wind tunnel construction. Then, the geometric scale of the model was chosen as 1/100. The size of this scaled model was comparable to that of the sectional models, so that the structural details of the truss girder and auxiliary members, including connection devices of each element, could be modeled with accuracy. For this program, a large boundary layer wind tunnel was especially constructed by Honshu-Shikoku Bridge Authority at the Public Works Research Institute, Ministry of Construction, to install a very long full aeroelastic bridge model of total length about 40 m.

WIND TUNNEL AND TURBULENCE GENERATION

In full aeroelastic model wind tunnel testings, the determination of test similarity is very important. For structures where the resistance to deformation is influenced by the action of gravity, such as suspension bridges on which the tensile force of main cables caused by dead load plays a primary role, F_r number shall be satisfied with because the acceleration of gravity must be equal in model and in prototype. Then, the scale of wind speed is the square root of the length scale. Actually, the maximum wind speed available in the facility was set to be 12 m/s by taking a margin for pressure loss in simulated boundary layer turbulent flow, which could allow sufficiently for the range of wind speed to verify flutter instability as required in the code (78 m/s in a smooth flow in this case). The primary dimensions of test section were chosen 41 m wide, 4 m high and 30 m long in order to install an about 40 m long full bridge model as well as to generate well simulated boundary layer turbulent flows which could allow for appropriate thickness over the bridge model in the rear. The wind tunnel is basically of the section Eiffel type, but the air is circulated within the building. Other properties and arrangements of the facility were specified by investigating the test results of a half model of 1/16 scaled pilot wind tunnel (Miyata et al 1992b; Miyata and Yamaguchi 1993).

To evaluate the aerodynamic stability in turbulent air flow, preliminary studies were carried out to examine how to generate a boundary-layer turbulence, also 1/100 in scale, simulating the case with natural wind prevailing at the bridge construction site. As an aid in generating the simulated boundary layer turbulence, it was actually decided to use spires and roughness blocks in alternately different combinations. The simulation process in the wind tunnel consisted, so to say, of trial and error, with reference to empirical formula (Irwin 1979). The boundary layer turbulence characteristics targeted to reproduce were, taking into account as basic requirements, the following parameters specified by the code requirement for this bridge; power exponent of vertical mean wind speed distribution α = 1/8 and intensity of along wind turbulence at the girder height I_u = 10% and 5%. In designing the spire, a pilot scheme was conducted using spires at the wind tunnel, 1/16 in scale, to decide on approximate size and shape to which to fabricate the spires for use in the large wind tunnel. Consequently, it was decided that generating a turbulent flow exceeding 3 m in boundary layer thickness would require the use of spires about as tall as the height (4 m) of the wind tunnel ceiling and that to ensure the spire structural integrity, each spire be fixed at its top to the ceiling. In the process of generating the turbulent flow, it was necessary to add the use of brick-blocks placed on the floor surface to get the required level (10%) of intensity (Miyata and Yamaguchi 1993). Finally, judging the first priority to be to simulate the intensity of turbulence at the girder height, the following two types of turbulent flow were provided to use for the test; I_u = 9.5% (I_w = 6.8%) and I_u = 6.5% (I_w = 5.5%), where the integral scale of along wind L_u^x was about 1 m long, the decay factor of correlation K_u^x =5~12 and power spectrum plotting in shape agreed fairly well with that of natural wind prescribed for the code requirement. It was found that the integral scale was slightly smaller than required, and some other parameters also slightly differed from the required. However, the difference was considered not so large, and adjustable for the evaluation of test results through the correction by analytical calculations.

FULL AEROELASTIC BRIDGE MODEL SIMULATION

In order to verify the detailed three-dimensional behaviors of a long-span truss stiffened suspension bridge, a lot of efforts to realize a good performance on structural as well as aerodynamic properties must be done as for the similarity of the model in configuration, mass distribution, stiffness distribution, damping capacity, aerodynamic interference between each element and so on. In this case of the Akashi Kaikyo Bridge, the geometric scale of full aeroelastic bridge model was chosen 1/100, and various analytical studies and preliminary model tests were carried out in advance.

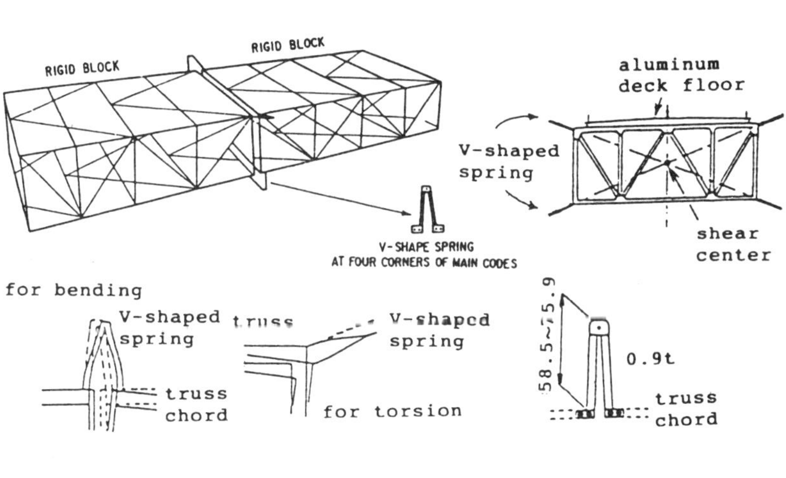

Fig. 4 Modeling of truss stiffening girder (unit:mm)

Modeling of the Stiffening Truss Girder

The stiffening girder model is shown in Fig. 4. The model is composed of rigid blocks of four truss panels connected by V-shaped springs, which represent expansion and contraction of the girder chords resulting in the simulation of vertical and lateral rigidities, while torsional stiffness is simulated by the out-of-plane bending of each member of a V-shaped spring, which means disparity of the chord axis. By directing the four V-shaped springs toward the center of the truss section, their large axial stiffness can guarantee a constant shear center of the cross section. In order to simulate the mass, polar moment of inertia and the center of gravity at the same time, the mass distribution of the model cross section must be analogous to that of the actual bridge. This requirement is satisfied by using light, rigid composite material (0.3 mm carbon fiber plate and foaming polystrol core) for the truss members, and an aluminum plate for the road deck floor. The other parts of the road deck and auxiliary facilities are made of a wood whose specific mass density is very small. The structural and aerodynamic performance of the stiffening girder model was investigated before assembling it to the full model. The result showed the discrepancies of bending stiffness and torsional stiffness, unavoidable nonlinearity taking place in the higher range of design wind speed though, with required values were within 10%, and those of mass, polar moment of inertia and center of gravity were within 2%. The sectional model tests, conducted with/without the V-shaped springs, showed only a minor effect on the aerodynamic properties of the cross section.

Modeling of the Main Tower

The stiffness of the tower has a significant effect on the entire behavior of the bridge. This modeling was carefully studied, and the tower model shown in Fig. 5 was finally selected. The model skeleton was consisted of a frame of steel bars on which a wood replica was mounted to simulate the geometric shape. The elastic modeling was based on how to select the bending rigidity appropriately for the skeleton members, as a result of which the longitudinal rigidity of the members was significantly greater than the simulation requirement. The diagonal members were disconnected at the intersection portions in order to reduce the unfavorable contribution to the tower stiffness by those rigidity. With the tower in free standing condition, the static deformation and natural vibration properties of the model showed a good simulation of the prototype. However, a FEM structural analysis for the entire bridge model revealed that the nonsimulated longitudinal rigidity of the members, especially that of tower post members, caused a little difference in the torsional displacement and natural frequency from those calculated by using the actual bridge dimensions.

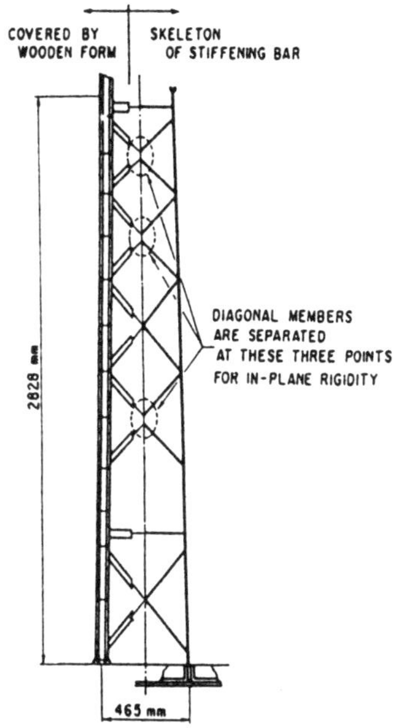

Fig. 5 Modeling of main tower (unit:mm)

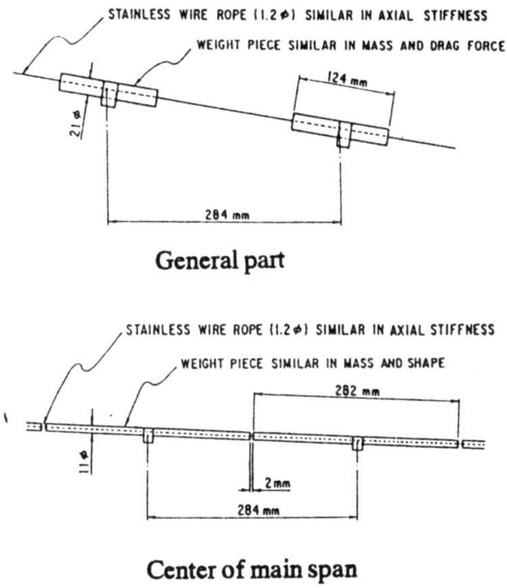

Fig. 6 Modeling of main cable (unit:mm)

Modeling of the Main Cable

In modeling the cable, simulation of mass, longitudinal rigidity and aerostatic drag was required at the same time. Since there was no material that satisfied such a requirement, a separate weight model shown in Fig. 6 was chosen. To obtain the proper longitudinal rigidity as well as sufficient rupture strength, stainless steel woven wire (1.2 mm Ø, 1×19) was used. The weight, configuration and its mounting rate were determined from the mass and aerostatic drag similarity. Regarding drag simulation, the effect of Reynolds' number and three-dimensional flow at the edge of the cylindrical weight was investigated by the wind tunnel test on the cable model. Results of the test indicated that the model had the proper overall drag coefficient effect. It was found out in the sectional wind tunnel tests that the aerodynamic properties were affected by the existence of the cables, and this influence could not be ignored unless the distance between the cables and the girder became nearly the same to the girder height. Considering the aerodynamic interference effects on the entire behavior of the bridge, it was chosen to use a geometrically simulated, mass also simulated though, cable for the center portion of the main span.

Table 1 Natural frequency and damping of 1/100-scaled full model

Predominant mode shape	Natural frequency (Hz)		Logarithmic decrement	
	Measured	Calculated*	Measured	Required*
1.vertical, sym. 1st 	0.647	0.638	0.032	
2.vertical, asym. 1st-1 	0.757	0.745	0.031	
3.vertical, asym. 1st-2 	0.855	0.835	0.035	0.030
4.vertical, sym. 2nd 	1.245	1.213	0.036	
5.vertical, sym. 3rd 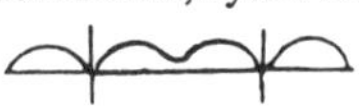	1.636	1.637	0.055	
6.torsional, sym. 1st 	1.560	1.497	0.031	0.020
7.torsional, asym. 1st 	2.075	2.077	0.030	
8.horizontal, sym. 1st 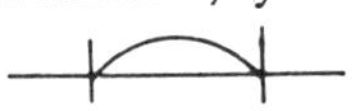	0.391	0.387	0.016	
9.horizontal, asym. 1st 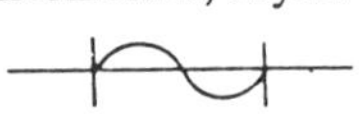	0.781	0.775	0.032	0.030
10.horizontal, sym. 2nd 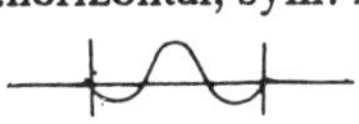	1.270	1.265	0.041	

*calculated; converted to the model from that of the prototype bridge

*required; prescribed in the code requirement

After assembling each part and element to the full suspension bridge model, the properties of static deformation and natural vibration were examined and assured of its performance. As for the static deformation with vertical, lateral and torsional loadings acted respectively, analytical calculations were compared with measured data, which showed rather good agreement. Natural frequency, vibration mode shape and structural damping are quite essential and important dynamic properties. On the bridge model, the various modes of vibration were excited initially by an electric exciter, and from subsequent decaying vibration, the above dynamic properties were measured. Table 1 shows the natural frequency and damping of various modes. The frequencies of vertical and horizontal modes were generally found to agree rather well with those calculated on the prototype bridge. Only the frequency of the first symmetric torsional mode was slightly higher than that of the actual bridge. With regard to structural damping, the logarithmic decrements of torsional modes were, in particular, slightly higher than those required in the code, although the decrements of vertical and horizontal modes were comparable with the required. However, those discrepancies were judged not so great and then adjustable in the final stage of interpretation to the actual bridge on the targeted aerodynamic behaviors. Investigating these measured properties of the full model, the performance of the model was considered rather well.

FULL MODEL WIND TUNNEL TEST RESULTS

Primary test results will be described to show how the full bridge model behaved in a smooth flow as well as in the prescribed turbulent flows. For the better understanding of basic characteristics of wind-induced deformation and flutter instability, first of all, the test started with a smooth flow. The tests were conducted on a couple of types of cross section, paying attention to the effects of such stabilizing devices as installed under and/or on the road deck at its center.

With the rise of wind speed, the truss girder was laterally deflected and twisted by wind loads as shown in Photo. 1. Each component of static displacement is such as illustrated in Fig. 7. Because of large drag force on a truss stiffening girder, the horizontal displacement reached nearly the scale of road deck width at the midpoint of main span under the equivalent design wind speed. Along with greater horizontal displacement, the large torsional one also took place at the same time as much as -4 degrees, windward-down or negative to the wind action at the midpoint, accompanied with vertical displacement lifted. It can be seen that such large deformation was primarily caused by large drag force on the truss girder. The drag directly made the horizontal displacement to take place. As for the torsional displacement, a kind of coupled moment effect, caused by a combination of drag force acting at the center of the truss girder and hanger reaction force at the top of it which acted in the opposite direction, played a primary role. For analytical prediction of those displacements, it was confirmed that a three-dimensional frame model method of load increment, considering the structural geometric nonlinearity due to the deformation as well as the variation of aerostatic forces on the girder twisted along the bridge axis, was necessary to get a good agreement with the test results.

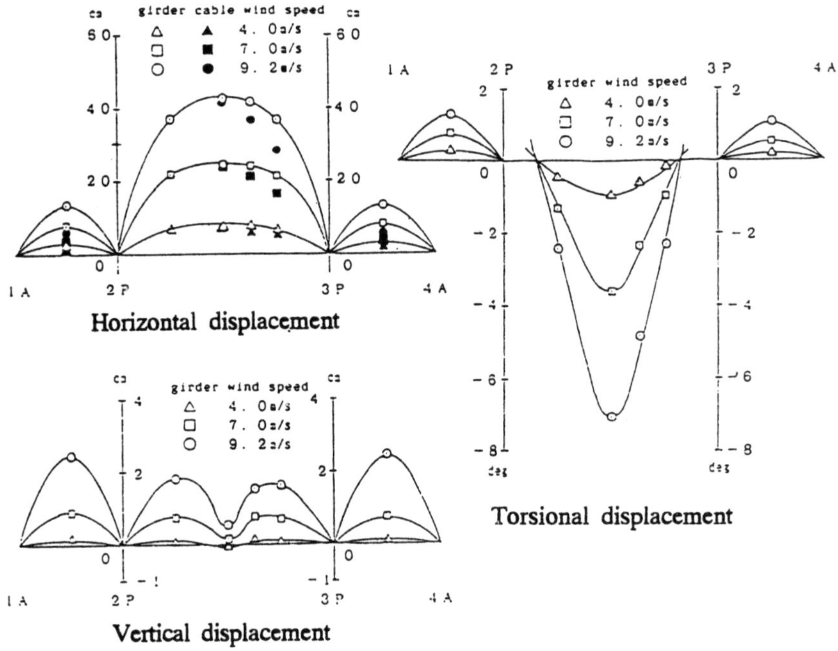

Fig. 7 Displacement of full model by wind load

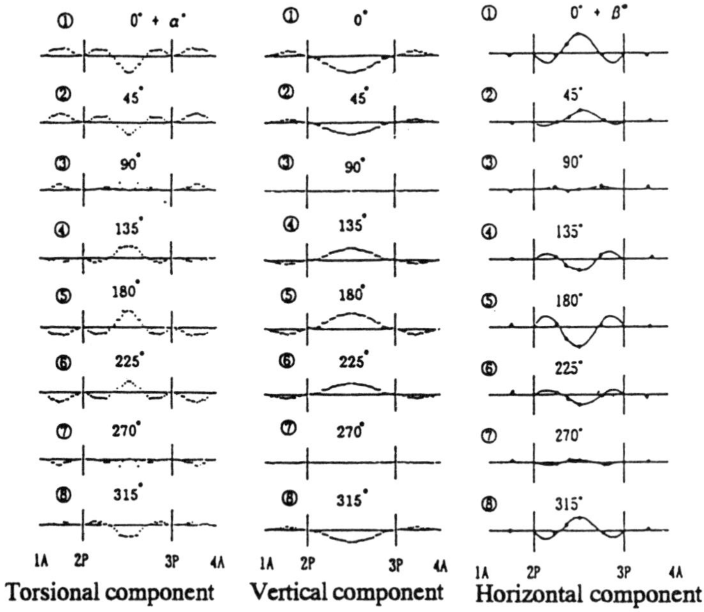

Fig. 8 Flutter mode shape at every 1/8 fraction of one period (α, β = phase lag in Fig. 9, basic section, U = 8.45 m/s)

Photo 1 Largely deformed model by aerodynamic forces at high wind speed

The twisted girder means the occurrence of additional relative angle of the girder to uniform wind action which really varied along the bridge axis. In fact, then, negative torsional displacement might result in the raise of onset wind speed of flutter instability, because the specified cross section of the truss girder came up with more stable aerodynamic characteristics at the negative angle of attack of flow incidence, looking at the results of sectional model wind tunnel tests carried out in advance.

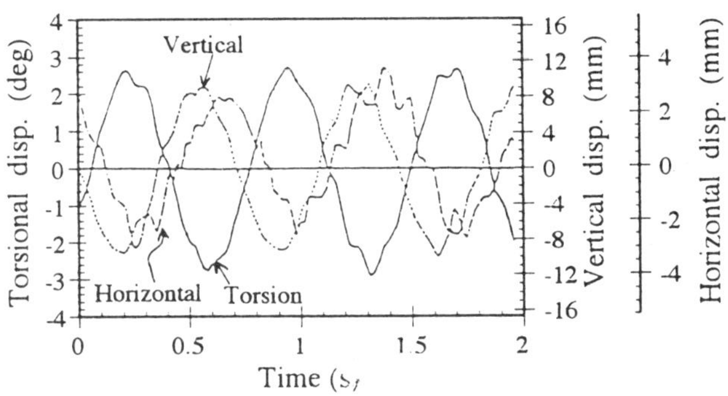

Fig. 9 Time history response during flutter at midpoint of main span (basic section, U = 8.45 m/s, α = 0 deg)

A coupled flutter instability occurred at higher wind speed in the full model test, although it was somewhat different from that estimated from the sectional model test. It can be seen from Fig. 8 that the observed coupled flutter had a particular mode shape of vibration with predominant torsional and vertical motions slightly accompanied by horizontal one. Fig. 9. shows a time history record at the time of flutter onset at the midpoint of main span, from which it can be understood that there were pretty large phase differences between each component of motion. Looking at the change of the center of rotation along the bridge axis, assumed at a point during one period, it can be also found out that the flutter mode shape was quite a complicated, three-dimensional one. The full model tests were conducted on a couple of cross sections; a) the basic section of a truss stiffening girder originally designed and b) the improved sections with stabilizing devices against flutter. The angles of flow incidence were implemented in a horizontal wind as well as a blow-up one of about 2.7 deg. at the girder height, because flutter instability was found quite susceptible to the angle of attack of air flow. Fig. 10 shows the onset wind speeds of flutter instability in the basic section and the improved one with stabilizing devices installed on entire spans, including those of sectional models as well. As for the angles of attack in the full model, a range of varied relative angle over the entire spans is indicated because the flutter at specified wind speed was accompanied by large torsional displacement caused by wind loads.

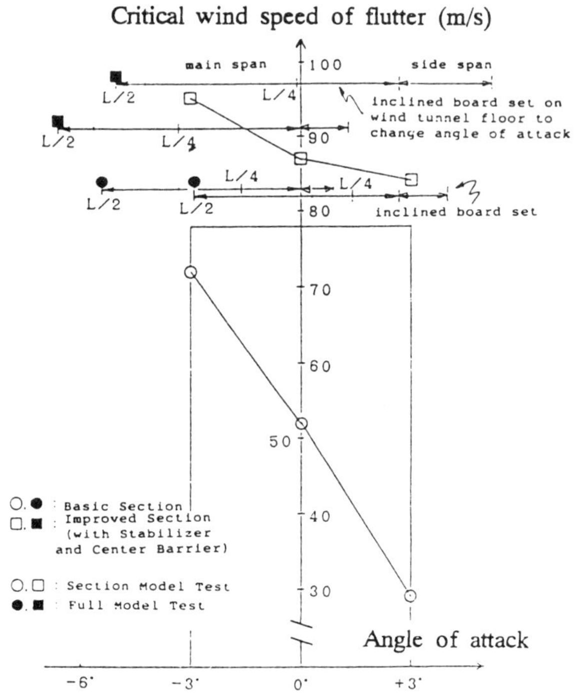

Fig. 10 Relation between flutter onset wind speed and angle of attack

It can be found from sectional model tests that: 1) the basic section was unstable in flutter, the onset wind speed of which was lower than the code requirement, although it was getting more stable with angle of attack reduced to the negative; and 2) the improved section was satisfied with the code requirement by installing such stabilizing devices as shown in Fig. 3, the onset wind speed of which was also slightly increasing in the negative angle of attack. Looking at onset wind speeds in the full model tests, not to speak of the improved section, those of the basic section were also higher than required in the code, although actually comparable with the prescribed after correcting some avoidable inconsistency in the test similarity of full model such as a slightly higher realization of the lowest torsional natural frequency. There was no difference in the basic section practically seen, while a reversal in the improved one, between two sets of angle of attack in 0 and 2.7 deg. Such complicated increase of onset wind speed in the full model was obviously caused by a combined result of higher occurrence in the negative angle of attack, characteristic of this truss-stiffened bridge girder, and large negative torsional displacement found in the midspan at higher wind speed as mentioned above. This means that the study on aerodynamic instability should be carried out with consideration of static wind-induced displacement at the same time.

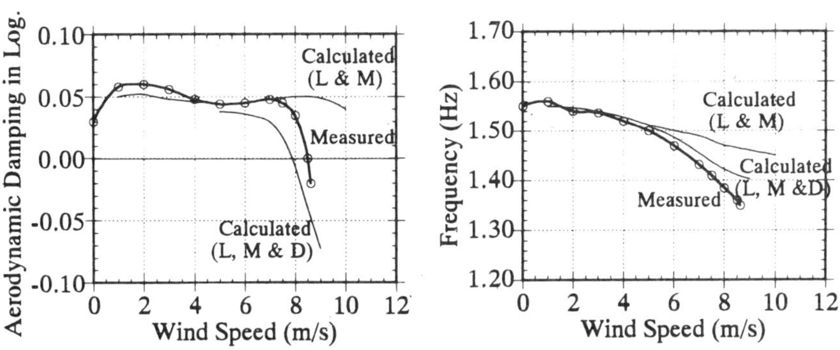

Fig. 11 Change of aerodynamic damping and frequency with increase of wind speed in full model (basic section, smooth flow, α = 0 deg)

For the better understanding of basic characteristics of the coupled flutter instability observed in the full model, of course in the sectional model as well, the change of aerodynamic damping and frequency with the rise of wind speed was measured at a specified amplitude from the damped vibration after excitation in case of positive damping, while from the divergent vibration developed from the rest in case of negative damping, at specified wind speeds. Fig. 11 (a) presents one of those measurements for the basic section in a smooth flow set at 0 deg. angle. The aerodynamic damping in logarithmic decrement can be traced in a slightly varied curve with the increase of wind speed, and dropping abruptly toward the negative in higher wind speed than about 7 m/s. The wind speed of 8.4 m/s (84 m/s for the actual bridge), in this case, is a critical one of flutter onset where the damping changes from positive to negative. The flutter occurred at the wind speed over 8.4 m/s was, in fact, of a coupled type showing a particular mode shape, as mentioned above. The torsional component was practically the same in shape as the first symmetric natural mode at no wind speed, while the vertical component was so complicated in shape that it was not similar to any of vertical natural modes.

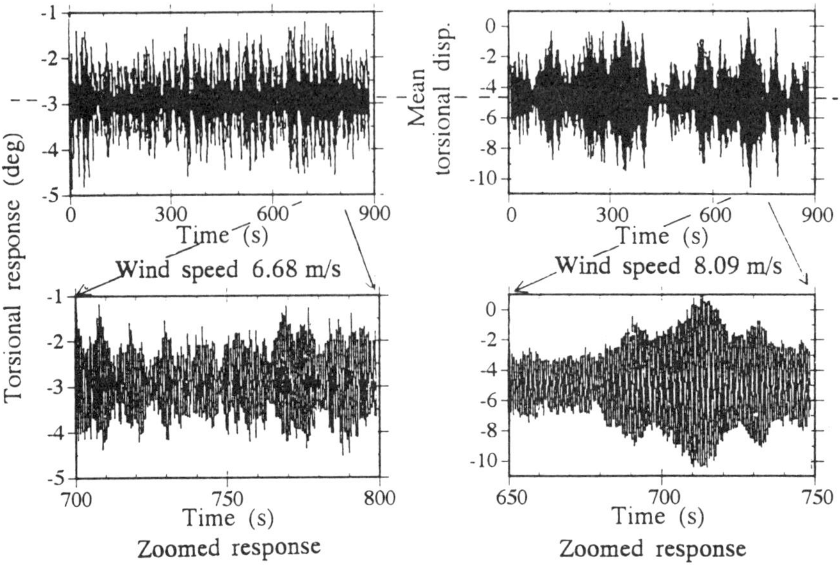

Fig. 12 Example of gust response in turbulent flow (basic section, I_u = 6.5 %)

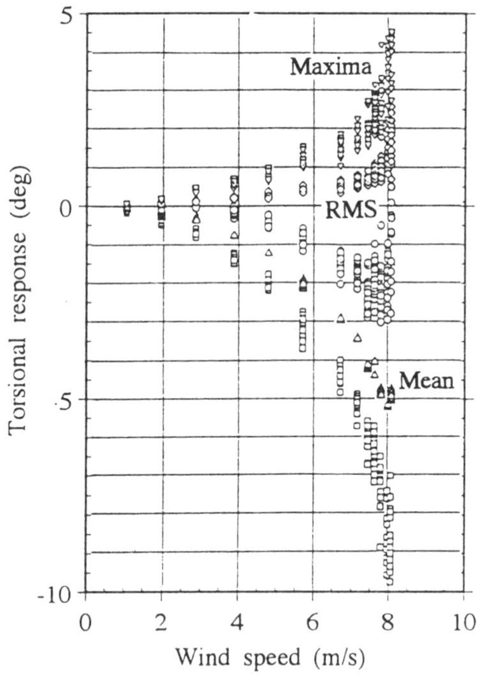

Fig. 13 Torsional gust response at midpoint of main span (basic section, I_u = 6.5 %)

Actually, those components of flutter mode shape should be also influenced by the specified displacement due to wind loads. As for the change of frequency specified in a particular mode branch of the first symmetric torsion, as indicated in Fig. 11 (b), natural frequency at no wind speed was reduced with the rise of wind speed and changed to about 90% of it at the flutter onset. This is typically because of stiffness effects of aerodynamic unsteady forces on this specified truss girder.

Following the smooth flow test, a couple of tests were carried out in the prescribed boundary layer turbulent flows. As expected, quite three- dimensional random oscillations took place, sometimes showing snake-like motions of a kind of wave propagation on the girder under statically deflected and twisted displacements caused by mean wind loads as in a smooth flow. Fig. 12 presents typical buffeting torsional responses measured at the midpoint of main span at two wind speeds, in which the random fluctuation of amplitude can be seen around the mean displacement. Taking the statistical properties averaged over 1 min. (10 min. for the actual bridge) measured at each wind speed, those of mean, root mean square, peak(maximum and minimum) values are scattered as plotted in Fig. 13. With the increase of wind speed, the scatterness of peak and root mean square becomes larger, resulting to make us recognize the possibility of flutter onset in the turbulent flow at about 8 m/s and much higher mean wind speed, quite close to that in a smooth flow. Significant random responses were similarly observed in vertical and horizontal motions as well. Actually, in comparison among a couple of gust effects observed in the tests, the horizontal buffeting response was pretty smaller than expected by conventional spectral analysis for individual motion, while the vertical and torsional responses were rather larger. As for one of reasons why the measured responses were different from the calculated on the basis of conventional spectral analysis, it may be likely caused by their complicated, three-dimensional fashion.

THREE-DIMENSIONAL FLUTTER INSTABILITY ANALYSIS

It is well known that a calculation of critical wind speed for heaving bending-torsion coupled flutter has been often made as a kind of effective index to identify wind-resistibility of long-span bridges, as shown in Fig. 1, on assumption of flat plate unsteady aerodynamic lift and pitching moment forces. In this calculation, a combination of one heaving-bending and one torsional natural modes, usually the lowest ones in similar mode shape, has been used to assume the flutter mode shape (Bleich et al 1950; Selberg 1964). This fact may be originated in conventional approach to the flutter problem of aircraft wings by means of a rigid sectional model of two-degrees-freedom of heaving and rotation, and then allow the use of sectional model mounted two-dimensionally in the tests on the bridge studies as well. However, looking at a complicated, three-dimensional mode shape of coupled flutter instability found in the present full model test, it can be seen that a single combination of the lowest two modes is not necessarily correct to describe the actual behavior of a full model, and also only a sectional model wind tunnel test of similar single combination is not good enough to verify the wind resistibility of a specified bridge.

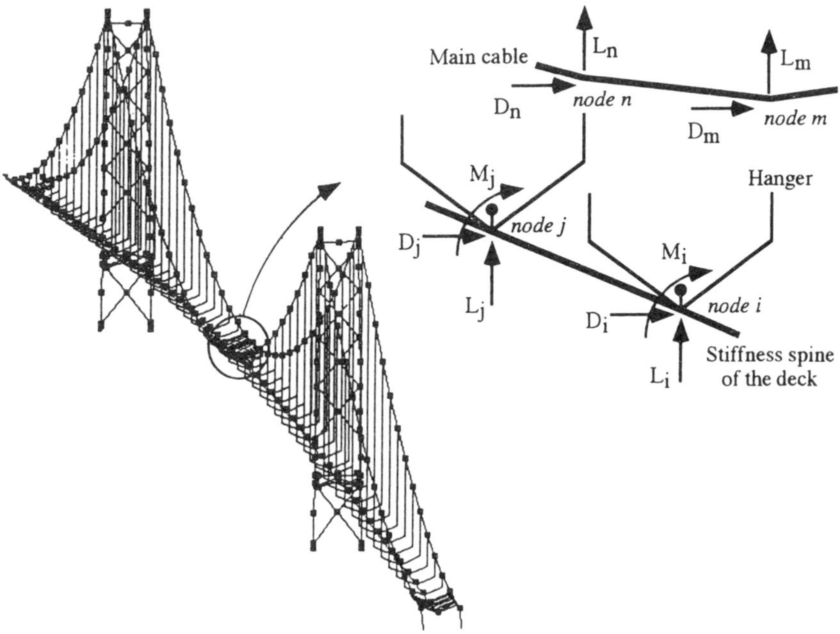

Fig. 14 Frame model for three-dimensional flutter analysis

In order to analyze such a complicated, three-dimensional flutter instability on the deflected and twisted truss girder as observed actually in the full model test, a new method was investigated (Miyata and Ohta 1986; Miyata and Yamada 1988) by the use of an equation motion formulated by applying the unsteady aerodynamic forces, or derivatives, to a three-dimensional structural frame model, as illustrated in Fig. 14, of the entire bridge structure deformed under specified wind loads. The equation of motion is described as follows;

$$[\mathbf{M}]\{\ddot{\mathbf{u}}\}+[\mathbf{C}]\{\dot{\mathbf{u}}\}+[\mathbf{K}]\{\mathbf{u}\}=[\mathbf{F}_A]\{\ddot{\mathbf{u}}\}+[\mathbf{F}_V]\{\dot{\mathbf{u}}\}+[\mathbf{F}_D]\{\mathbf{u}\} \tag{2}$$

where $\{\mathbf{u}\}$: deformed displacement vector = $[\mathbf{u}_S \;\; \mathbf{\Phi}_S \;\; \mathbf{v}_S \;\; \mathbf{u}_c \;\; \mathbf{v}_c \;\; \mathbf{v}_h]^T$,

$[\mathbf{M}]$: deformed mass matrix,

$[\mathbf{C}]$: deformed structural damping matrix,

$[\mathbf{K}]$: deformed stiffness matrix,

and [F] : unsteady aerodynamic force matrices which are defined as functions of the reduced frequency $k=\omega b/U$, measured experimentally in sectional model wind tunnel tests under specified harmonic motions on particular configurations of the girder at specified angles of attack and with or without the simulated cables equipped. Those unsteady aerodynamic forces, or derivatives, proportional to displacement, velocity and acceleration of motion, the last is usually neglected though, are written in the form;

$$[\mathbf{F}]=\begin{bmatrix} \boxed{L_s(u)\;\; L_s(\Phi)} & L_s(v) & 0 & 0 & 0 \\ \boxed{M_s(u)\;\; M_s(\Phi)} & M_s(v) & 0 & 0 & 0 \\ \underline{D_s(u)\;\; D_s(\Phi)} & \boxed{D_s(v)} & 0 & 0 & 0 \\ 0 \quad\; 0 & 0 & \boxed{L_c(u)} & 0 & 0 \\ 0 \quad\; 0 & 0 & 0 & \boxed{D_c(v)} & 0 \\ 0 \quad\; 0 & 0 & 0 & 0 & D_h(v) \end{bmatrix} \tag{3}$$

where L, M and D are lift, pitching moment and drag forces respectively, u, Φ, and v are vertical, torsional and horizontal displacements respectively and the subscripts s, c and h mean stiffening girder, main cable and hanger cable respectively. The reduced frequency $k=\omega b/U$ is defined in a harmonic motion of $u=X\exp(i\omega t)$, where ω is the circular frequency, b half width of the girder and U wind speed.

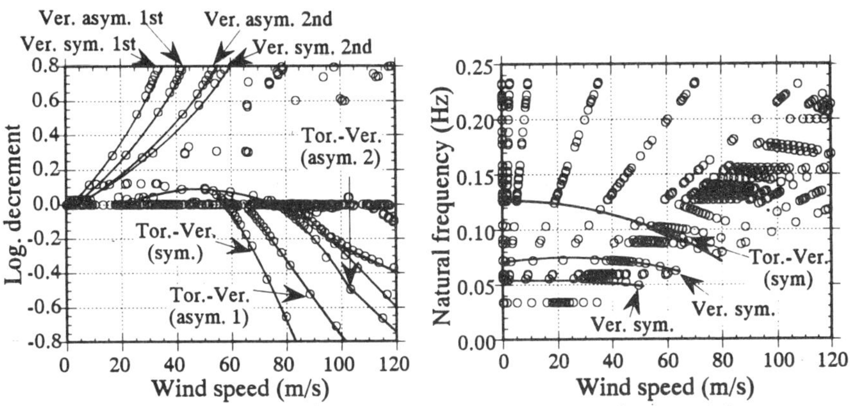

(a) Change in damping due to wind speed

(b) Change in natural frequency due to wind speed

Fig. 15 Root traces by three-dimensional flutter analysis

Making the complex eigenvalue analysis for Eq. (2) with a specified reduced frequency k, a set of complex frequencies $\omega=\omega_R+i\omega_I$ and complex mode vectors

$\mathbf{X}=\mathbf{X}_R+i\mathbf{X}_I$ are directly obtained as much as the degrees of freedom of the three-dimensional structural frame model, from which the aerodynamic damping decrement can be derived in the form: $\delta_a = \omega_I / \sqrt{\omega_R^2 + \omega_I^2}$ at a specified wind speed $U=\omega_R b/k$, along with the complex mode shape. Carrying out similar calculation by changing the reduced frequency over an appropriate range concerned, the change of aerodynamic damping and frequency with the rise of wind speed is to be described in the form of such root traces as shown in Fig. 15, connected with adjacent points of similar mode shape at different wind speeds. Among a lot of significant branches of those root traces, some branches illustrate a particular change of aerodynamic damping from the positive to the negative sign where the zero crossing means a critical wind speed of flutter instability in the specified mode shape. The branch to give the lowest critical wind speed must be of the present concern.

Along with the direct analysis on Eq. (2), an approximate method is available by assuming its flutter mode shape by means of a combination of natural mode shapes at the deflected stage in specified wind speed or the non-deflected stage of zero wind speed. It can be seen that the combined number of natural mode shapes may be appropriately limited to make an easier analysis in case of measured data of flutter instability in hand as a target, while not so simply in general case without a target in the beginning of calculation.

Meanwhile, applying a simplified method to analyze the present flutter behavior observed in the full model wind tunnel test, it was expected for the first time that a good agreement might be seen between measured and calculated aerodynamic damping changes, on the past understanding of coupled flutter where the vibration was to be governed by coupled unsteady lift and pitching moment due to vertical and torsional motions of the stiffening girder, and also by some additional unsteady forces such as drag due to horizontal motion of the girder as well as drag and lift due to horizontal and vertical motions of the main cables, all boxed in Eq. (3). In fact, as shown in Fig. 11 (a), the agreement was pretty good in lower wind speed range up to about 7 m/s. However, the larger changes in higher wind speed range beyond it could not be successfully followed by this idea. Further investigations have been done to find out a key factor to present the abrupt drop of aerodynamic damping to the negative in the higher wind speed, resulting in a discovery of significant contribution of unsteady drag forces due to torsional and vertical motions of the girder, indicated by dotted and solid lines in Eq. (3). That is, looking at a rather large change of averaged static drag with the increase of negative angle of attack, it was recognized that the effect of unsteady drag due to change of torsional displacement was not so small, particularly on the windward-down twisted stiffening girder around the center of main span in higher wind speed as mentioned in the preceding. In reality, applying those additional terms of unsteady force, the targeted abrupt drop of aerodynamic damping could be realized in the calculation as attached also in Fig. 11 (a), still certain difference found from the measured though. As for the change of frequency with the rise of wind speed, there happened also better agreement with the measured as shown in Fig. 11 (b). Such a complicated behavior of flutter instability might be caused by the truss stiffening girder itself in this case, because the drag force on it was essentially large. It is concluded that the effect of drag on coupled flutter was one of new findings from the full model wind tunnel tests.

ANALYSIS OF GUST RESPONSE BEHAVED THREE-DIMENSIONALLY

The random gust responses caused by turbulence of natural wind take place at any time on any structural system. Analytical description of those gust responses is normally done by conventional spectral approach in the frequency domain, making use of appropriately determined power spectrum, correlation functions of turbulence and aerodynamic admittances along with the aerodynamic damping defined by the quasi-steady assumption, particularly only for a single-dimensional motion concerned. As already mentioned in the preceding, those gust responses observed in the full model wind tunnel tests were actually taking place quite three-dimensionally, and pretty smaller in horizontal motion, while rather larger in vertical and torsional motions than calculated by conventional spectral approach although a couple of factors associated with those responses were separately investigated, for instance, by the use of directly measured data on the site.

In order to reflect such three-dimensional responses as observed in the full model tests, a trial of time domain approach may be useful to analyze the entire random response at the same time, although quite complicated and troublesome in actual treatment (Diana et al 1986; Scanlan 1984; Scanlan 1988; Kovacs et al 1992; Santos et al 1993; Santos 1994). The most simplified approach in the time domain normally starts to formulate appropriately following two basic variables; instantaneous relative wind speed as well as effective angle of attack which combines the wind speed fluctuations and the structural motions behaved three-dimensionally. On the basis of quasi-steady assumption, the aerodynamic forces can be expressed in the global directions for the entire structure at a particular instant time as shown in Fig. 16;

$$F_Z(t) = -D\cos\psi + L\sin\psi \quad (4a)$$

$$F_Y(t) = D\sin\psi + L\cos\psi \quad (4a)$$

$$M_X(t) = -M \quad (4c)$$

where the aerodynamic drag, lift and pitching moment in the relative wind axes are given by

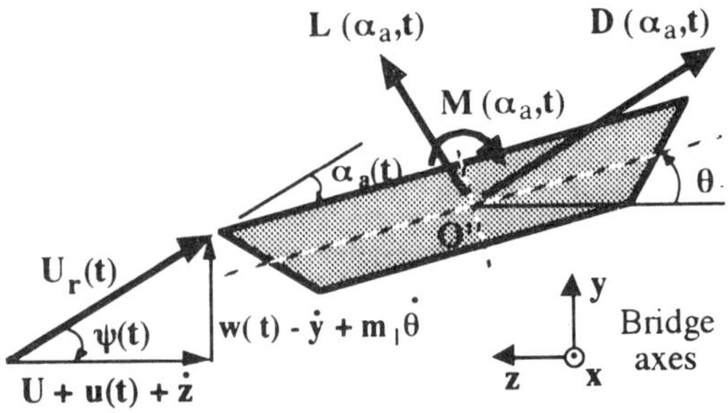

Fig. 16 Relative wind speed and angle of attack on a bridge girder

$$D(\alpha_a, t) = \frac{1}{2}\rho U_r^2 A_n C_D(\alpha_a) \tag{5a}$$

$$L(\alpha_a, t) = \frac{1}{2}\rho U_r^2 B C_L(\alpha_a) \tag{5b}$$

$$M(\alpha_a, t) = \frac{1}{2}\rho U_r^2 B^2 C_M(\alpha_a) \tag{5c}$$

The instantaneous relative wind speed, using velocities $\dot{z}$, $\dot{y}$ and $\dot{\theta}$ of horizontal, vertical and torsional motions of the girder along with a kind of distance m_ℓ assumed to describe the vertical effect of the last torsional motion $\theta(t)$, may be written as

$$U_r(t) = \sqrt{(U+u(t)+\dot{z})^2 + (w(t)-\dot{y}+m_\ell\dot{\theta})^2} \tag{6}$$

and then the effective angle of attack is expressed in the form of

$$\alpha_a(t) = \psi(t) - \theta(t) \tag{7}$$

in which U = mean wind speed; and $u(t)$ and $w(t)$ = along-wind and vertical fluctuating wind speeds.

In this simplified quasi-steady model of the forces concerned, by fitting those averaged static coefficients C_D, C_L and C_M with the polynomials of effective angle of attack and also simulating time series of wind fluctuations by an appropriate stochastic method, ARMA model, the three-dimensional random responses come to be available in the time integration. Actually, in order to carry out a calculation, a measure of distance m_ℓ associated with torsional motion must be assumed appropriately, say to be equal to $B/4$, that is, taking account of the wind loads on the girder concentrated at the quarter chord point, although this is quite a gross assumption which may not realistically represent the location of loading of wind forces and may lead to an overestimate of the aerodynamic damping since there are indications that the value of m_ℓ must be frequency-dependent and vary with the entire bridge, because of those unsteady aerodynamic forces, or derivatives, really coupled as seen in Eq. (3) in three-dimensional fashion. Fig. 17 presents one of comparisons among a couple of gust response analyses with the measured data in the full model wind tunnel tests mentioned in the preceding (Santos et al 1993; Santos 1994). It can be seen that the calculated results by present simplified approach in the time domain are much closer to the measured than the calculated by conventional spectral analysis for a single-dimensional motion concerned. In reality, this may be a result of introducing the effects of three-dimensional instantaneous coupled motion more inclusively into the evaluation of mode mass as well as mode damping.

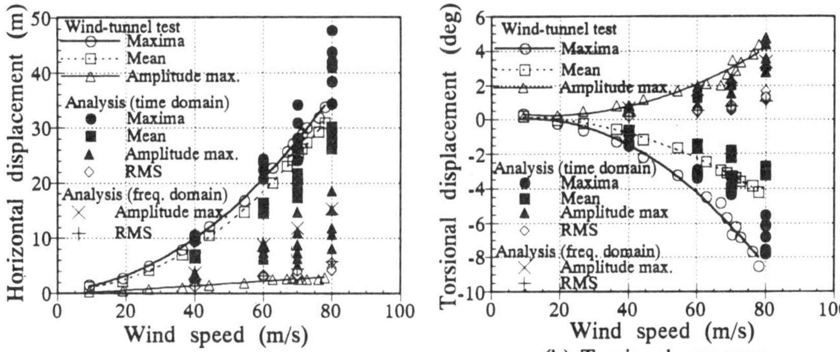

(a) Horizontal response (b) Torsional response

Fig. 17 Comparison of gust response at midpoint of main span

A numerical approach to perform the three-dimensional response of a truss-stiffened suspension bridge under gusty wind seems quite effective, which focused on the time domain to look closely into the instantaneous coupled response considering the time-space variation in wind speeds and aerodynamic forces. Actually, however, difficulties were encountered in expressing the wind-force-response relationship in the time domain so that much simplification had to be made. Therefore, further efforts to advance in the numerical prediction of gust response are needed, particularly in more realistic description of instantaneous aerodynamic forces, for instance, by the use of indicial functions to consider the effect of time delay between wind fluctuation, structural motion and corresponding aerodynamic force. Such indicial functions in the time domain are to correspond to self-excited unsteady aerodynamic forces, or derivatives, caused by coupled structural motions as well as to aerodynamic admittances used to bring wind fluctuations to random exciting forces, which are all defined in the form of frequency-dependence.

TOPOGRAPHICAL EFFECTS ON ENTIRE BRIDGE RESPONSES

The description of wind is normally dependent on site terrain and location of the structure concerned. As for the structure stretching over a long distance, the wind structure at one point may be different from that at another point because of surrounding topography. Those influences of topography on the responses of long-span bridges have been indicated on several occasions (Davenport and King 1993). Some of general conclusions, derived from topographical wind tunnel testings as well as field observations of natural wind, are as follows;

(a) Decreases in the strength of mean wind are frequently accompanied by increase in the level of turbulence, which can increase the amplitude of gust response but decrease the static mean response and suppress the vortex excitation.

(b) By screening the wind from directions of a prevailing storm, the risk of severe winds or the safety of specified objects can be significantly altered.

(c) Investigations of site meteorology through the measurements of full scale wind conditions have proven quite useful in the calibration of topographical wind tunnel model studies.

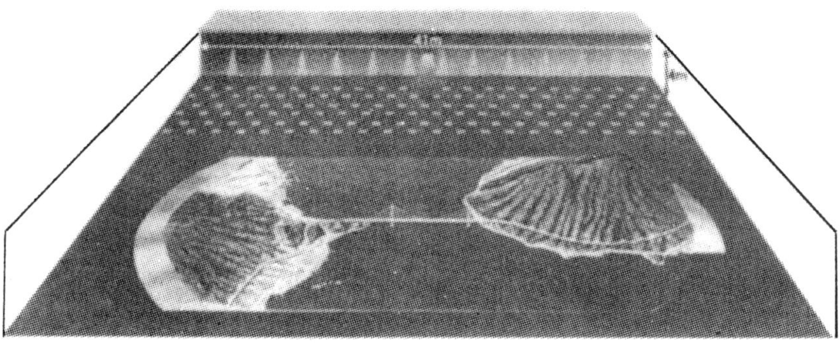

Fig. 18 Sitting of full Tatara bridge and surrounding topography models

The sitting of a very long-span cable-stayed bridge, the Tatara Bridge, was recognized to be relative to significant topographic features because it was to be constructed spanning over a strait between an island and another, both standing in excess of 400 m above sea level as shown in Fig. 18 (Miyata et al 1994c). With the wind of the direction in which the bridge was hidden behind the islands, for instance, the flow might probably make a sharp turn in the direction of bridge axis, resulting in the possibility of deflected wind loads or increased gust responses on the bridge posing more seriously under the construction stage that was still much unstable structurally as well. To make a rough survey of wind pattern prevailing at the bridge site, a 1/2000-scale relief model wind tunnel tests, representing the topography of a region within the radius of 6 km, were conducted in advance, with due reference to vertical distribution of wind speeds and turbulence intensities measured at observation towers on the shore.

On evaluating the aerodynamic stability of this very long-span cable-stayed bridge, it involves the consideration of structural and aerodynamic problems, the former being the problem of the stay cables and deck or pylons vibrating in mutual interference and the latter that of particular site wind pattern resulted from the local topography. In addition, the wind resistance is warranted not only after completion but also during erection. Actually, however, it was very difficult to devise a workable wind tunnel test plan capable of satisfying all the requirements at the same time to build a scale bridge as well as topography models. In other words, while producing the stay cable structural properties (weight distribution or tension) precisely requires preferably a largest possible model over 1/70 in scale, the scale of boundary layer turbulent flow to be generated in the large wind tunnel, mentioned in the preceding, requires the model to measure 1/200 and under in scale. Therefore, in this case, it has been judged that those contradictory requirements should be met by carrying out two different kinds of wind tunnel test, one using a large bridge model with an emphasis on precise simulation of structural properties and the other using a small model with an emphasis on necessary production of surrounding wind pattern influenced by topography. Between both results taken in two models, there must be careful investigations and arguments to be interpreted.

CONCLUDING REMARKS

Focusing on how to approach to full model wind tunnel testings and how to get better findings from those tests, some of governing factors interactions and influenced behaviors of wind effects on very long-span bridges are illustrated and discussed with particular reference to the Akashi Kaikyo Bridge and partly to the Tatara Bridge, for which construction is well under way. Recent other wind tunnel testings by means of full model of very long-span bridges can be referred to in a couple of cases such as the Great Belt East Bridge (Reinhold et al 1993), the Messina Straits Bridge (Diana 1993) and the Normandie Bridge (Larose and Livesey 1993). It goes without saying that individual circumstances should have been taken into account in carrying out those full model wind tunnel testings in the sense of the choice of targeted behaviors under wind, the similarity of entire structural systems concerned and their dimensions, the reliability of test results and their analytical explanations, and so on.

In process of practical experiments on the wind-induced behaviors of bridge structures, generally speaking, the sectional model wind tunnel tests are still effective in verifying and measuring the single-dimensional vibrations and aerodynamic forces. Particularly, it is quite so in case of finding out desirable configurations of bridge deck proposals. As described in the preceding, however, the full model wind tunnel tests are quite essential in those cases of possibility of such as complicated three-dimensional behaviors, interacted between associated factors and also under turbulent flows. In this sense, it is an important key point to prospect their possibility in advance by appropriate means. Further efforts are needed to rebuild more effective approach for this purpose by expanding the ideas of conventional sectional model wind tunnel testings. Of course, a detailed full model testing must be conducted appropriately for a kind of calibration.

Nowadays, a tool of numerical computation is quite powerful in analytical explanation of wind tunnel test results. Even so, the computer model actually realized is only a model assumed in certain definition and appropriate size, which often results in a cause of some inconsistency with the full model in wind tunnel, although the full model also has its own defect and limitation in modeling. In the argument of a triangular relationship of those three models, here including the prototype structure as a kind of conceptual model because it is not existing in reality at the stage of investigation before its completion, it is quite natural to keep a careful insight on the comparison and interpretation among three in order to minimize a likely inconsistency taking place essentially.

ACKNOWLEDGEMENT

The author would like to express his sincere gratitude to Honshu-Shikoku Bridge Authority for the use of those data associated with full model wind tunnel testings. All the members of the committee for wind resistant design of Honshu-Shikoku bridges and its subcommittee for the wind tunnel project, Bridge and Offshore Engineering Association, would also be acknowledged for their cooperation.

REFERENCES

Astiz,M.A. (1993), "Wind Resistant Design of the Gibraltar Straits Crossing Bridge", Proc. Int. Sem. on Utilization of Large Boundary Layer Wind Tunnel, Tsukuba, Japan, Dec., pp. 203-221.

Billington,D.P., and Deodatis,G. (1991), "Form and Aesthetics in Cable-Stayed Bridges", Cable-Stayed Bridges: Recent Developments and their Future, pp. 35-55.

Bleich,F., et. al. (1950), "The Mathematical Theory of Vibration in Suspension Bridges", U.S.Gov.Print. Office.

Davenport,A.G., and King,J.P.C. (1993), "The Influence of Topography and Storm Structure on Long Span Bridge Behavior in Wind", Proc. Int. Sem. on Utilization of Large Boundary Layer Wind Tunnel, Tsukuba, Japan, Dec., pp. 11-21.

Diana,G., et.al. (1986), "Wind Effects on the Dynamic Behavior of a Suspension Bridge", Tech. Rept., Dep. di Meccanica, Politech. di Milano ,Jan..

Diana,G. (1993), "Analytical and Wind-Tunnel Simulations for the Aeroelastic Design of the Messina Straits Bridge", Proc. Int. Sem. on Utilization of Large Boundary Layer Wind Tunnel, Tsukuba, Japan, Dec., pp. 183-202.

Endo,T., et.al. (1991), "The Technical Challenge of a Long Cable-Stayed Bridge-Tatara Bridge", Cable-Stayed Bridges; Recent Developments and their Future, Elsevier, pp. 417-436.

Honshu-Shikoku Bridge Authority (1976), Wind Resistant Design Code for Honshu-Shikoku Bridges.

Honshu-Shikoku Bridge Authority (1980), Wind Tunnel Testing Manual for Honshu-Shikoku Bridges.

Honshu-Shikoku Bridge Authority (1990), Wind Resistant Design Code for Akashi Kaikyo Bridge.

Irwin,H.P.A.H., (1979) "Design and Use of Spires for Natural Wind Simulation", LTR-LA-223 , NRC Canada.

Kovacs,I., et.al. (1992), "Analytical Aerodynamic Investigation of Cable-Stayed Helgeland Bridge", J.Struct. Engrg., ASCE, Vol.118, No.1, pp. 147-168.

Larose,G.L., and Livesey,F.M. (1993), "Wind-Tunnel Study of the Pont de Normandie during Construction: Aeroelastic Model Tests", prepared for Monberg& Thorsen A/S, DMI Rept. 92159.03, Denmark, Jun..

Leonhardt,F., and Zellner,W. (1991), "Past, Present and Future of Cable-Stayed Bridges", Cable-Stayed Bridges: Recent Developments and their Future, Elsevier, pp.1-33.

Miyata,T, and Ohta,H. (1986), "Considerations for Flutter and Gust Response of Very Long-Span Suspension Bridge", Proc. Nat. Symp. Wind Engrg., Dec., pp.253-258 (in Japanese).

Miyata,T., and Yamada,H. (1988), "Coupled Flutter Estimate of a 'Suspension Bridge", J. Wind Engrg. (Japan), No.37, Oct., pp. 485-492.

Miyata,T. (1991), "Design Considerations for Wind Effects on Long-Span Cable-Stayed Bridges", Cable-Stayed Bridges: Recent Developments and their Future, pp. 235-256.

Miyata,T., Matsumoto,H., and Yasuda,M. (1992a), "Circumstances of Wind Resistant Design Examination for Very Long Suspension Bridge", J. Wind Engrg. & Ind. Aero., 41-44, pp. 1371-1382.

Miyata,T., Yamada,H., Yokoyama,K., et.al. (1992b), "Construction of Boundary Layer Wind Tunnel for Long-Span Bridges", J. Wind Engrg. & Ind.Aero., 41-44, pp. 885-896.

Miyata,T., Yokoyama,K., Yasuda,M., and Hikami,Y. (1992c), "Akashi Kaikyo Bridge:Wind Effects and Full Model Wind Tunnel Tests", Aerodynamics of Large Bridges, Balkema, pp. 217-236.

Miyata,T., and Yamaguchi,K. (1993), "Aerodynamics of Wind Effects on the Akashi Kaikyo Bridge", J. Wind Engrg. & Ind.Aero., 48, pp.287-315.

Miyata,T., Tada,K., et.al. (1994a), "New Findings of Coupled-Flutter in Full Model Wind Tunnel Tests on the Akashi Kaikyo Bridge", Proc. Int. Conf. on Cable-Stayed and Suspension Bridges, Deauville, France, Oct..

Miyata,T., Kitagawa,M., Yamada,H., et. al. (1994b), "Design Considerations on 3-Dimensional Gust Response in the Akashi Kaikyo Bridge", Proc. Int. Conf. on Cable-Stayed and Suspension Bridges, Deauville, France, Oct..

Miyata,T., Sato,H., et. al. (1994c), "Full Model Wind Tunnel Tests for a Very Long-Span Cable-Stayed Bridge: Tatara Bridge", Proc. Int. Conf. on Cable-Stayed and Suspension Bridges, Deauville, France, Oct..

Reinhold,T., Larsen,A., Damsgard,A., and Svensson,E. (1993), "The Role of Full-Bridge Model Tests in Wind Engineering Studies for the Great Belt East Suspension Bridge", Proc. Int. Sem. on Utilization of Large Boundary Layer Wind Tunnel, Tsukuba, Japan, Dec., pp.163-181.

Santos,J.C., Miyata,T., and Yamada,H. (1993), "Gust Response of a Long Span Bridge by the Time Domain Approach", Proc. 3rd Asia-Pacific Symp. Wind Engrg., Hong Kong, Dec., pp.211-216.

Santos,J.C. (1994), "Time Domain Approach in Gust Response for Long-Span Bridges", Doctoral Thesis, Yokohama Nat. Univ., Japan, March.

Scanlan, R.H. (1984), "Role of Indicial Functions in Buffeting Analysis of Bridges", J. Struct. Div., ASCE, Vol.110, No.7, pp.1433-1446.

Scanlan,R.H. (1988), "On Flutter and Buffeting Mechanisms in Long-Span Bridges", Probabilistic Engrg. Mech., Vol.3, No.1, March, pp. 22-27.

Selberg,A. (1964), "Oscillation and Aerodynamic Stability of Suspension Bridge", ACTA Polytechnica Scandinavia, Ci 13.

STATE OF THE ART VOLUME

SUPPRESSION OF WIND-INDUCED VIBRATIONS OF STRUCTURES

MANABU ITO

Saitama University, Japan

SUMMARY

Design of very flexible structures such as long span steel bridges and high rise towers is frequently governed by wind-induced vibrations. The measures controlling these dynamic responses are classified into aerodynamic and mechanical means. The first step is to select aerodynamically stable cross section under acceptable levels of functional and economic requirements for the structure, in particular for the case of long span cable-supported bridges. When the original design is found to be unacceptable in view of ultimate, serviceability or fatigue limit states, aerodynamic appendages or mechanical means shall be considered. In general, the former takes precedence over the latter in bridge design. As to the mechanical means, a variety of damping devices are now available, and various passive type dampers have been increasingly implemented on bridges, buildings and towers. On the other hand, active control means, both mechanical and aerodynamic, has also attracted interests of engineers these days.

INTRODUCTION

With increase of light, flexible, low-damping and long-span or high-rise structures, considerations of dynamic effects of wind on these structures have gained more importance. When wind-induced vibrations of a structure or structural member are likely to occur below the design wind speed or when their response amplitude is expected so large as to cause structural damage, functional trouble or human discomfort, some preventive measures have to be taken.

A distinctive feature of wind-induced vibrations of a flexible structure is their variety as shown in Table 1. Although the phenomena to be considered are different for respective structure, there is a possibility of multiple failure or unserviceability modes for a specific structure at different wind speeds, and some of the phenomena can occur concurrently.

Table 1. Classification of dynamic responses to wind

type of motion		typical structures
limited-amplitude response	gust response (buffeting)	all flexible structures
	vortex excitation	pylons, stacks, bridge girders, buildings etc.
divergent-amplitude response	galloping (cross-wind instability)	pylons, girders, cables
	torsional flutter	cable-supported bridge girders
	coupled flutter	

The design of a structure shall be made firstly in accordance with structural and functional requirements under primary loadings, but consideration on aerodynamic stability at preliminary design stage is very essential for such a flexible structure as long span cable-supported bridge and high rise tower located at the site where strong wind is anticipated. Japanese practice of bridge deck design in these cases, for example, has been to provide with such torsional stiffness that the fundamental natural frequency in torsion exceeds at least twice that in bending and also to select aerodynamically stable cross section.

When the original design is found to be unacceptable in view of the aerodynamic limit states of the structure, some appropriate countermeasures described later shall be taken to meet either of the following requirements:

1. pushing out the critical wind speed at which the vibration starts, above the design wind speed, or
2. reducing the maximum response amplitude below the acceptable limit.

The former requirement must be applied to the divergent amplitude responses such as galloping and flutter, whereas the latter is usually considered for the limited amplitude oscillations such as gust response and vortex excitation.

Measures suppressing wind-induced vibrations are classified into aerodynamic and mechanical means as shown in Fig.1 [Tsuji (1984)]. The state-of-the-art of these means is going to be reviewed in what follows [cf. Ito (1987), Fujino & Tamura (1990)].

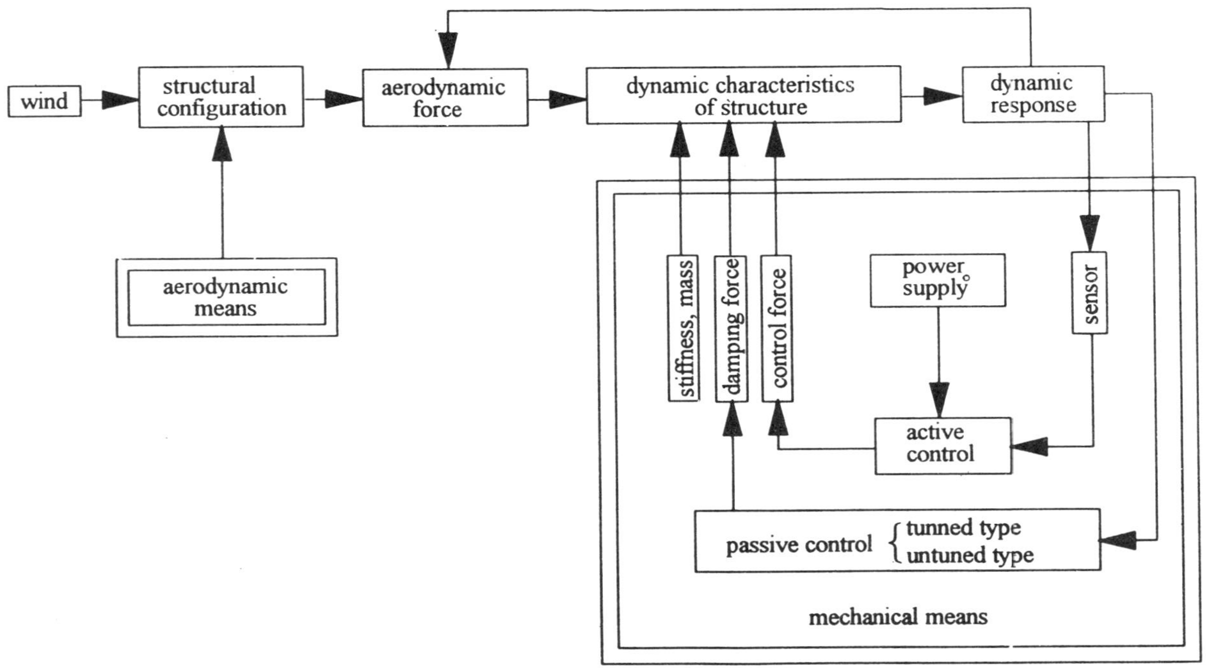

Fig.1 Classification of preventive measures for wind-induced vibrations [Tsuji (1984)]

AERODYNAMIC MEANS

In principle, the aerodynamic countermeasures utilize change of air flow around the structure by geometrical modifications of its cross-sectional shape or by small attachments to the structure.

In almost all of the cable-supported bridges in Japan, wind tunnel tests are conducted and better geometrical cross sections including aerodynamic countermeasures are sought whenever unacceptable behaviors are observed in the test. In the case of tall buildings, not only the wind-induced structural response but also the change of wind environment around the building are often investigated with wind tunnel model. As an example, the NEC Building in Tokyo was provided with the wide opening at its middle height not to deteriorate local wind condition.

Selection of Basic Cross Section

The geometrical shape plays an important role in the wind effects on structures. Although the cross-sectional shape and dimensions are primarily determined in accordance with structural, functional and, in some cases, aesthetic requirements, the selection of aerodynamically stable cross section often results in the most economical design for the girder of long span cable-supported bridge.

Streamlining is one of the solutions and approaches to an airfoil-like shape which is free from separated-flow instabilities, including vortex excitation, under horizontal wind. In this extreme case, the classical flutter theory may be applicable and its critical wind velocity is usually high. As a practical matter, however, the cross section of bridge deck is usually bluff, so some aerodynamic appendages mentioned later may have to be added to the basic structure section. Use of shallow trapezoidal or hexagonal box girder is now popular on long span cable-stayed bridges in Japan. Twin box design gave better performance in some cases. In the case of very long span suspension bridge, air gaps at the center of bridge deck to reduce the pressure difference may be useful.

Another approach to attain wind-resistant design is the use of truss structure with enough opening. Although a truss is generally considered disagreeable from aesthetical viewpoint, it is used even in cable-stayed bridges for other design reasons such as double deck layout. It is known that the stiffening truss of very long span suspension bridge becomes more stable under wind action when provided with gratings on the roadway as well as the openings between the roadway and chord members. It must borne in mind, however, that existence of solid handrails, thick curb stones, and other small attachment may cause unfavorable effects on overall stability of the basic cross section.

The cross sectional shape of bridge tower is sometimes modified by aerodynamic reason. Since stay cables can hardly contribute to the constraint of lateral bending of a cable-stayed bridge tower, the trapezoidal and octagonal (corner cut on rectangular shape) sections are adopted for the tower shafts of the Tsurumi Bridge and the Meikoh-Chuo Bridge, respectively. A horizontal strut was added to the inverse V shape tower of the Meikoh-Higashi Bridge, while a vertical slit was provided on the Sugawara-Shirokita Bridge pylon. Suspension bridge towers have been considered aerodynamically stable after the completion of bridge, but in case of the Akashi-Kaikyo Bridge having the world's longest span, the corner cut on tower shafts was made to prevent wind-induced vibrations, together with installation of mechanical dampers.

Even on high rise buildings wind-induced vibrations may become the primary concern for structural safety and occupant's comfort. Care should be taken on the selection of the cross-sectional shape of the super high-rise buildings that will be contemplated in the future, from this viewpoint [Hayashida & Iwasa (1990)].

Aerodynamic Appendages

When the substantial change of the basic cross section is not allowed for design reasons, some aerodynamic control devices are adopted for wind-sensitive structures with bluff cross sections. Zdravkovich (1981) classified a wide variety of aerodynamic appendages suppressing vortex excitation into the following three categories in accordance with the phenomenological mechanism of vortex shedding.

1) surface protrusions (strakes, helical wires, fins and studs, etc.), which destroy or reduce the coherence of the shed vortices by affecting separation lines and /or separated shear layers
2) shrouds (perforated, gauze, axial rods and axial slats, etc.), which affect the entraint layers, and

3) near wake stabilizers (splitter and saw-tooth plates, guiding plates and vanes, base-bleed, slits cut along the cylinder, etc.), which prevent interaction of entraint layers.

The most typical structure where the above means are applied is a circular cylinder such as steel stacks and light poles. Most means in the first two categories mentioned above are effective irrespective of the direction of wind, whereas some means in 1) and all in 3) are unidirectional. So it depends on the given case which means is effective. Increase of drag coefficient in some of the above means should be also borne in mind.

Similar investigations have been done on tall buildings. As an example, Kwok and Bailey (1987) compared three different modifications to square tower model, namely those with small fins, vented fins and slotted corners, respectively at the four corners. They reported that the overall performance of the slotted corners was very encouraging. The tall pylons of cable-supported bridges are the structures in this category. The examples used in Japan are corner deflectors on the Katsushika Harp Bridge and corner cuts on the Higashi-Kobe Bridge and the Akashi suspension bridge. Web plate of slender tension members with H-section is sometimes perforated to control vortex excitation. Successful effect of openings in reducing the lift response is also reported on building models [Isyumov et al. (1989)].

A wide variety of aerodynamic appendages are used on slender bridge girders. Examples are farings (triangular or semicircular) or wind nose, flaps, splitter plates, corner deflectors, soffit plates, baffle plates and so forth. Sometimes the intermittent arrangement of the appendages is effective due to the reduction of spanwise correlation of exciting force. The aerodynamic appendages are occasionally provided even on truss girders. Triangular farings are attached to the upper chord of the truss of the Yokohama Bay Bridge, a cable-stayed bridge with a main span of 460m, since this upper chord is a shallow box section which works also as the bridge deck. In case of the Ohnaruto and the Akashi suspension bridges, center barrier was attached to the stiffening truss.

Although preventive measures for the wind-induced vibrations of cables are mostly mechanical, a few aerodynamic means have been employed in cable-stayed bridges. The polyethylene sheaths with protuberances along the cable axis were used on the Higashi-Kobe Bridge.

When the aerodynamic appendages are contemplated, attention must be paid on the trade-off with the possibility of increasing cross-wind gust response, which may be also the case of streamlining, and spoiling the appearance of the structure.

Active Control

For extremely long span or high rise structures it may be necessary to provide special facilities. Thus active aerodynamic control to suppress wind-induced vibrations has recently gained interests [Ostenfeld & Larsen (1992), Kobayashi (1992), Kubo (1992)]. Various means such as actively controlled rotating flaps, farings or rod and use of jet nozzles are being proposed, though not yet realized in civil engineering structures. As compared with the mechanical active control referred to later, active aerodynamic control may have to rely heavily on the experimental verification, while it seems to be more economical from the viewpoint of required energy [Fujino (1993)].

MECHANICAL MEANS

When any aerodynamic countermeasures are not effective, when aerodynamic appendages spoil the appearance of the structure, or when the countermeasures are required only during erection stage, mechanical control means are considered to be used. Mechanical means aim at controlling the dynamic response of the structure to wind by adjusting the mass distribution, increasing the stiffness, adding the structural damping, or applying the control force to the structure (cf. Fig.1). Different from the aerodynamic countermeasures, the mechanical control means may be doubly beneficial as the preventive measures for the vibrations induced by other sources such as earthquakes and vehicle traffic, if they are properly designed.

Table 2 shows the examples of mechanical means used on various structures. Although the data are not necessarily exhaustive, one can recognize several interesting trends in the use of the mechanical means. Some of the facts are as follows:

1. The tuned mass dampers (TMD) are widely used on a variety of structures.
2. Various alternative means are used particularly for towers and cables of cable- stayed bridges.

3. Mechanical control means have been not so often used on the girders of flexible bridges because aerodynamic countermeasures seem to be given priority so far.
4. Active control means have started to come into practical use in recent years.

Table 2. Mechanical means implemented in wind sensitive structures

structure / means	bridge				building	tower	stack	antenna
	girder	tower	cable	member				light pole
added mass				1				
tie rope or member		2	6	2		1		
spacer or space damper			2					
friction damper		5	1		2			
damper block		5						
viscous damper			6		2			
hydraulic damper		1	5					
impact damper								2
TLD	1	3				3		
TMD	4	11	4	2	2	5	2	2
active or hybrid TMD		5			3			
active mass damper		1						

Notes : 1) include the means during erection of bridge girder and tower
2) exclude the means for transmission lines
3) exclude strengthening of structural member after causing damage
4) TMD : tuned mass damper, TLD : tuned liquid damper

Added Mass and Stiffness

The increase in mass and rigidities of the structure, respectively reduces the wind induced response when other parameters are kept constant. However, the increase in mass results in the increase in the natural period and the dead weight of the structure, so the application seems very limited to such a case that the steel pipe posts of a deck type arch bridge are partially filled with concrete or gravel. On the other hand, the increase in rigidity of the structure is doubly beneficial in the case of vortex excitation in particular, because the response amplitude reduces and the resonance wind speed increases by increasing the natural frequency. In general, however, the increase of rigidities results in uneconomical design. It may be not wise to put too much materials into the structure for wind design alone.

A simple way of increasing structural rigidity is to enlarge the cross sectional dimensions, and in the case of vortex excitation this brings on a double benefit again ; that is, the increase in resonance wind speed can be attained by the increase of both natural frequencies and cross sectional size of the structure. Alternative way to contribute to stiffness is to provide additional members or to give constraint to the members susceptible to wind-induced vibrations. The increase of torsional stiffness in particular is very effective to raise the critical wind velocity for flutter. Tie wires sometimes used on multi-stay cables of cable stayed bridges or the vertical hangers of through-type arch bridges. These tie wires contribute to both stiffness and damping of the original members, but it has been shown that the tie wires on stay cables are subject to high repeating stress and fatigue failures at the connection occurred in some of the bridges.

Passive Damping Devices

Direct energy absorbers and energy dissipaters are most prevailing to damp the response of the structure, prevent resonant energy buildup and sometimes augment the critical wind velocity of dynamic instability. As seen in Table 2, various damping devices have been developed and put to practical use. These " dampers " are generally classified into passive and active types. As compared with active control devices that will be described later, passive systems rely only on behavior of their materials or their mechanism and, given proper design and adequate quality of fabrication, they will act reliably with little dependence on maintenance activities or external systems. The passive dampers are further assorted by whether tuned or not.

The most simple idea of artificial damper is realized by inserting viscous or viscoelastic materials to the members concerned or their connections or by connecting such members with tie ropes or inserting spacers. Although the latter may primarily aim at increasing constraint or keeping the positions of the members, they also result in increase of the structural damping of connected members due to the constraint given to the member and the interaction between the motions of the members connected each other. The use of tie ropes in cable stayed bridges and the associated problem have been just mentioned in the previous section. On the other hand, the shear type viscous dampers are provided near the ends of stay cables of the Sakitama Bridge in Japan. Other examples of the adoption of viscoelastic damper are on the frame of tall buildings such as the World Trade Center Towers in New York and the Columbia Seafirst Center in Seattle. Care should be given to the selection or the material properties, in particular their temperature dependence and aging effects.

During the period from 1960's to the mid-1980's, sliding blocks or viscous dampers connected to the top of suspension bridge towers at free standing stage were widely employed. Submerged block or mass spring system may be substituted for the sliding block. But since the tuned mass damper (TMD) of pendulum type was used on the Meikoh-Nishi cable stayed bridge tower in 1984, a variety of TMDs have been developed and employed to suppress the wind-induced vibrations of slender bridge towers during erection. The compact TMDs with long natural period are now available. These TMDs are, if necessary, left even after completion of the bridge as permanent fixtures. TMDs are also used on tall buildings to reduce their motions insofar as they affect occupant comfort.

The natural frequency of TMD is tuned to that of the structure. The necessary mass of TMD may be only several percent of that of the structure, whereas the optimal damping of TMD depends on the type of vibration concerned [Fujino et al. (1990)]. It should be reminded that TMD is effective only for the vibration mode whose frequency is tuned.

The tuned liquid damper (TLD) [Fujino et al. (1988)] which relies on liquid motion in a container is similar to TMD. Tuned frequency and appropriate own damping are required. Damping control of the liquid damper is made by utilizing boundary layer on the bottom surface in shallow water case, by generating turbulence through nets or rods placed in the liquid in deep water case, or by other methods. Another type of liquid damper is the tuned liquid column damper (TLCD) utilizing the motion of liquid in U-shape tube [Sakai et al. (1989)]. Frequency adjustment can be done by controlling air pressure in the tube, while the orifice in the tube servers for damping control. TLCD can be applied not only to horizontal motion but to rotating motion of the structure in the vertical plane.

The advantages of the liquid damper are a) simple and economical, b) applicable to very small amplitude. c) multi directional (in case of TLD), d) reliable, and e) easy to transfer in general, though it may occupy rather large space according to situations.

Very simple and small TMDs represented by the classical Stockbridge damper have been prevailing technique to prevent excessive vortex excitation of transmission power lines. It aims at suppressing mainly the last half-wave generated by the cable oscillation. Other vibration problems of long span power lines are associated with full-span galloping and sub-span wake-induced galloping. TMDs may also be provided at the center of cable span to alleviate the galloping of power lines. Countermeasures for sub-span wake-induced galloping occurring in bundled conductors have included the special spacers to detune the various cables in a bundle from each other and the energy dissipating spacers, though they are not always fully effective.

The similar techniques to the above mentioned have been applied to the vibrations of stay cables or diagonal hangers of cable-supported bridges. Among various preventive measures, use of cable-to-cable ties, viscous dampers and aerodynamic means were already mentioned. Although the Stockbridge dampers were employed in several European bridges, they are not popular for aesthetic reason. The most prevailing recently is the installation of viscous or hydraulic (oil) dampers near the cable anchorage.

The impact damper is generally categorized in TMD. The vibration energy of the structure is absorbed by the impact between the buffer fixed to the structure and the moving block, ball or chain. The block suspended by wires or the suspended chain itself moves as a pendulum which has own natural frequency. They are applied to light, vertical structures such as light pole.

The application of a gyroscope to suspension bridge flutter [Murata & Ito (1971)] and building motions was once investigated. It is one of the passive type devices to counter the external action induced by wind, and the experiments proved its effect.

Active Control Means

Active devices depend on a control loop that provides a force to counter those responses induced by external disturbances. More concretely, the control force estimated from the dynamic response of the structure, which is detected by a sensor, is generated by driving an actuator and applied to the structure to suppress the dynamic response. The advantage of the active system is that, if the correct control algorithm is developed for the structural system in which it is installed, the precisely correct response of the structure can be commanded by the actuator, while disadvantages lie in the cost, reliability, and maintenance cost.

Active mass dampers which have recently been put into practical use are classified as seen in Fig. 2 , the respective features of which are as follows.

a) active type : The mass is driven by actuator, and its reaction acts controlling force suppressing the response of the structure. The mechanism is simple, but the actuator expends much energy.

b) hybrid type : Passive type damper (TMD) is driven by actuator, and its reaction acts as controlling force. Although the system is more complex, the TMD works even if power failure occurs.

c) semi-active type : The actuator controls the dynamic characteristics of the structure so that the vibration of the structure is suppressed or the effect of the control system is optimized.

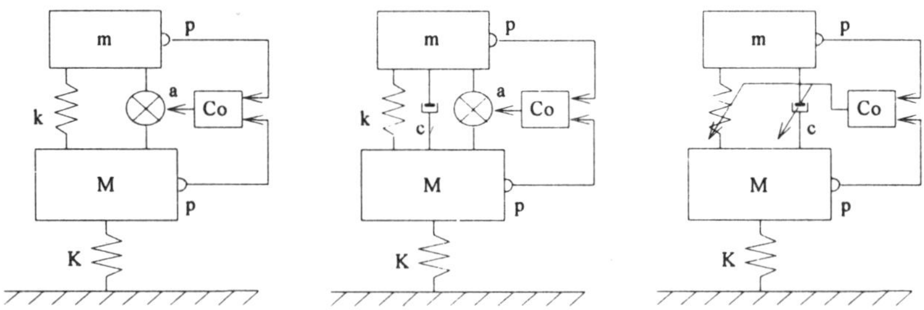

(a) active type (b) hybrid type (c) semi-active type

Fig.2 Classification of active mass damper
(p: sensor, a: actuator, Co: controller)

The active mass dampers, in particular hybrid or semi-active type, were employed during the erection of a few cable-supported bridge towers and on some buildings in Japan [JSCE (1991)].

CONCLUDING REMARKS

In principle, the aerodynamic means shall be firstly explored, and when it is found difficult, the use of mechanical means shall be considered as an alternative or additional measure. In some cases, the countermeasures may be prepared to be used only if any malfunction is observed on the completed structure.

A variety of damping devices are now available, and the use of passive type dampers represented by TMD has been increasing. Active control means have also started to be put into practical use. These situations are much owed to very vigorous research and development in this field. More efforts to find more efficient and more reliable countermeasures are expected so as to achieve more rational design of wind-sensitive structures and to realize extremely long span or high rise structures in the future.

REFERENCES

Fujino, Y (1993), " Recent research and developments on control of bridges under wind and traffic excitation in Japan", Proc. International Workshop on Structural Control, Honolulu, August 1993.

Fujino, Y., Ito, M. and Pacheco, B. (1988), "New dampers for tower of long span suspension bridge ", Preprint of the 13th Congress of IABSE, Helsinki, June 6-10, 1988, pp.923-928.

Fujino, Y. and Tamura, Y. (1990), "Control of wind-induced vibrations in civil engineering structures and buildings (in Japanese)", Journal of Wind Engineering, JAWE, No.44, July 1990, pp.53-84.

Hayashida, H. and Iwasa, Y. (1990), "Aerodynamic shape effects of tall building for vortex induced vibrations", Bluff Body Aerodynamics and Its Applications, 1990, pp.237-242, Elsevier.

Isyumov, N. et al. (1989), "Aerodynamic methods for mitigating wind-induced building motions", Structural Design, 1989.

Ito, M. "Measures against wind-induced vibrations of bridges", Proc. Structures Congress '87, Bridges and Transmission Line Structures, ASCE, August 17-20, 1987, pp.129-139.

Japan Society of Civil Engineers (1991), Text of the Colloquium on Control of Structures (in Japanese), July 1991, JSCE.

Kobayashi, H. and Nagaoka, H. (1992), "Active control of flutter of a suspension bridge", Journal of Wind Engineering and Industrial Aerodynamics, Vol.41, 1992, pp.143-151.

Kubo, Y. , Modi, V. J., Yasuda, H. and Kato, K. (1992), "On the suppression of aerodynamic instabilities through the moving boundary layer control", ditto. , pp. 205-216.

Kwok, K. C. and Bailey, P. A. (1987), "Aerodynamic devices for tall buildings and structures", Journal of Engineering Mechanics, ASCE, Vol.113, No.3, March 1987, pp.349-365.

Murata, M. and Ito, M. (1971), "Suppression of wind-induced vibration of a suspension bridge by means of a gyroscope", Proc. 3rd International Conference on Wind Effects on Buildings and Structures, Tokyo, Sept. 1971, pp.1057-1066.

Ostenfeld, K. and Larsen, A. (1992), "Bridge engineering and aerodynamics", Aerodynamics of Large Bridges, 1992, pp.3-22, Balkema.

Sakai, F., Takaeda, S. and Tamaki, T. (1989), "Tuned liquid column damper : New device for suppression of building vibrations", Proc. International Conference on High Rise Buildings, Nanjing, March 1989.

Tsuji, M. (1984), "Review of various mechanical means for suppressing wind-induced vibration of structures (in Japanese)", Journal of Wind Engineering, JAWE, No.20, June 1984, pp.81-102.

Zdravkovich, M M. (1981), "Review and classification of various aerodynamic and hydrodynamic means for suppressing vortex shedding", Journal of Wind Engineering and Industrial Aerodynamics, Vol.7, 1981, pp.145-189.

BUFFETING RESPONSE ANALYSIS AND CONTROL OF LONG-SPAN BRIDGES

HAIFAN XIANG
Tongji University, Shanghai, China

SUMMARY

Since 1977, China entered a new stage with reform and open policy, more than 60 cable-stayed bridges have been built in past 20 years. After the successful construction of two remarkable cable-stayed bridges in Shanghai, the Chinese bridge engineers have full confidence in building extra long-span cable-stayed bridges with a main span beyond 600 m, and modern suspension bridges. 4 suspension bridges are being constructed simultaneously in China, and some other suspension bridges are under design or planning.

The passive aerodynamic countermeasures (fairings, air gap, etc.) is effective in providing high critical wind speed of flutter even for very long-span suspension bridges. However, the buffeting response, although it is not a catastrophic phenomenon, is becoming increased concern from traffic discomfort, unsafety and the fatigue of structure as the principal wind problem of long-span bridges.

In this paper, the state of the art of long-span cable-supported bridges in China are summarized. Some innovative developments on the basis of buffeting response theory proposed by Professors Davenport and Scanlan are introduced.

BACKGROUND [Xiang, 1993]

The revival of modern cable-stayed bridges is one of the greatest achievements in bridge engineering after World War II.

Cable-stayed bridges were first introduced into China in the early sixties. Two trial bridges, the Xinwu Bridge with a main span of 54m in Shanghai and the Tangxi Bridge with a span of 78.5m in Yuyang, Sichuan Province — both are reinforced concrete cable-stayed bridges — were completed in 1975.

Table 1. Major cable-stayed bridges in China (L > 400m)

	Name	Location	Main Span	Year of Completion	Deck Type
1	Nanpu Bridge	Shanghai	423m	1991	comp.
2	Yangpu Bridge	Shanghai	602m	1993	comp.
3	Yunxian Bridge	Hubei	414m	1994	p.c.
4	Wuhan 2nd Bridge over Yangtse River	Hubei	400m	u.c. (1995)	p.c.
5	Tongling Bridge over Yangtse River	Anhui	436m	u.c. (1995)	p.c.
6	Chongqing 2nd Bridge over Yangtse River	Sichuan	444m	u.c. (1995)	p.c.
7	Xupu Bridge	Shanghai	590m	u.c.(1996)	comp.
8	Qingzhoulu Bridge	Fujian	605m	u.d.	comp.
9	Zhanjiang Bay Bridge	Guangdong	480m	u.d.	p.c.
10	Zhujiajian Bridge	Zhoushan, Zhejiang	450m	u.p.	p.c.
11	Zhengyang Bridge	Jiangsu	850m	u.p.	comp.
12	Orient Bridge(W1)	Zhoushan, Zhejiang	600m	u.p.	comp.
13	Orient Bridge(W2)		900m	u.p.	steel
14	Nanjing 2nd Bridge over Yangtse River	Jiangsu	608m	u.p.	comp.

Since 1977 when China entered a new era with its reform and open policy, with the traffic development, Chinese bridge engineering has made great strides in the construction of long-span cable-stayed bridges. The Jinan Bridge across the

Yellow River with a main span of 220m, which was completed in 1982, can be regarded as a successful achievement of the stage — the learning stage — in the history of Chinese cable-stayed bridge construction.

In the eighties, the construction of cable-stayed bridges developed rapidly over a wide area in China. More than 30 bridges of various types were built in different provinces and municipalities, for example, the Yonghe Bridge in Tianjin, which has a span as long as 260m, and the Dongying Bridge in Shandong Province, whose span reaches 288m, the latter being Chinese first steel cable-stayed bridges.

Table 2. Major Modern Suspension Bridges in China

	Name	Location	Main Span	Year of Completion	Deck Type
1	Shantou Bay Bridge	Guangdong	452m	u.c. (1996)	P.C. Box
2	Xiling Bridge over Yangtse River	Sichuan	900m	u.c. (1996)	Steel Box
3	Humen (Tigergate) Bridge	Guangdong	888m	u.c. (1997)	Steel Box
4	Jiangying Bridge over Yangtse River	Jiangsu	1385m	u.c. (1998)	Steel Box
5	Lingding Bridge	Zhuhai, Guangdong	920m	u.d.	Steel Box
6	Haicang Bridge	Xiamen, Fujian	690m	u.d.	Steel Box
7	Zhenyang Bridge over Yangtse River	Jiangsu	1300m	u.p.	Steel Box
8	Orient Bridge(E1)	Zhoushan, Zhejiang	1650m	u.p.	Steel Box

At the beginning of this decade, with the completion of Nanpu Bridge in Shanghai in 1991, a new high tide of construction of cable-stayed bridges began to surge in China. The most outstanding one is the Yangpu Bridge with a record-breaking span of 602m, another composite deck cable-stayed bridges in Shanghai. Now, a large number of cable-stayed bridges with a span of over 400m are under design and construction(see Table 1). It can be predicted that by the turn of this century the construction of cable-stayed bridges in China will have attracted world attention not only for its large scale but for its speedy development as well.

Comparatively speaking, the construction of modern suspension bridges in China started rather late. In the sixties, some flexible suspension bridges with spans less than 200m were built in the mountain areas of southwestern China, the Chaoyang Bridge in Chongqin, Sichuan province being the most famous one. However the Dazi Bridge in Tibet completed in 1984 has a span as long as 500m.

The upsurge of traffic engineering construction in nineties led to a new stage of modern suspension bridges. The Bay Bridge in Shantou, Guangdong Province will have a 452m concrete stiffening girder; the Tigergate Bridge will be a steel box girder suspension bridge with a span of 888m; the Xiling Bridge over the Yangtse River will be a 900m steel box deck suspension bridge; the Jiangying Bridge over Yangtse River, whose main span will be 1385m. The simultaneous construction of the four suspension bridges is a convincing evidence that Chinese bridge science and technology will soon catch up with the advanced world level(see Table 2).

BUFFETING THEORIES

Introduction

The wind-induced vibration of long-span cable supported bridges (cable-stayed bridge and suspension bridge) behaves mainly as the flutter and buffeting response of deck as well as the wind-induced vibration of cables. Flutter is a self-excited divergent vibration, but a well-designed cross section of deck may have large enough critical wind speed in order to avoid the occurrence of such catastrophic failure. The local wind-induced vibration of cables can also be suppressed through some effective measures for vibration reduction.

With the increase of span length, the bridge structure tends to be more flexible. The excessive buffeting response under the action of near-ground turbulent wind, although it is not destructive, may cause fatigue problems due to high frequency of occurrence and traffic discomfort, even lead to the unsafety of passing vehicles with high speed.

The method developed by Davenport and Scanlan are generally adopted in the buffeting response analysis. Davenport introduced the aerodynamic admittance function and joint acceptance function in order to modify the buffeting force spectrum, but considered simply the self-excited forces[Davenport, 1962, 1963]. Oppositely, Scanlan considered completely the coaction of self-excited forces. These two theories have the same assumption that the buffeting response is linear, and the superposition of modal responses can be used[Scanlan, 1974, 1977, 1978, 1984, 1990, 1992].

Refinement of Buffeting Theory [Scanlan, 1993; Lin, 1979, 1983, 1993]

The buffeting response of actual bluff bridge deck under turbulent flow is a complicated problem, because both the approach flow turbulence and the body signature turbulence interacted with the motion of structure will influence the buffeting force to a greater or lesser extent.

The conventional frequency-domain buffeting analysis framework is based on the quasi-steady assumption with a consideration of aerodynamic admittance for the non-steady effect. Although the frequency-domain approach is successfully applied to the buffeting analysis of many long-span bridges, the refinement of buffeting theory for more accurate response prediction is necessary in following aspects:

(1) Flutter derivatives, especially the aerodynamic damping terms, strongly influence the buffeting response. So the effect of turbulence on flutter derivatives should be further studied.

(2) Aerodynamic admittance functions for commonly used bridge deck should be identified and classified into several categories and, if possible, given approximately by formulae.

(3)Correlation index λ in wind speed spectrum is a important value for the accuracy of buffeting response prediction. Field measurement of space correlation of wind speed on the existed long-span bridges should be done.

(4) Interaction between turbulence components and motion components of bridge deck should be investigated through theoretical analysis for assessing their effects on buffeting response.

(5) Coupling effect between motions and vibration modes should be analyzed especially at the erection stage.

(6) Finally, the time-domain analysis would be a useful approach in considering both aerodynamic and structural non-linear effects.

BUFFETING RESPONSE SPECTRUM METHOD [Chen, 1993; Xiang, 1994]

Introduction

The theoretical analysis and field measurement show that the influence of aerodynamic coupling between modes is negligible. Therefore, for the case of vertical bending, lateral bending and mainly torsional mode, we can take each mode for estimating separately their buffeting response. This situation is similar to the seismic analysis by using the response spectrum method.

In order to make the estimation of buffeting response of long-span bridges simple and clear, and to assess easily the aerodynamic behaviour of cross section from the viewpoint of buffeting response in the preliminary design stage, it is possible to establish a practical calculation method for the buffeting response spectrum.

Fundamental Formulae for Buffeting Response Spectrum

Based on the random vibration theory, and by ignoring the aerodynamic coupling between the different motion components, and the cross spectra between horizontal fluctuating wind $u(x,t)$ and vertical one $w(x,t)$, the RMS response spectrum of all three motions can be written as follows:

$$\left.\begin{aligned} R_h(\omega_h,\zeta_h) &= \left|\Phi_h(x)\right|_{\max}(\rho UB)\sqrt{\int_0^\infty \left|\tilde{H}_h(\omega,\tilde{\omega}_h,\tilde{\zeta}_h)\right|^2 \left|J_h(\omega)\right|^2 S_L(\omega)d\omega} \\ R_\alpha(\omega_\alpha,\zeta_\alpha) &= \left|\Phi_\alpha(x)\right|_{\max}(\frac{1}{2}\rho UB^3)\sqrt{\int_0^\infty \left|\tilde{H}_\alpha(\omega,\tilde{\omega}_\alpha,\tilde{\zeta}_\alpha)\right|^2 \left|J_\alpha(\omega)\right|^2 S_M(\omega)d\omega} \\ R_p(\omega_p,\zeta_p) &= \left|\Phi_p(x)\right|_{\max}(\rho UB)\sqrt{\int_0^\infty \left|\tilde{H}_p(\omega,\tilde{\omega}_p,\tilde{\zeta}_p)\right|^2 \left|J_p(\omega)\right|^2 S_D(\omega)d\omega} \end{aligned}\right\} \quad (1)$$

in which, the torsional response spectrum has been multiplied by $B/2$ (half width of bridge deck) for the unity of dimension in order to make comparison with vertical displacement h and lateral one p; $S_L(\omega)$, $S_M(\omega)$ and $S_D(\omega)$ are the buffeting force spectra for three direction considering the aerodynamic admittance function respectively; $|J_h(\omega)|^2$, $|J_\alpha(\omega)|^2$ and $|J_p(\omega)|^2$ are the joint acceptance functions corresponding to three generalized coordinates for the basic mode shapes $h(t)$, $\alpha(t)$ and $p(t)$ respectively, which reflect the space correlation of fluctuating wind, $|\tilde{H}_h(\omega,\tilde{\omega}_h,\tilde{\zeta}_h)|^2$, $|\tilde{H}_\alpha(\omega,\tilde{\omega}_\alpha,\tilde{\zeta}_\alpha)|^2$ and $|\tilde{H}_p(\omega,\tilde{\omega}_p,\tilde{\zeta}_p)|^2$ are the aerodynamic transfer functions for three displacement degrees of freedom, where the effects of self-excited forces(aerodynamic derivatives) are included in frequency $\tilde{\omega}$ and damping ratio $\tilde{\zeta}$.

If only the resonance part is taken into account, the calculation formulae for buffeting response spectrum expressed by Eqs.(1) can be simplified as follows:

$$\left.\begin{aligned} R_h(\omega_h,\zeta_h) &= \frac{(\rho UB)\cdot|\Phi_h(x)|_{\max}}{\tilde{\omega}_h^{\,2}}\cdot\sqrt{\frac{\pi\tilde{\omega}_h}{4\tilde{\zeta}_h}\cdot|J_h(\tilde{\omega}_h)|^2\cdot S_L(\tilde{\omega}_h)} \\ R_\alpha(\omega_\alpha,\zeta_\alpha) &= \frac{(\rho UB^3)\cdot|\Phi_\alpha(x)|_{\max}}{2\tilde{\omega}_\alpha^{\,2}}\cdot\sqrt{\frac{\pi\tilde{\omega}_\alpha}{4\tilde{\zeta}_\alpha}\cdot|J_\alpha(\tilde{\omega}_\alpha)|^2\cdot S_M(\tilde{\omega}_\alpha)} \\ R_p(\omega_p,\zeta_p) &= \frac{(\rho UB)\cdot|\Phi_p(x)|_{\max}}{\tilde{\omega}_p^{\,2}}\cdot\sqrt{\frac{\pi\tilde{\omega}_p}{4\tilde{\zeta}_p}\cdot|J_p(\tilde{\omega}_p)|^2\cdot S_D(\tilde{\omega}_p)} \end{aligned}\right\} \tag{2}$$

Practical Calculation Formulae for Buffeting Response Spectrum

In the conventional seismic analysis of bridges, the response spectrum method is often used. In order to make the estimation of buffeting response simple and clear such as the calculation of seismic response, it is necessary to neglect some secondary factors, so as to simplify further the fundamental formulae. The buffeting response spectrum of vertical bending mode is taken as an example for explanation

(1) The dimensionless reduced frequency $k=(\omega\cdot B)/U$ is introduced as a self-variable, and then the RMS buffeting response spectrum for $h(t)$ can be rewritten as follows:

$$R_h(k_h)=\frac{\sqrt{\pi}\cdot(\rho B^2)\cdot|\Phi_h(x)|_{\max}}{2k_h}\cdot\sqrt{\frac{|J_h(k_h)|^2\cdot S_L(k_h)}{\zeta_h Z_H\omega_h}} \tag{3}$$

where,

$$Z_H = 1 - \frac{\rho B^2}{2\zeta_h m} \cdot H_1^* \tag{4}$$

$$\frac{S_L(k_h)}{\omega_h} = \frac{1}{4}(C_L' + \frac{A}{B} \cdot C_D)^2 \cdot \frac{S_w(k_h)}{\omega_h} \cdot [1 + \varepsilon_h \cdot \eta(k_h)] \cdot |\gamma(k_h)|^2 \tag{5}$$

$|J_h(k_h)|^2$ has the same meaning as before; $|\gamma(k_h)|^2$ is the aerodynamic admittance function; $k_h = (\omega_h \cdot B)/U$

(2) By noticing that, for the majority of bridge cross section, $C_L' >> C_L$, and $C_L' >> \frac{A}{B} C_D$, the value of $\varepsilon_h = \left[2C_L \big/ (C_L' + \frac{A}{b} C_D) \right]^2$ in Eq.(5) should be small, generally less than 0.04. If the term $\varepsilon_h \cdot \eta(K_h)$ in the squared bracket is neglected, Eq.(3) can be further simplified.

(3) The joint acceptance function $|J_h(k_h)|^2$ is a dual integration. If the first mode shape of deck can be described approximately by an equivalent sine function, the expression can be obtained as follows:

$$|J_h(k_h)|^2 = \frac{4}{m^2 \cdot |\Phi_h(x)|^2_{\max}} \cdot \frac{1}{p^2 + \beta^2} \cdot \left(\beta + \frac{2p^2}{p^2 + \beta^2} \right) \tag{6}$$

in which, $\beta = \lambda L k_h / (2\pi B)$, and for symmetrical mode shapes $p = \pi$; for asymmetrical one $p = 2\pi$.

According to the analysis and simplification mentioned above, the practical calculation formula for RMS buffeting response spectrum of vertical bending modes can be obtained as follows:

$$R_h(k_h) = C_{wh} \cdot C_{Ah} \cdot C_{Sh} \cdot \psi(\beta_h) \cdot \mu(k_h) \cdot \phi_w(f_h) \tag{7}$$

in which, $C_{wh} = 0.346\rho / \ln(z / z_0)$, related to the wind speed spectrum and topography; $C_{Ah} = C_L' + \frac{A}{B} C_D \approx C_L'$, related to the aerodynamic force coefficients and their slope; $C_{Sh} = B^2 \cdot \sqrt{z \cdot B} \Big/ \left[m \cdot \sqrt{\left(1 - \frac{\rho B^2}{2m\zeta_h} \cdot H_1^*\right) \cdot \zeta_h} \right]$, related to the dimension of structure, inertia, damping ratio and the aerodynamic derivative as well, which reflects the total damping level of structure; $\psi(\beta)$ is a coefficient related to the joint acceptance function. It can be calculated by equivalent length of sine function L, see Fig.1; $\mu(k_h)$ is a coefficient related to the aerodynamic admittance function, see Fig.2; $\phi_w(f) = 1/(1+4f)$, $(f = n \cdot z / U)$ is a coefficient related to the vertical wind spectrum, see Fig.3;

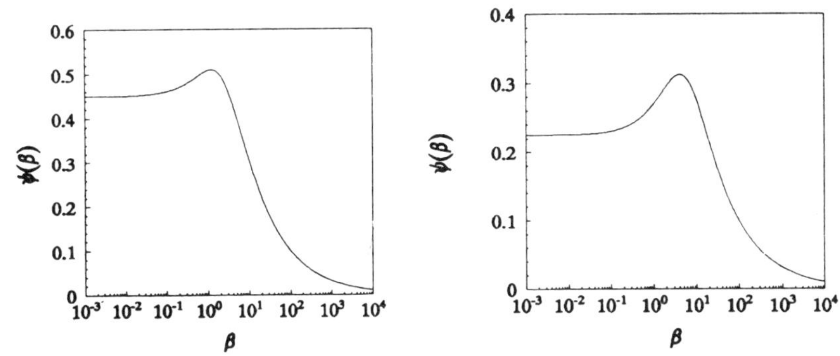

(a) $\psi(\beta)$ for symmetric mode shapes (b) $\psi(\beta)$ for asymmetric mode shapes

Fig.1 Curve of $\psi(\beta)$ function

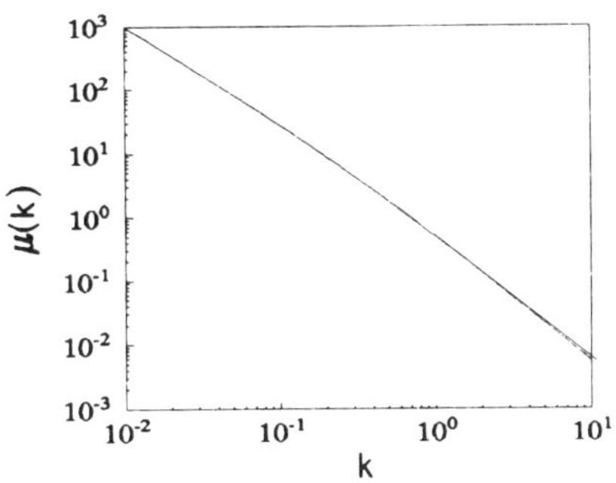

Fig.2 Curve of $\mu(k)$ function

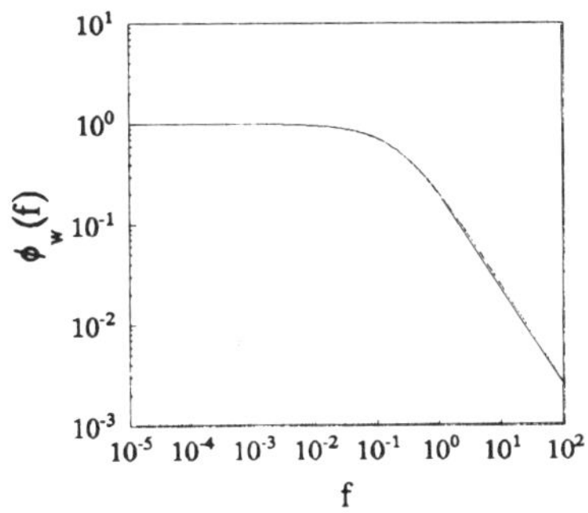

Fig.3 Curve of $\phi_w(f)$ function

Similarly, the practical calculation formulae from RMS buffeting response spectrum of torsion and lateral bending can be written as:

$$\left.\begin{aligned} R_\alpha(k_\alpha) &= C_{w\alpha}\cdot C_{A\alpha}\cdot C_{S\alpha}\cdot \psi(\tilde{\beta}_\alpha)\cdot \mu(\tilde{k}_\alpha)\cdot \phi_w(\tilde{f}_\alpha) \\ R_p(k_p) &= C_{wp}\cdot C_{Ap}\cdot C_{Sp}\cdot \psi(\beta_p)\cdot \mu(\tilde{k}_p)\cdot \phi_u(f_p) \end{aligned}\right\} \tag{8}$$

where, $\tilde{\beta}_\alpha$, $\tilde{k}_\alpha$ and $\tilde{f}_\alpha$ are parameters considering the effect of aerodynamic derivatives; $C_{w\alpha} = \dfrac{0.172\rho}{\ln(z/z_0)}$; $C_{wp} = \dfrac{4.0\rho}{\ln(z/z_0)}$; $C_{A\alpha} = C_M{}'$; $C_{Ap} = \dfrac{A}{B} C_D$;

$$C_{S\alpha} = B^4\sqrt{z\,B}\Bigg/\left[I\sqrt{\left(1-\frac{\rho B^4}{2I\zeta_\alpha}A_2^*\right)\zeta_\alpha}\right];$$

$$C_{Sp} = B^2\sqrt{z\,B}\Bigg/\left[m\sqrt{\left(1-\frac{\rho B^2}{2m\zeta_p}P_1^*\right)\zeta_p}\right];$$

$\phi_u(f) = 1\big/(1+50f)^{5/6}$ is related to lateral wind spectrum, which is used for calculating the lateral buffeting, see Fig.4.

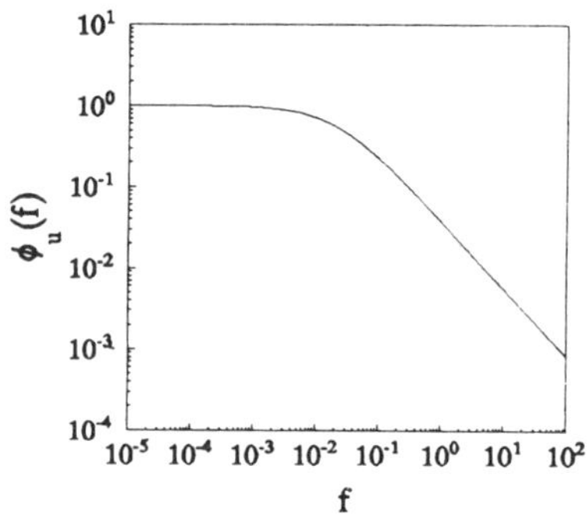

Fig.4 Curve of $\phi_u(f)$ function

Buffeting-based Selection of Deck Section [Gu, 1993]

Aerodynamic selection of deck cross section shape is very important in the preliminary design stage of a long-span bridge. In the past years, this selection mainly aimed at flutter-based selection, but it is better to further choose one section according to the buffeting characteristics in several cross sections which are all satisfactory from flutter-based selection.

It can be assumed that the difference of deck cross section shapes have only effects on the aerodynamic loads acting on each deck, while have no effects on the structural dynamic behaviour, such as eigenfrequencies, mode shapes, etc.

As a comparative study, a basic reference cross section is first chosen from the provided cross sections in buffeting-based selection, and others are compared with it. Let the RMS buffeting responses of the reference one to be $(R_{h_1})_b$, $(R_{p_1})_b$ and $(R_{\alpha_1})_b$, and the others to be $(R_{h_1})_i$, $(R_{p_1})_i$ and $(R_{\alpha_1})_i$, "comparable buffeting coefficient" are defined and derived as follows:

$$\eta_{h_1} = \frac{(R_{h_1})_i}{(R_{h_1})_b} = \sqrt{\frac{\eta_{sh}}{\eta_{zh_1}}} \tag{9}$$

$$\eta_{p_1} = \frac{(R_{p_1})_i}{(R_{p_1})_b} = \left[\frac{(C_D)_i}{(C_D)_b}\right] \Big/ \sqrt{\eta_{zp_1}} \tag{10}$$

$$\eta_{\alpha_1} = \frac{(R_{\alpha_1})_i}{(R_{\alpha_1})_b} = \sqrt{\frac{\eta_{s\alpha}}{\eta_{z\alpha_1}}} \tag{11}$$

where,

$$\eta_{sh} = \frac{[S_{h_1}(\omega_{h_1})]_i}{[S_{h_1}(\omega_{h_1})]_b} = \frac{(C_L)_i^2 \eta_{uw}(\omega_{h_1}) + \frac{1}{4}[(C_L')_i + \frac{A}{B}(C_D)_i]^2}{(C_L)_b^2 \eta_{uw}(\omega_{h_1}) + \frac{1}{4}[(C_L')_b + \frac{A}{B}(C_D)_b]^2} \tag{12}$$

$$\eta_{s\alpha} = \frac{[S_{\alpha_1}(\omega_{\alpha_1})]_i}{[S_{\alpha_1}(\omega_{\alpha_1})]_b} = \frac{(C_M)_i^2 \eta_{uw}(\omega_{\alpha_1}) + \frac{1}{4}(C_M')_i^2}{(C_M)_b^2 \eta_{uw}(\omega_{\alpha_1}) + \frac{1}{4}(C_M')_b^2} \tag{13}$$

$$\eta_{zh_1} = \frac{\zeta_{h_1} - \rho B^2 (H_1^*)_i/(2\overline{m})}{\zeta_{h_1} - \rho B^2 (H_1^*)_b/(2\overline{m})} \tag{14}$$

$$\eta_{zp_1} = \frac{\zeta_{p_1} - \rho B^2 (P_1^*)_i/(2\overline{m})}{\zeta_{p_1} - \rho B^2 (P_1^*)_b/(2\overline{m})} \tag{15}$$

$$\eta_{z\alpha_1} = \frac{\zeta_{\alpha_1} - \rho B^4 (A_2^*)_i/(2\overline{I})}{\zeta_{\alpha_1} - \rho B^4 (A_2^*)_b/(2\overline{I})} \tag{16}$$

$S_u(\omega)$ and $S_w(\omega)$ here are taken as Simiu and Panofsky formulae, respectively. Hence

$$\eta_{uw}(\omega) = \frac{200}{6} \times \left[(1 + \frac{2Z}{\pi U}\omega)^2 \Big/ \sqrt[3]{(1 + \frac{25Z}{\pi U}\omega)^5} \right] \tag{17}$$

where Z is the bridge deck level.

For further simplification of the method, η_{sh} and η_{sa} can be expressed as:

$$\eta_{sh} = (C_L')_i^2 / (C_L')_b^2 \tag{18}$$

$$\eta_{sa} = (C_M')_i^2 / (C_M')_b^2 \tag{19}$$

as the condition of $\eta_{uw} = 2\sim4$, $C_L' >> C_L$, $C_M' >> C_M$ and $C_L' >> (A/B)C_D$ for many bridges. It means that, for a relative comparison, the slope of the aerodynamic force coefficient curve of the deck section C_L' and C_M' are the predominant parameters in the buffeting-based selection.

BUFFETING RESPONSE ANALYSIS IN TIME DOMAIN

Introduction

In order to investigate the response characteristics of a long span bridge, such as the effects of instantaneous relative velocity and the effective angle of attack, and structural nonlinearity, some researchers have utilized the time domain approach. J. C. Santos(1993) and I. Kovacs(1992) have presented solution method in their papers. Wind forces are modeled by using the quasi-steady formulation which takes instantaneous effective angle of attack and time-space variation of wind velocity fluctuation into account.

In the time-domain analysis, turbulent components $u(x,t)$ and $w(x,t)$ and buffeting forces are transformed into time history using computer simulation, in which time-space variation of wind velocity fluctuation is considered. The self-excited forces appear to be in the form of convolution between impulse function and response, whose parameters are determined by fitting aerodynamic derivatives obtained in bridge section model wind tunnel test. Furthermore, the instantaneous change of drag, moment and lift coefficients with the effective angle of attack, especially the wind velocity turbulence effect on aerodynamic loads are considered, which can not be performed in the frequency domain.

Simulation of Wind Turbulent Components

The wind velocity at one point is assumed to compose of a mean part U, and a fluctuating part $u(x,t)$ for the along wind direction and a fluctuating part $w(x,t)$ for the vertical direction. When performing the time domain analysis of long span bridge under wind action, the time-space variation of wind velocities along the bridge axis is concerned, i.e., turbulent components $u(x,t)$ and $w(x,t)$ are modeled as multi-correlated random process. Simulation of multi-correlated random processes can be done in several ways, such as methods based on the summation of trigonometric function (wave superposition) and methods based on the convolution of white noise with a kernel function or integration of a differential equation driven by white noise (digital filtering). These techniques vary in their applicability, complexity, computer storage requirement, and computing time. Due to its good versatility, wave superposition method based on a prescribed correlation function and power spectral density function is obtained here.

To comply with the coherence criteria, the turbulent part of the momentary wind velocity in point x is generated as the following sum:

$$v(x,t)=\sum_{j=1}\left\{\sqrt{\frac{4S(n_j)\Delta n}{\sum_k C^2(\xi_k,n_j)}}\cdot\sum_k C(x-\xi_k,n_j)\cdot\cos\left[(\omega_j+\delta\,\omega)t+\phi_{kj}\right]\right\} \tag{20}$$

where, $C(\xi,n)=e^{-2kn|\xi|/\bar{U}}$; k is decaying coefficients; $S(n)$ is target spectrum; ξ_k is generation spot; ω_j is circular frequency; $\delta\omega_1$ is circular frequency to avoid periodicity of generated random process, $\delta\omega\in(-\Delta\omega'/2,\Delta\omega'/2)$, $\Delta\omega'<<\Delta\omega$; ϕ_{kj} is starting phase, uniformly distributed between 0 and 2π ; n is frequency (Hz).

For the longitudinal component $u(x,t)$ of the wind velocity fluctuation, the turbulence spectrum is defined, according to Simiu, as

$$\frac{nS_u(n)}{u_*^2} = \frac{200f}{(1+50f)^{5/3}} \tag{21}$$

where, $f = n\,z/U(z)$, is reduced frequency; $u_* = KU(z)/\ln[(z-z_d)/z_0]$; K is Karman constant($=0.4$); $z_d = 0$, in open zone; z_0 is roughness length of the ground surface (0.003m); z is height; $U(z)$ is mean velocity at height z.

On the other hand, for the vertical component of velocity fluctuation, the Panofsky-McCormick spectrum is expressed as

$$\frac{n\,S_w(n)}{u_*^2} = \frac{6f}{(1+4f)^2} \tag{22}$$

where, $f = n\,z/U(z)$, the remaining symbols are the same as the Eq.(21).

For the Shantou Bay Bridge, the turbulent components $u(x,t)$ of a mean wind speed $U = 30$m/s are simulated along the bridge span, shown in Fig.5. Comparison between the target spectra and the spectra obtained from the simulated velocities is shown in Fig.6. It can be seen that velocity fluctuation can be simulated satisfactorily using Eq.(20).

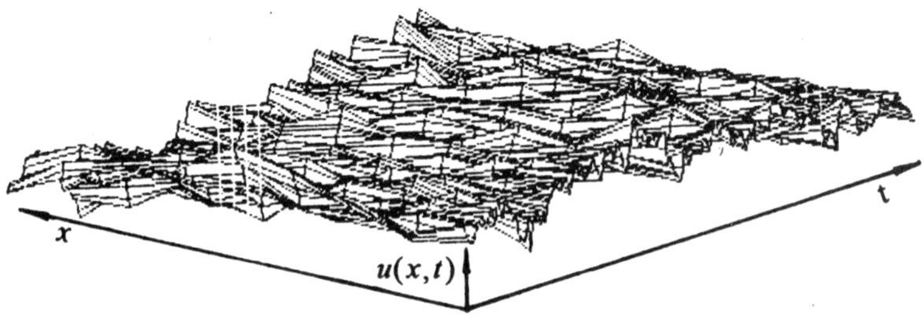

Fig.5 Generated wind turbulence along Shantou Bay Bridge axis

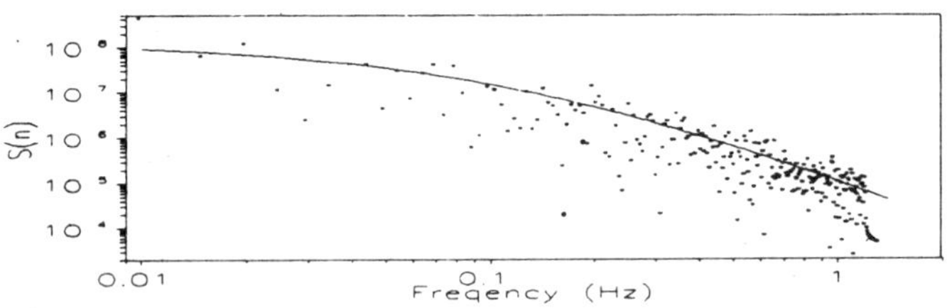

Fig.6 Target spectrum and simulated spectrum

Simulation of Buffeting Forces

At a particular time instant, buffeting forces derived from the quasi-steady formulation are expressed as:

$$\left.\begin{aligned} L_b(x,t) &= -\frac{1}{2}\rho \bar{U}^2 B\left\{2C_L(\alpha_0)\frac{u(x,t)}{\bar{U}} + \left[C_L'(\alpha_0)+\frac{A}{B}C_D(\alpha_0)\right]\frac{w(x,t)}{\bar{U}}\right\} \\ M_b(x,t) &= -\frac{1}{2}\rho \bar{U}^2 B\left\{2C_M(\alpha_0)\frac{u(x,t)}{\bar{U}} + C_M'(\alpha_0)\frac{w(x,t)}{\bar{U}}\right\} \\ D_b(x,t) &= -\frac{1}{2}\rho \bar{U}^2 B\left\{\frac{A}{B}C_D(\alpha_0)\cdot 2\frac{u(x,t)}{\bar{U}}\right\} \end{aligned}\right\} \quad (23)$$

where, A is cross-wind projected area (per unit span) normal to $\bar{U}$; α_0 is the effective angle of attack under the wind velocity; C_D, C_L, C_M are instantaneous drag, lift and moment coefficients per unit span, respectively; C_L' , C_M' are instantaneous slope of C_L and C_M at angle instant of attack; ρ is air density; B is bridge deck width.

Taking aerodynamic admittance and correlation function to Eq.(23), and making the Fourier transform, the buffeting forces spectra are obtained as following:

$$\left.\begin{aligned} S_L(n) &= \rho^2 U^2 B^2 \gamma^2(n)\left\{C_L^2(\alpha_0)S_u(n) + \frac{1}{4}\left[\left.\frac{dC_L}{d\alpha}\right|_{\alpha=\alpha_0} + \frac{A}{B}C_D(\alpha_0)\right]^2 S_\omega(n)\right\} \\ S_M(n) &= \rho^2 U^2 B^4 \gamma^2(n)\left\{C_M^2(\alpha_0)S_u(n) + \frac{1}{4}\left[\left.\frac{dC_M}{d\alpha}\right|_{\alpha=\alpha_0}\right]^2 S_\omega(n)\right\} \\ S_D(n) &= \rho^2 U^2 B^2\, C_D^2(\alpha_0)\, S_u(n)\, \gamma^2(n) \end{aligned}\right\} \quad (24)$$

Assuming $[S_L]$, $[S_M]$, and $[S_D]$ are buffeting force – lift, moment and drag – spectral density matrix respectively, they can be expressed as the following:

$$\left.\begin{aligned} [S_L] &= [S_1]\cdot S_L(n) \\ [S_M] &= [S_1]\cdot S_M(n) \\ [S_D] &= [S_1]\, S_D(n) \end{aligned}\right\} \quad (25)$$

where

$$[S_1] = \begin{bmatrix} 1 & e^{-\lambda n|\Delta_{12}|/\bar{U}} & e^{-\lambda n|\Delta_{13}|/\bar{U}} & \cdots & e^{-\lambda n|\Delta_{1n}|/\bar{U}} \\ & 1 & e^{-\lambda n|\Delta_{23}|/\bar{U}} & \cdots & e^{-\lambda n|\Delta_{2n}|/\bar{U}} \\ & & 1 & \cdots & e^{-\lambda n|\Delta_{3n}|/\bar{U}} \\ & symetric & & \ddots & \vdots \\ & & & & 1 \end{bmatrix}$$

According to R. H. Scanlan and Y. K. Lin, self-excited loads are assumed to be generated by linear mechanisms and expressed in terms of convolution integrals as follows:

$$\left.\begin{aligned} M(t) &= M_{\alpha}(t) + M_h(t) = \int_{-\infty}^{t} f_{M_{\alpha}}(t-\tau)\alpha(\tau)d\tau + \int_{-\infty}^{t} f_{M_h}(t-\tau)h(\tau)d\tau \\ L(t) &= L_{\alpha}(t) + L_h(t) = \int_{-\infty}^{t} f_{L_{\alpha}}(t-\tau)\alpha(\tau)d\tau + \int_{-\infty}^{t} f_{L_h}(t-\tau)h(\tau)d\tau \end{aligned}\right\} \tag{26}$$

where, $f_{M_{\alpha}}(t-\tau), f_{M_h}(t-\tau)$ are impulse response function.

Assuming that the surrounding air flow provides a set of filter-like devices in generating the buffeting forces on a bridge, the transfer function $F_{M_{\alpha}}$ may be approximated as follows for the purpose of fitting the experimental data:

$$F_{M_{\alpha}}(\omega) = \rho B^2 U^2 \left[c_1 + \frac{B}{U} i\,\omega\, c_2 + \frac{c_3\, i\,\omega}{c_5 \frac{U}{B} + i\,\omega} + \frac{c_4\, i\,\omega}{c_6 \frac{U}{B} + i\,\omega} \right] \tag{27}$$

where, $c_1, c_2, c_3, c_4, c_5, c_6$ are real dimensionless coefficients to be inferred from measured flutter derivatives; U is instantaneous total wind speed $(\bar{U} + u)$

Time History Analysis of Buffeting Response

The equations governing the motion of a bridge deck is as follows:

$$
\left.\begin{aligned}
m_h\left(\ddot{h}+2\zeta_h\,\omega_h\,\dot{h}+\omega_h^2\,h\right) &= L_{se}(t)+L_b(t)\\
I_\alpha\left(\ddot{h}+2\zeta_\alpha\,\omega_\alpha\,\dot{\alpha}+\omega_\alpha^2\,\alpha\right) &= M_{se}(t)+M_b(t)\\
m_p\left(\ddot{p}+2\zeta_p\,\omega_p\,\dot{p}+\omega_p^2\,p\right) &= D_{se}(t)+D_b(t)
\end{aligned}\right\} \qquad (28)
$$

where, $L_b(t), M_b(t), D_b(t)$ are generalized fluctuating components-induced lift, moment and drag, respectively; h, α, p are generalized vertical bending, torsional and swaying displacements, respectively; $\zeta_h, \zeta_\alpha, \zeta_p$ are structural damping ratios; $\omega_h . \omega_\alpha, \omega_p$ are structural circular frequencies; m_h, I_α, m_p are generalized mass; $L_{se}(t), M_{se}(t), D_{se}(t)$ are self-excited force.

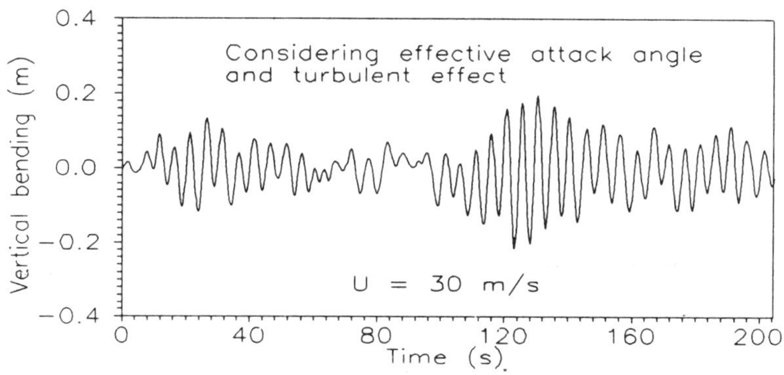

Fig.7 Vertical bending response of center point of mid-span of Shantou Bay Bridge

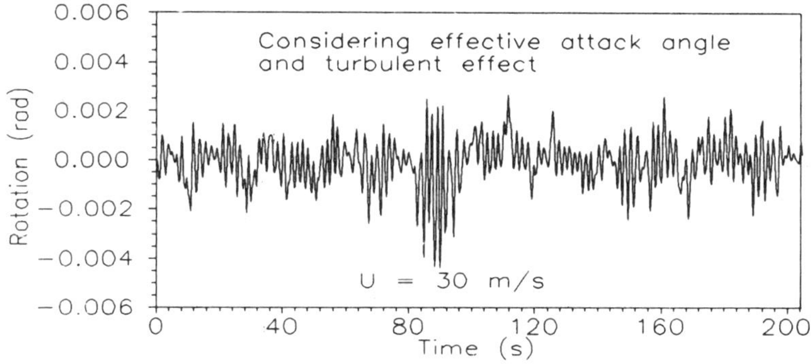

Fig.8 Rotation response of center point of mid-span of Shantou Bay Bridge

In the analysis, the wind fluctuation along the bridge axis are generated using Eq.(20) derived from the given Simiu spectrum, Panofsky-McCormick spectrum and Davenport coherence function. At each time increment, the drag, lift and moment coefficients vary with the effective angle of attack.

The Shantou Bay Bridge's coupled buffeting response of vertical bending and torsion motions corresponding to fundamental vibration mode are calculated as a result, shown in Fig.7 and Fig.8. Fig.9 demonstrates a good accordance among the results of buffeting responses calculated in time domain and frequency domain, and measured by 3D full scale wind tunnel test.

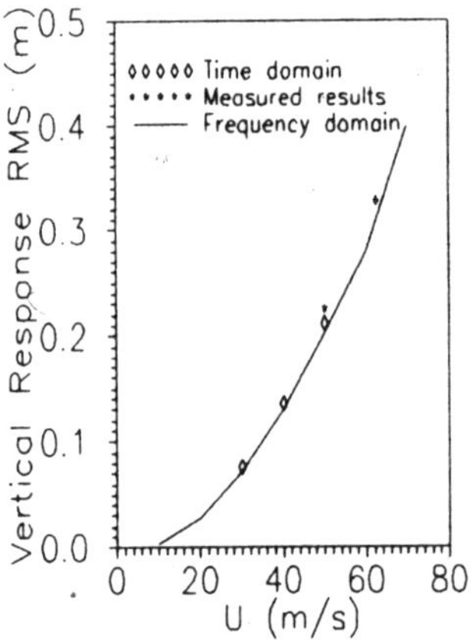

Fig.9 Comparison among the results of buffeting responses obtained by three ways

PASSIVE TMD CONTROL OF BUFFETING

Introduction

At present, the measures which are adopted for increasing the critical flutter wind speed of bridges include improving the aerodynamic behaviour mainly by modifying the cross section shape, and increasing the first torsional frequency by using inclined cable-planes. However, buffeting is still a problem during erection and operation. Many engineers and researchers are concerned about how to suppress the buffeting of bridges effectively. TMD (tuned mass damper) devices with the advantages of reasonable cost and convenience, and effectiveness in usage, have been mounted on some bridges[Wardlaw, 1992; Huffmann, 1987]. TMD for increasing the flutter and galloping critical wind speed[Nobuto, 1988; Fujino, 1985] as well as suppressing buffeting[Mcnamara, 1977; Gu, 1994, 1992] of long- span bridges have been theoretically and experimentally studied.

Control of Vertical Bending Buffeting

For bridges with a higher first torsional frequency, the contribution from torsional modes to the total buffeting response may be neglected. Therefore, it can be assumed that the total buffeting response only takes place in vertical bending modes, and further in the first vertical bending mode, so $\Phi_1(x)$, generally speaking, occupies a dominant position in the total response. The mechanical model for analysis is shown in Fig.10(b). The girder displacement and the TMD displacement relative to the girder can be written as $Y_1(x,t) = \Phi_1(x)\xi_1(t)$ and $Y_2(t) = \Phi_1(x_0)\xi_2(t)$ respectively. The equations governing the motions of a bridge deck and TMD are, respectively, as follows:

$$M_1(\ddot{\xi}_1 + 2\zeta_1\omega_1\dot{\xi}_1 + \omega_1^2\xi_1) = \int_0^{L_s} F_{ae}(x,t)\Phi_1(x)dx + \int_0^{L_s} F_b(x,t)\Phi_1(x)dx + M_1(2\mu\zeta_2\omega_2\dot{\xi}_2 + \mu\omega_2^2\xi_2) \tag{29}$$

$$\ddot{\xi}_2 + 2\zeta_2\omega_2\dot{\xi}_2 + \omega_2^2\xi_2 = -\ddot{\xi}_1 \tag{30}$$

where ξ_1 and ξ_2 are the first mode time-dependent vertical generalized coordinates of the bridge deck and TMD, respectively; $\Phi_1(x_0)$ is the value of $\Phi_1(x)$ at the site where TMD is mounted; M_1, ω_1 and ζ_1 are the first mode generalized mass, circular frequency and damping ratio of bridge, respectively; ω_2 and ζ_2 are the circular frequency and the damping ratio of TMD, respectively; $\mu = m_2\Phi_1^2(x_0)/M_1$ is the generalized mass ratio of TMD to the bridge; m_2 is the mass of TMD; L_S is the total length of the main bridge; $F_{ae}(x,t)$ and $F_b(x,t)$ are the aeroelastic force and buffeting force acting on the bridge deck proposed by Scanlan,

From the Eqs. (29) and (30), the mean square displacements of the bridge deck and TMD can be easily derived respectively. Then

$$\eta = \frac{R_y(\bar{U},x_0) - R_{y_1}(\bar{U},x_0)}{R_y(\bar{U},x_0)} = 1 - \frac{R_{y_1}}{R_y} \tag{31}$$

is defined as the reduction ratio, where R_{y_1} and R_y denote the RMS displacements of the bridge deck with and without TMD respectively.

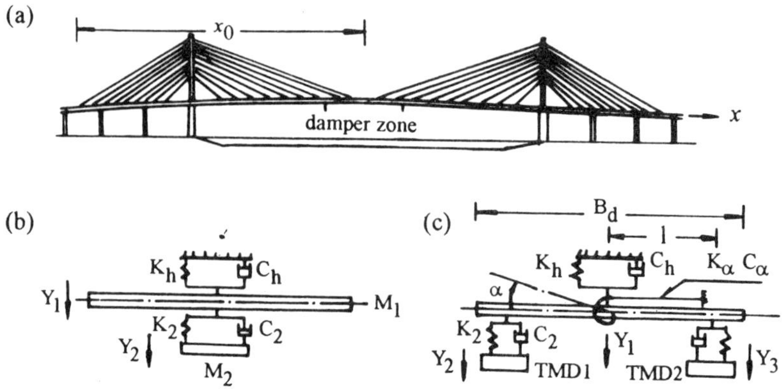

Fig.10 Mechanical model for analysis

Eq.(31) is here treated as the objective function for optimization of parameters of TMD. The equations which describe how the optimal damping ratio(ζ_{20}) and frequency(f_{20}) depend on the generalized mass ratio of TMD to bridge(μ) and ζ_t are given as follows:

$$\zeta_{20} = 0.5\sqrt{\mu} \tag{32}$$

$$\frac{f_{20}}{f_1} = \frac{1}{1+\mu} + 3\mu^2 - 0.2\zeta_t^2 \tag{33}$$

where f_1 is the vertical bending eigenfrequency of structure which need to be controlled, ζ_t is the total damping ratio, the sum of structural one and aerodynamic one, which is a function of mean wind speed $\overline{U}$

Eqs.(32) and (33) are the same as the classical equations for very small ζ_t. But when ζ_t is much larger than the inherent structural one, the classical equation will affect the optimal frequency of TMD to a certain extent.

By using the Least Square Method, an expression for the optimal reduction ratio corresponding to the optimal parameters from Eqs.(32) and (33) is derived as follows:

$$\eta_0 = \frac{0.0158}{0.1277+\zeta_t}\ln\mu + \frac{0.0496}{0.0405+\zeta_t} \tag{34}$$

or

$$\mu = \exp\left[\left(\eta_0 - \frac{0.0496}{0.0405+\zeta_t}\right)\cdot\frac{0.1277+\zeta_t}{0.0158}\right] \tag{35}$$

It should be noted that η_0 may be much larger at allowable wind speed than at a high one because it is inverse proportional to ζ_t, which depends directly on wind speed. For these reasons, it is suggested that in designing TMD for a given bridge, designers first should select a "TMD design speed", at which the parameters of TMD are optimized.

When engineers design a TMD for a given bridge and a desired reduction ratio, they can determine a value of μ (mass of TMD) of a selected "TMD design speed" by using Eq.(35) at first. Then, from Eqs.(32) and (33), the optimal parameters of TMD can be easily found.

The Yangpu Bridge[28] is a cable-stayed bridge with a composite deck, include cable planes and a center span of 602m. The first vertical bending frequency is 0.305Hz. The inherent structural damping ratio is taken as 0.01. According to the results of buffeting analysis for the bridge before closure, the required reduction ratio is at least 35% at "TMD design wind speed" of 25m/s. The calculated total mass of TMD using Eq.(35) is about 40 tons, and the optimal frequency and damping ratio using Eqs.(32) and (33) are 0.298 and 0.072 respectively.

Control of Buffeting Response with Vertical Bending — Torsional Modes Interaction

For some cases that the vertical buffeting and torsional one are of the same significance, the control of buffeting response should take the interaction between two modes into account. Fig.10(c) shows a mechanical model for the control of interaction buffeting, in which Y_1 and α denote vertical displacement and torsional angle of the bridge deck respectively; Y_2 and Y_3 denote vertical displacements of TMD1 and TMD2 relative to Y_1 respectively. TMD1 and TMD2 have the same mass, spring stiffness and damping ratio, that is, $M_2 = M_3$, $K_2 = K_3$ and $C_2 = C_3$ ($\zeta_2 = \zeta_3$).

The Objective function for optimization of the TMDs' parameters are taken as the mean square displacement (R_B^2) comprising vertical value (R_b^2) and torsional angle (R_t^2) times $(B/2)^2$, where B is the width of bridge deck, that is,

$$\min f(\mu, \zeta_2, \omega_2) = R_B^2 = R_b^2 + R_t^2 (B/2)^2 \tag{36}$$

The amplitudes of TMD can be taken into consideration of the constraint equations. If the bridge vibrates only in vertical motion, the vibration amplitudes of TMD1 and TMD2 are the same. But the existence of torsional motion leads to the situation that the amplitudes of TMD1 and TMD2 are not the same. Larger RMS (root mean square) displacement between TMD1 and TMD2 can be found and used to check the engineering constraint condition.

In general, the optimal frequency of TMD is near one frequency on which the buffeting energy concentrates mainly. When the frequency of TMD varies apart from the optimal value between the vertical and torsional eigenfrequencies, the reduction ratio decreases slowly. While it is away from the optimal value outside the two frequencies, the reduction ratio falls down quickly.

Constraint Condition

It is important to establish the engineering constraint conditions and to take them into consideration in designing TMD. The equation for checking simultaneously the fatigue strength and static strength as well as the space required for the motion of TMD is given as follows:

$$\min\left\{\frac{[\tau]\pi C_0 H_0}{2gKG}, \frac{\tau_0 \pi C_0^2 H_0}{(n_F + 0.75 + n_F g - 0.75g)Kg}, \frac{H_B - H_0 - \Delta}{2g}\right\} - R_{y_2} > 0 \tag{37}$$

where $[\tau]$ and G are the allowable shearing stress and the shearing elasticity of the spring material respectively; τ_0 is the shearing fatigue limit of the spring material under pulsation shearing stress; n_F is the safety factor for fatigue check; H_0 is the original length of each spring; H_B is the depth of the bridge deck; g is the peak factor; R_{y_2} is the RMS displacement of TMD, K, C_0, etc. are all coefficients of spring.

In Eq.(37), the two left items in brackets are both directly proportional to H_0, while the third item is inversely proportional to it. This means that when

$$\min\left\{\frac{[\tau]\pi C_0 H_0}{2gKG}, \frac{\tau_0 \pi C_0^2 H_0}{(n_F + 0.75 + n_F g - 0.75g)Kg}\right\} = \frac{H_B - H_0 - \Delta}{2g}, \tag{38}$$

$$\min\left\{\frac{[\tau]\pi C_0 H_0}{2gKG}, \frac{\tau_0 \pi C_0^2 H_0}{(n_F + 0.75 + n_F g - 0.75g)Kg}, \frac{H_B - H_0 - \Delta}{2g}\right\} \tag{39}$$

takes the maximum value, which can be bring the constraint Eq. (37) into full play. So, H_0 can be appropriately determined from Eq.(38).

CONCLUSIONS

(1) With the increase of span-length, the buffeting problem becomes more important compared with flutter phenomenon, which can be solved through some aerodynamic measures even for extra long-span suspension bridges.

(2) Two basic buffeting theories developed by A. Davenport and R.H. Scanlan independently have been close each other by considering fully and rationally all important factors concerned.

(3) For the buffeting response analysis of long-span bridges, the approximate calculation formulae including some factors similar to the seismic response spectrum method can be established through the rational simplification to the theory of buffeting response spectrum.

(4) The buffeting-based selection of deck section is important in the preliminary design stage together with the conventional flutter-based selection. As a relative comparison, the slope of lift force curve of deck section is a predominant parameter in the buffeting-based selection.

(5) A method for the buffeting response in time-domain is useful in considering the non-linear factor both structurally and aerodynamically, as well as the effect of additional dynamic attack angle, which makes the buffeting analysis more to reality.

(6) The buffeting response of long-span bridges with strong enough torsional stiffness is mainly contributed by the vertical bending mode, and the TMD set is effective and convenient in suppressing vertical buffeting.

MAIN REFERENCES

Chen, W. (1993), "Study on Buffeting Response Spectrum of Long-span Bridges", Ph D Dissertation, Tongji Univ.

Davenport, A.G. (1962), "Buffeting of a Suspension Bridge by Storm Wind", J. Struct. Div., Proc. ASCE, Vol.88, ST3, pp1353-1375.

Davenport, A.G.(1963), "The Buffeting of Structures by Gusts Proceedings", Symposium on Wind Effects on Buildings and Structures, Vol.1, Teddington U.K.

Fujino, Y. et al (1985), "Suppression of Galloping of Bridge Tower Using Tuned Mass Damper", J. Fac. Eng., The Univ. of Tokyo(B), 38(2), pp.49.

Gu. M. & Xiang, H.F. (1992), "Optimization of TMD for Suppressing Buffeting Response of Long-span Bridges", J. Wind Eng. Ind. Aerodyn. 42, pp.1383.

Gu, M. et al (1994), "A Practical Method of Passive TMD for Suppressing Wind-induced Vertical Buffeting of Long-span Cable-stayed Bridges and its Application, J. of WEIA, 51, pp.203-213.

Gu, M. et al (1993), "Study on Buffeting-based Selection of Bridges", Recent Advances and Applications in Wind Effects on Structures, Tongji University Press, pp185-189.

Huffmann, G. et al (1987), "Dao Khanong Cable-stayed Bridge Tuned Dampers in Deck and Pylon", Proc. Int. Conf. on Cable-stayed Bridges, Bangkok, pp.1485.

Ioi, T. & Ikeda, K. (1978), "On the Dynamic Vibration Damped Absorber of the Vibration System", Bull. JSME, pp.64.

Kovacs, I. et al (1992), "Analytical Aerodynamic Investigation of Cable-stayed Helgelang Bridge, J. Struct. Eng., ASCE, Vol.118, No.1, pp.147-168.

Lin, Y.K. (1979), "Motion of Suspension Bridge in Turbulent Winds", J. Eng. Mech., ASCE, 105(6).

Lin, Y.k. & Yang, J.N.(1983), "Multi-mode Bridge Response to Wind Excitation", J. Eng. Mech., ASCE, 109(2).

Lin, Y.K. & Li, Q. C. (1993), "New Stochastic Theory for Bridge Stability in Turbulent Flow", J. Eng. Mech., ASCE, Vol.119, pp.113-127.

Mcnamara, R.J. (1977), "Tuned Mass Dampers for Buildings", J. Struct. Div. ASCE, 103, pp.1785.

Nobuto, J. et al (1988), "A Study on the Effectiveness of TMD to Suppress a Coupled Flutter of Bridge Deck", J. JSCE, 413(in Japanese).

Santos, J.C. et al (1993), "Gust Response of Long-span Bridge by the Time Domain Approach", Third Asia-pacific Symposium on Wind Engineering, Dec. 13-15, Hong Kong, pp.211-216

Scanlan, R.H. & Gade, R. (1977), "Motion of Suspended Bridge Spans under Gusty Wind", J. Struct. Div. Proc. ASCE, Vol.103, ST9, pp.1867-1883.

Scanlan, R.H.(1978), "The Action of Flexible Bridges under Wind: II: Bufffeting Theory", J. Sound and Vibration, Vol.60.

Scanlan, R.H. (1984), "Role of Indicial Functions in Buffeting Analysis of bridges", J. Struct. Eng. , Vol.10, No.7.

Scanlan, R.H. (1992), "Wind Dynamics of Long-span Bridges", Aerodynamics of Large Bridges.

Scanlan, R.H. & Jones N.P. (1990), "Aeroelastic Analysis of Cable-stayed Bridges", J. Struct. Eng., ASCE, 116(2).

Scanlan, R.H et al (1974), "Indicial Aerodynamic Functions for Bridge Decks", J. Eng. Mech. Div., ASCE, Vol.100,.pp.657-672.

Scanlan, R.H. (1993), "Problematics in Formulation of Wind-force Models for Bridge Deck", J. Eng. Mech., Vol 119, No.7, pp.1353-1375.

Simiu, E (1974), "Wind Spectra and Dynamic Alongwind Response", ASCE, ST9, Vol.100, pp.1897-1910.

Wardlaw, R.L. (1992), "The Improvement of Aerodynamic Performance, in: A. Larsen (Ed), Aerodyn. of Large Bridges", Balkema, Rotterdam, pp.59.

Xiang, H.F et al (1993), "Bridges in China", Tongji University Press/A&U Publication (HK) Ltd.

Xiang, H.F et al (1994), "Practical Calculation Method for Buffeting Response Spectrum of Long-span Bridges", Selected Paper of Tongji Univ

Xiang, H.F et al (1992), "Wind Resistant Design of Yangpu Bridge", Research Report (in Chinese), Dept. of Bridge Eng., Tongji University.

NON-LINEAR EFFECTS IN THE BUFFETING PROBLEM

H.J. NIEMANN, Prof. Dr.-Ing. and R. HOFFER, Dipl.-Ing.
Building Aerodynamics Laboratory,
Ruhr-Universität, Bochum, Germany

SUMMARY

Modern design procedures for wind-sensitive structures aim on ultimate limit states in which non-linearity is dominating. This review focusses on the treatment of geometrical and physical non-linear structures. Contributions coming from the structure-flow interactions are also considered. Approximate methods in the frequency domain as well as in time domain are discussed.

INTRODUCTION

The classical approach to modelling the stochastic wind loading process and to calculating the structural response it induces relies on several linearizations. This applies particularly to the buffeting, along-wind and cross-wind, loads which are imposed by wind gustiness and by signature turbulence due to flow separation. Linear models have also been developed for strongly non-linear phenomena such as flutter, galloping, and vortex resonance. These shall not be considered in what follows. The first linearization concerns the wind speed-wind load transfer. It is assumed that the load fluctuations are related linearly to the velocity fluctuations; higher order terms contained in the basic fluid dynamic equations are neglected. In other words, the aerodynamic admittance is modelled as a linear filter. As well, the structural behaviour is represented by a linear, differential equation, in which stiffness and damping are constants. Additional interactive forces are caused by the displacement velocity of the structural movements, and are accounted for as linear aerodynamic damping and stiffness within the scope of the linearized theory. The linear theory was developed, starting in the early sixties, by the fundamental work of Alan G. Davenport. It provided for the first time a realistic design tool for wind effects on structures, and has become the backbone of modern wind loading codes.

The linearizations make the theory particularly efficient for two reasons. Firstly, the calculations can be performed in the frequency domain using Fourier Transforms and making use of a well-developed mechanical theory, the spectral approach. The second reason is that a system exposed to a Gaussian input typical for turbulent velocity fluctuations, will respond with a Gaussian output. This allows to predict statistically the extremes of the linear structural response.

On the other hand it is clear that most structures behave strongly non-linear in their ultimate limit state immediately before they fail. Modern design methods aim at calculating this state using higher order mechanical methods. It seems necessary to inspect wheter the linear wind loading theory is appliable within this new type of design calculations and to review more appropriate methods. The following contribution is an attempt in this respect.

LINEARIZATIONS

Elastic non-linear restoring forces

Guyed masts respond to wind excitations with forced vibrations of the mast, forced and parametric excited vibrations of single ropes or bundles of ropes, and oscillations of the guy-mast system. The restoring forces of guy-mast systems are non-linear because of the sags of the tilted rope bundels. The degree of non-linearity is determined by the amount of initial stress, the length, and the weight of a rope. The dynamic behaviour of a guy rope is governed by a non-linear differential equation of motion (DEQM), which contains square and cubic non-linear terms, and terms of parametric excitations. Tonis (1989) tested spatial oscillations of a tilted rope, which undergoes a forced excitation by the periodical, longitudinal motion of its upper support, like it is imposed from a periodically moving mast. He stated that linearized DEQM cannot describe the spatial movement. The number of response frequencies of a rope to a monofrequent forced excitation is drastically reduced. Cai and Chen (1994) found similar results for the forced excitation of the upper support in orthogonal directions.

However, guyed masts are often modelled as an elastically supported beam, which means that the guy supports are treated as linear. The linearization of the non-linear equation of motion is performed around the deflected position under average wind conditions or for the initial undeflected state. To reduce the computational effort, Gerstoft (1986) proposed a simplified procedure, in which bending moments of the mast are calculated as the background response using influence lines from the linearized system and applying patch loads to the mast. The dynamic response is estimated by multiplying the background response by a dynamical amplification factor. This is possible, since the deformations, which result from applying specific patch loads, are in affinity to the important mode shapes. For design purposes, simplifying approximative equations are given in the mentioned paper to determine the amplification factor. The author limits the applicability of the procedure to certain ranges of mass and stiffness properties and 'slackness' ratios for the guys. The latter ratio only treats the initial stress of the guys in the undeflected position. Kahla (1994) pointed out that lowest oscillatory frequencies for the mast have been found for slack guys while the structure vibrates about its deflected position under average wind conditions. Large oscillations of leeward guys are possible accompanied by large guy tension which likely would cause the collapse of the structure, either by rupture of guys or guy connectors, or by local buckling of some elements of the mast. Peil (1993) proposes to linearize the restoring force around the operating point of the loaded guy-mast system. The stiffness of the guy support depends on frequency, since the mass of the ropes is taken into account. The mean wind loads on the ropes are included into the computation of the operating point. It is found that the mean wind loads have to be included for the computation of the support's stiffness. The calculated eigenfrequencies and the spectral location of measured peaks are in good accordance, but the amplitude spectrum of e.g. the tip deflection is considerably overestimated. To improve the linear approximation, Peil (1993) adopts an equivalent damping ratio, which is determined from the peaks of the non-linearly and the linearly calculated amplitude spectra. The difference of the amplitude spectra is explained in Petersen (1992) as a consequence of the permanent detuning of the stiffness of the support, which acts as a damper. A comparison of the amplitude spectra of a non-linear guyed mast and a linearized system is shown in Fig. 1, where the stiffness of the linear oscillator is the tangential stiffness of the original non-linear system in the undeflected position, the damping ratio is the same in both cases. It is stated in Petersen (1992) that the non-linearity leads to a jump phenomenon with a considerably reduced maximum amplitude, which is not affected

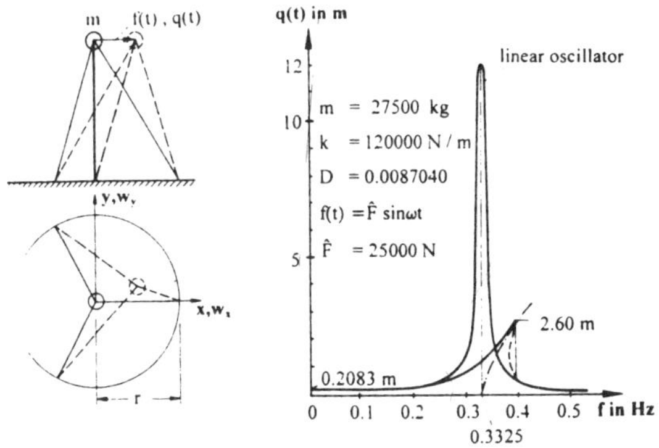

Fig. 1: a. system sketch
b. response amplitude spectra of the non-linear system and the linearized oscillator (after Petersen (1992))

from viscose damping. Chaotic oscillations are detected. He found, that the response of a guyed mast in turbulent wind is limited to much smaller amplitudes as those of a linear system with the tangential stiffness of the guyed mast in undeflected position.

Statistical equivalent linearization

The equivalent linearization technique is an efficient analytical tool for approximating the response of systems with weak and in some cases even strong non-linearities, provided the structure is deterministic. A certain drawback of the method is that the probability distribution of the response must be known to some extend. A DEQM of a linearized oscillator is constructed, which satisfies the non-linear response in a mean sense. In the field of random vibrations, the method was first applied to non-linear SDOF-systems under stationary excitations. The approach were later extended to MDOF-systems and to hysteretic structures.

Given the DEQM of a non-linear multi-degree-of-freedom (MDOF) system

$$\mathbf{g}(\ddot{\mathbf{q}},\dot{\mathbf{q}},\mathbf{q}) = \mathbf{f}(t) \tag{1}$$

where $\mathbf{g}(\ddot{\mathbf{q}},\dot{\mathbf{q}},\mathbf{q})$ is the vector of the deterministic non-linear restoring forces and $\mathbf{f}(t)$ is the vector of the external excitation, and

$$\mathbf{M}\,\ddot{\mathbf{q}} + \mathbf{D}\,\dot{\mathbf{q}} + \mathbf{K}\mathbf{q} = \mathbf{f}(t) \tag{2}$$

where **M**, **D**, **K** are the equivalent mass, damping, and stiffness matrices of the linearized system with the entries m_{ij}, d_{ij}, k_{ij}, $i,j=0,1,2,\ldots,N$, N is the number of DOF.

The minimization in the mean square sense

$$E\,\{\mathbf{e}^T \cdot \mathbf{e}\} \tag{3}$$

of the error

$$\mathbf{e} = \mathbf{g}(\ddot{\mathbf{q}},\dot{\mathbf{q}},\mathbf{q}) - \mathbf{M}\ddot{\mathbf{q}} - \mathbf{D}\dot{\mathbf{q}} - \mathbf{K}\mathbf{q} \tag{4}$$

serves as the criterion for choosing the equivalent system properties in eq. 2. Atalık and Utku (1976) proved that the matrix entries

$$m_{ij} = E\left\{\frac{\partial g_i(\ddot{q},\dot{q},q)}{\partial \ddot{q}_j}\right\},\quad c_{ij} = E\left\{\frac{\partial g_i(\ddot{q},\dot{q},q)}{\partial \dot{q}_j}\right\},\quad k_{ij} = E\left\{\frac{\partial g_i(\ddot{q},\dot{q},q)}{\partial q_j}\right\} \tag{5}$$

satisfy the criterion eq. 3 for a Gaussian probability density function (PDF) of the response quantities. The derivation inside the brackets in eq. 5 means the calculation of a tangential mass, damping, and stiffness properties from $\mathbf{g}(\ddot{\mathbf{q}},\dot{\mathbf{q}},\mathbf{q})$ for each DOF After the ensemble average is performed, a single matrix entry contains the expected value of the tangent property, which is similar, but generally not the same as the partial tangent property at the (discrete) operating point. The mean and r.m.s response of eq. 2 is an approximation of the statistical response parameters of the original non-linear system and can be calculated by use of usual methods. The spectral method is advantageous for turbulent wind excitations.

In the case of a considerable deviation of the real non-Gaussian response from the Gaussian assumption, the accuracy of such statistical equivalent linearization is not adequate, since an inexact PDF in eq. 5 introduces an error into the first two moments. Moreover, a reliability analysis is very sensitive to the tail of the PDF of the response. Pradlwarter et al. (1988) extended the approach to hysteretic MDOF-systems under non-stationary excitations and, as the essential point, to a response PDF of arbitrary shape. Therfore, a vector transformation

$$\mathbf{t}(\mathbf{q}) = \mathbf{F}_t^{-1}\left(\mathbf{F}_q(\mathbf{q})\right) \tag{6}$$

is established, $\mathbf{t}(\mathbf{q})$ can be looked at as a realization of the response vector of the non-linear response, and $\mathbf{F}_t$ is the distribution function of the non-linear response. The shape of $\mathbf{F}_t$ is constructed, e.g. by physical consideration based on prior knowledge, or by a Monte-Carlo-simulation over a small number of samples. The ensemble averages for the evaluation of the equivalent linearized system properties, like those in eq. 5, are obtained by integration of the transformed random variables $\mathbf{t}(\mathbf{q})$. It is proved in Pradlwarter (1990) that the mean and the r.m.s. response of the linearized and the original non-linear system are identical.

As an application, the gust response factor (GRF) of concrete chimneys is investigated in Pradlwarter et al. (1991). The structure is subjected to loads which have a long-term character and to loads, such as wind turbulence, which are predominantly of short-term character. The CICIND Model Code for Chimneys (1987) accounts for the visco-elastic behaviour of dynamically stressed, reinforced concrete by adopting a linear-elastic stress-strain relationship up to the ultimate limit stress of concrete, which is valid for the action of dynamic wind. The ultimate limit stress f_{ck}/γ_{cu} for dynamic actions decreases with increasing long-term strain towards the long-term limit, but Young's modulus remains constant for dynamic effects. The material law for reinforcement steel contains the elastic-ideal plastic behaviour of steel, the tension stiffening effect due to the bond between cracked concrete and steel bars is superposed in the stress-strain relationship P-Δ-effects are included into the analysis. Linear damping is assumed. The non-linearity of the concrete chimney contains the softening of the structure caused by cracks which are assumed to develop due to gust action and not from fatigue. The wind loads are calculated based on the linearized quasi-stationary theory (cf. to 3.3). The GRF's are determined for the stationary states of the structural oscillations. It has been found out that the GRF's near the ultimate limit state are up to 20 % larger in the non-linear analysis than the respective values based on linear dynamic analysis. This effect is explained by the 'softening' of the structure when it enters the highly non-linear range. The eigenfrequencies then decrease and move towards the frequency of the maximum ordinate of the load spectrum, which lets the response increase (Fig. 2).

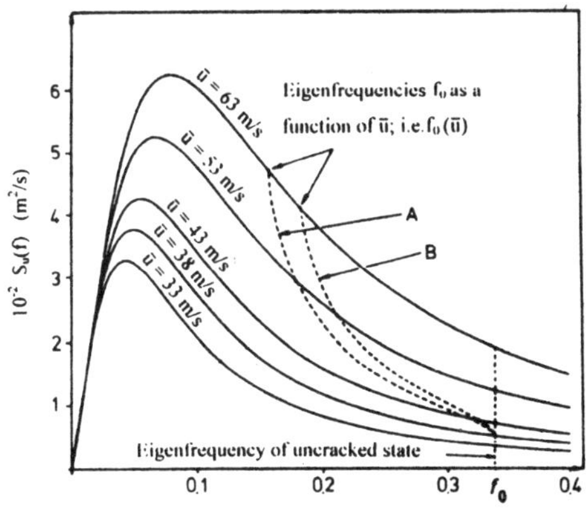

Fig. 2: Decrease of eigenfrequencies with increase in loading
A, B: without, with consideration of tension stiffening (after Pradlwarter et. al. (1991), minor changes)

Hysteretic response

A nominally elastic construction material, such as construction steel, follows Hooke's law, when it is loaded within its elastic range and if the loading is slow so that the material can establish the deformed state proportional to the load intensity (quasi-static loading). When the stress is imposed or released as a step function the strain lags behind its elastic value (creep). In the case of a harmonic load, such behaviour becomes apparent in the time domain as a phase lag of the strain behind the stress. The stress-strain relationship shows a dynamic hysteresis loop which is at small stress amplitudes slim and elliptical in shape. The term linear damping refers to this type of dynamic hysteresis. It can be modeled as the result of the combination of a linear spring and a viscous dashpot. The elastic part of the total energy, fed to the system, can be restored. The dissipated part is usually represented as the loss factor ψ of a material or construction component

$$\psi = \frac{w_{dis}}{2\pi w_{def}}, \tag{7}$$

where $w_{dis} = \oint \sigma(\varepsilon)\, d\varepsilon$, and $w_{def} = \int_0^{\varepsilon_0} E\varepsilon\, d\varepsilon$ are the rate of energy loss per unit volume and per loading cycle (w_{dis}), and the rate of specific deformation energy (w_{def}) per unit volume and cycle respectively, $\sigma(\varepsilon)$ is the stress due a strain ε, E is Young's modulus, and ε_0 is the maximum strain the material is subjected to. When the stress amplitude of the harmonic load exceeds a certain limit even within the elastic range - e.g. 1/20 of the fatigue strength of construction steel - experimental - hysteresis loops show a lancet-like shape, indicating a non-linear damping. The loss factor ψ tends to increase with increasing strains. It is often assumed based on experimental results that ψ increases proportional to the square of the deformations. Petersen (1971) proposed to retain the linear model, applying an equivalent loss-factor determined from the relation

$$\psi = \left(\frac{\varepsilon_0}{\varepsilon_{0r}}\right)^2 \cdot \psi_r \tag{8}$$

for stationary oscillations. The non-linear relationship eq. 8 just controls the lateral width of the loop, the index r stands for reference conditions, i.e. a reference strain amplitude ε_{0r} and a corresponding loss factor ψ_r obtained e.g. from an experiment. That means that the hysteretic loops do not retain affinity with increasing ε_0, as it is the case for strictly linear oscillations, but become fatter. The differential equation of motion can then be written as

$$\ddot{q} + \left(\frac{q_0}{q_{0r}}\right)^2 \left(\frac{\psi_r}{2\pi}\right)\frac{\omega_0^2}{\omega}\dot{q} + \omega_0^2\, q = \frac{Q}{m}\cdot \sin\omega t, \tag{9}$$

where ω_0 is the eigenfrequency for zero damping, q is the generalized displacement, q_0 is the (stationary) amplitude, q_{0r} is the reference amplitude for determining ψ_r, and Q/m is the generalized load amplitude normalized by the mass m.

The stationary solution for the amplification function V is given as

$$\frac{q_0}{q_{st}} = \frac{1}{\sqrt{\left[1-\left(\frac{\omega}{\omega_0}\right)^2\right]^2 + \left[\left(\frac{q_0}{q_{0r}}\right)^2 \frac{\psi_r}{2\pi}\right]^2}} = V\left(\frac{\omega}{\omega_0}, \frac{q_{0r}}{q_{st}}\right) \tag{10}$$

where q_{st} is the displacement under static load application.

The resonant behaviour is of major interest. Fig. 4 illustrates the change of the resonant response variances $\widetilde{R}^2$ of a weakly damped steel structure (i.e. $\psi_r < 0.4$) under buffeting wind load for damping proportional to the squared strain amplitude $(\widetilde{R}^2)$ against the strictly linear approach $(\widetilde{R}_0^2)$, following the procedure in Niemann (1990) for convenience. The ratio q_{0r} / q_{st} can be determined a-priori by static methods and is set to 1.5 here. Values for $V(\omega/\omega_0=1)/V_0(\omega/\omega_0=1)$ are taken from Petersen (1971), where V_0 is the amplification factor for $\psi = \psi_r$. It is obvious, that consideration of non-linear damping characteristics diminishes drastically the overall dynamic response of flexible steel structures under buffeting loads.

A linear, viscous model is no longer appropriate when the construction material is loaded at very high levels or beyond its elastic limit. The hysteresis loops become fat and sharp edged and can degenerate into shapes similar to those for Coulomb frictional damping. Fat hysteric loops occur also, when composite materials, such as reinforced concrete, is exposed to cyclic loading. For such cases Bouc (1967) and Wen (1976) proposed a model, in which the hysteretic force z is governed by the differential relationship

$$\dot{z} = -\gamma|\dot{q}|\,z|z|^{n-1} - \nu\dot{q}\,|z|^n + A\cdot\dot{q}, \tag{11}$$

where γ, ν, A, n are parameters which control the shape of the loop. A wide variety of hysteretic constitutive laws which can be expressed in this manner as was shown by Suzuki and Minai (1987). Such differential model of hysteresis can be combined with the system's differential equations to yield an overall differential model, which can be analysed by using techniques such as statistical equivalent linearization. The latter approach is applied extensively in earthquake engineering. The application of the method and of other stochastic approaches on hysteretic structures is reviewed in Roberts and Spanos (1991) and Schuëller (1991). Only few works exist about the treatment of wind excitated structures in the highly non-linear range, which is dominated by restoring forces with fat hysteretic loops, such as degraded structures. Chen and Ahmadi (1992) studied the sensitivity of the displacements of a base isolated three-storeyed building to wind loading. A quasi-stationary model was used for modeling the longitudinal turbulent wind action. The variation of the wind speed with height was neglected. The building is considered as a rigid body. The results mainly may serve for the comparison of the effect of different isolation types in severe storms and hurrican conditions.

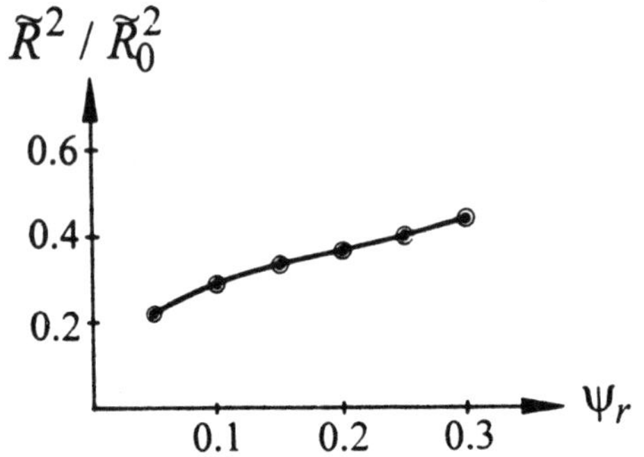

Fig. 3: Ratio of resonant response variances due to a non-linear damping law for various loss factors

PROCEDURES IN THE TIME DOMAIN

General aspects

Ultimate limit states of structures usually include large displacements and strong non-linearities of the structural behaviour. Parametric excitations and stochastic uncertainties in the structural parameters can be included. Nowadays, computer power allows for calculations in the time domain, even if the excitation is multi-dimensional and stochastic, such as the wind loading process. An important feature is, that flow-structure interaction and buffeting response are treated simultaneously. The calculations can be extended over a sufficient time intervall in order to ensure the statistical relevance of the results (Kareem (1987)).

In engineering practice, non-linear systems are mostly solved incrementally. Starting point is the non-linear condition of equilibrium (t), $\mathbf{g}(\ddot{\mathbf{q}},\dot{\mathbf{q}},\mathbf{q}) = \mathbf{f}(t)$, where $\mathbf{g}(\ddot{\mathbf{q}},\dot{\mathbf{q}},\mathbf{q})$ contains all structural properties and the load function $\mathbf{f}(t)$, both the purely aerodynamic forces, depending only on time, and aeroelastic components, self-excited by the current structural response. A suitable incremental iteration algorithm is presented in detail in Beem et al. (1994). The FE-code FEMAS is applicable to physically and geometrically non-linear truss, beam and shell structures, arbitrary stress-strain relations can be treated.

Load vector processes $\mathbf{f}(t)$ are required which can be applied either as nodal forces or nodally-defined element loads. The load vector processes may be obtained from measurements in full scale experiments or in a boundary layer wind tunnel, or ,more universally, from mathematical transformations of artificial wind time histories into loads, or direct generation of load processes.

Wind speed records

For time-history calculations, the loading process may either be obtained from direct load simulations, or by transforming time histories of the wind speeds into loads through an adequate wind load model. One of these is the quasi-stationary model in which is assumed that the stationary aerodynamic coefficients are valid for the fluctuating loads. Although the model is not very meaningful beyond some frequency limit, it has found wider application (e.g. Yang et al. (1981), Borri and Zahlten (1991), Kovacs et al. (1992)), and surprisingly good results can be achieved. One of its merits is, that the flow-structure interaction can be included principally revealing linear and non-linear contributions of aerodynamic damping and stiffness. More sophisticated models are based on the aerodynamic force response produced by a step change of the wind velocity, e.g. Küssner and Wagner functions. However, the first step in any of these approaches is to provide time-histories of the wind speed.

The application of measured wind time histories in full scale experiments suffers from the high experimental effort for recording long enough multi-dimensional wind vector processes at a dense mesh of monitoring points. Moreover, such wind speed records include eg. instationarities, which occur as trends without any visible period on the one hand. Sudden changes of the frequency content cannot be countered with trend removal techniques on the other hand. However, natural wind speed records have been transformed to load histories, (e.g. Lu et. al. (1994)).

An advantageous method to collect wind time histories is the construction of a multivariate random field. Basic generation techniques simulate the fluctuations of a single velocity component by superposing N cosinusoids per frequency f, N is the order of the model, each weighted with respect to the spectral ordinate of the target spectral density function. For multiple wind velocity histories, the Shinozuka II method (Shinozuka and Jan (1972)) yields the requested number of velocity processes, which match the target cross-spectral density functions. Numerically optimized versions of such methods are successfully applied for the analysis of the load-bearing capacity of large structures under buffeting excitation (Kovacs and Andrä (1994)).

It is generally accepted that the probability distribution of turbulence is Gaussian. While the target spectrum is usually well respresented within the procedure mentioned above, it cannot be taken for granted that higher moments than of second match the Gaussian values. The statistical inaccuracy may be misleading in estimating the fatigue damage, since it generally varies in accordance with higher moments (Lutes et. al. (1984)).

The filtering of Gaussian white noise, e.g. by an autoregressive (AR) filter, lead to more accurate statistical properties of the load. AR methods have been adopted for the simulation of wind turbulence (Iwatani (1982)). Extended autoregressive filter types have been applied, eg. ARMA (Maeda and Makino (1992)), for generating wind velocity fields. The wind vector $\underline{u}(t) = [u_1(t), u_2(t), \ldots, u_{2k}(t)]^T$ may consist of the fluctuating parts of the longitudinal $u_1(t), u_2(t), \ldots, u_k(t)$ and orthogonal components $u_{k+1}(t), u_{k+2}(t), \ldots, u_{2k}(t)$, $l=1,2, \ldots, k$ are the identifiers of the points at which the processes are required. These are usually the structural nodes on which aerodynamic loads are imposed. A general 2k-dimensional AR-model is given by

$$u(t) = \sum_{m=1}^{M} A(m) \cdot u(t - m \cdot \Delta t) + z(t) \tag{12}$$

where $\underline{z}(t) = [z_1(t), z_2(t), \ldots, z_{2k}(t)]^T$ is the vector of independent Gaussian white noise processes. A(m) is a 2k x 2k-matrix comprising the coefficients of the AR-model which can be calculated solving the Yule-Walker-equations

$$C(\tau) = \sum_{m=1}^{M} A(m) \cdot C(\tau - m \cdot \Delta t) + D(\tau) \quad , \tau = l \cdot \Delta t,\ l = 0,\ 1,\ \ldots,\ M \tag{13}$$

$\mathbf{C}(\tau)$ is the matrix of covariance functions (CVM), τ is the time lag, M is the depth of regression of the AR-model, and $D(\tau)$ the (diagonal) covariance matrix (CVM) of the white noise processes. When the longitudinal $u_l = u_l'$, $l=1,2,\ldots,k$ and the vertical $u_{k+l} = w_l'$ components of turbulent wind are considered, $\mathbf{C}(\tau)$ is not banded as a CVM of the u' and lateral v' components is, since it follows from the condition of continuity that the u' and w' components are coherent signals. Suitable models for the coherence can be found in Kristensen et al. (1989).

The model-order selection process is rather empirical, since criteria, such as the Akaike-FPE, PAIC, BIC, or AIC unfortunately give contrary predictions (Schrader (1993)). The natural frequencies of linear structures are often found beyond the spectral peak in the inertial subrange. In such cases, model orders of about 3 to 5 can be employed in structural anlysis. The interesting frequency range of non-linear vibration cannot be limited to some range around the eigenfrequencies of the linearized system (e.g. Kahla (1994), Pradlwarter et al. (1991)), the spectral properties of the AR-model should rather match the target cross-spectral density functions down to low frequencies. Higher orders are required then for the basic AR-model, since autoregressive methods suffer from inconvenience for the representation of peaks. (Höffer et al. (1993)) found an excellent fit to the production zone, the peak range, and the inertial subrange of v.Karman target spectra at an order of 90.

Effect of non-linear contributions to the dynamic pressure

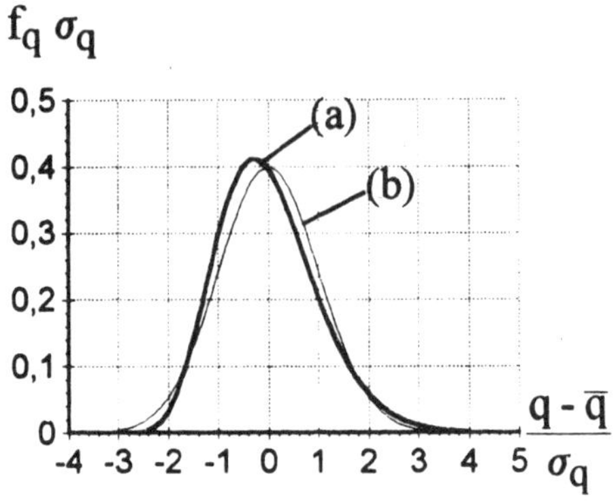

Fig. 4: Normalized PDF of the non-linear (a) and a linearized (b) load q

0.09	0.502	0.303	0.156
0.06	0.581	0.202	0.109
0.03	0.770	0.083	0.051
0.01	1.246	0.016	0.015

Table : Statistic of a SDOF oscillator to non-normal loading (after Peinelt (1989))

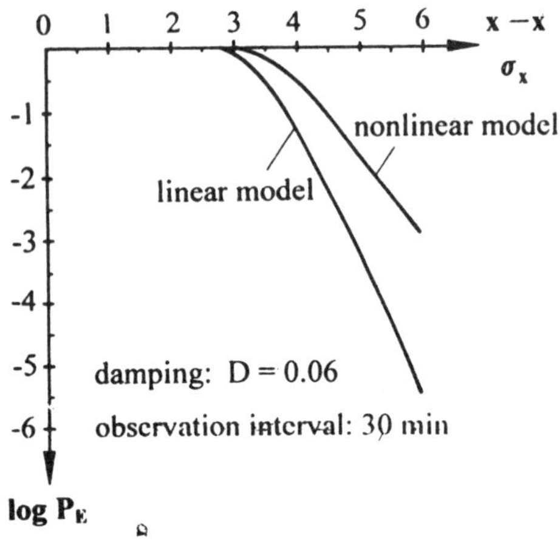

Fig. 5: Exceedance P_e probabilities of the response of a SDOF oscillator under a linearized and non-linear wind force (after Peinelt (1989))

When the buffeting forces are calculated using a quasi-stationary approach, linearizations of the mathematical model are not requested, since the dynamic pressure q(t) can be computed explicitly by

$$q(t) = \frac{\rho}{2}\left[(u + u'(t))^2 + v'^2(t)\right] , (14)$$

ρ is the air density. Eq. 14 yields a non-Gaussian dynamic pressure, since it is non-linear in u' and v', the wind speeds are usually supposed to be Gaussian distributed. In fig. 4 the non-Gaussian shape of the probability density function (PDF) of q after eq. 14 is compared to a linearized approximation, for which the square terms of turbulence are neglected. In the example, the turbulence intensity is 0.172, and Iu/Iv=0.8 is assumed. The exceedance probability of an extreme value of the load increases with the right-sided skewness of the PDF of q. The non-normality of the PDF is specified by the skewness coefficient γ_1 and the kurtosis γ_2. These statistical parameters are defined in the table. Neglecting lateral turbulence, the coefficient of variation (c.o.v.) of q is 0.3368, the skewness is 0.3368, and the kurtosis is 0.3442 for the example mentioned above.

Peinelt (1989) considers the response x of a linear SDOF oscillator under a non-Gaussian load due to eq. 3.3-1. The eigenfrequency of the oscillator is 1 rad/s. The damping ratio is stepwise decreased from 0.09 to 0.01. The variation of damping is found to alter the higher statistical moments (cf. table). For zero damping the response approaches Gaussian properties.

The exceedance probabilities P_e of multiples of the r.m.s. value of the response are compared in fig. 5. Curve (a) denotes the P_e due to the non-linear force, curve (b) represents the exceedance probabilities of a Gaussian load of same c.o.v.. The Gaussian excitation yields considerably smaller P_e for thresholds larger than three times σ_x.

Non-linear dependence of the aerodynamic coefficients on the angle of wind attack

In the frame of the classical quasi-stationary approach an aerodynamic coefficient c is solely a function of the angle of attack of the flow. It is obtained from measurements in smooth flow. The non-linear relation between the coefficient and the angle of wind attack must be known explicitly, in order to provide a function value $c(\varphi)$ for each φ within an interesting range. The precise mathematical incorporation of tabulated pairs of points c-φ into a formula presents little difficulties, e.g. for the representation of a curve with sharp peaks. Singular points of discontinuity in smooth flow set a limit to the general acceptance of the representation as a Taylor-type series, since the differential quotient $dc/d\varphi$ is not defined in any cases. Furthermore, the fluctuations of the angles of attack of the flow typically cover a range of about $\pm 20°$, so that $(dc/d\varphi)\cdot\varphi$ would result in significantly erroneous estimates of the actual coefficients. The consideration of the fluctuation $\Delta(\varphi'(t))$ instead of $(dc/d\varphi)\cdot\varphi'$ is a more exact description (Höffer et al. (1993)). $\Delta(\varphi'(t))$ is the deviation of a momentary coefficient $c(\bar{\varphi} + \varphi'(t))$ from $c(\bar{\varphi})$. To $\Delta(\varphi'(t))$ belongs the PDF $f_\Delta(\Delta)$.

The statistical properties of the non-linear transformed coefficients can be approximated without performing a Monte-Carlo-Simulation. The dynamic pressure and the angle of attack of the flow are assumed to be uncorrelated for this approximation. One obtains $f_\Delta(\Delta)$ from a nonlinear transformation of $f_\varphi(\varphi')$ considering the nonlinear interrelation between Δ and φ' The inverse function $\varphi' = g^{-1}(\Delta)$ may be ambiguous, so that

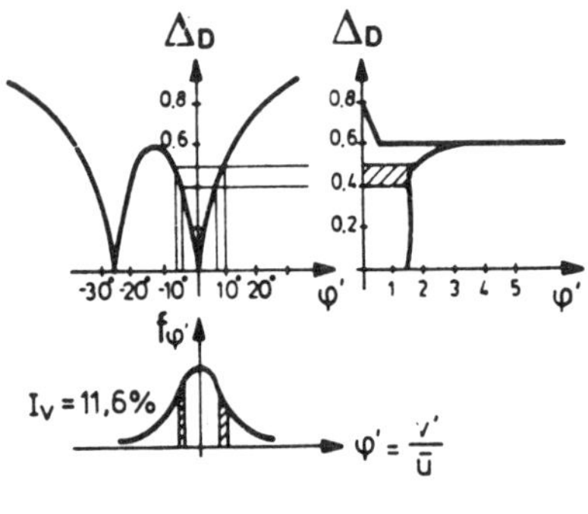

Fig.6: Construction of the turbulence induced increment of force coefficients

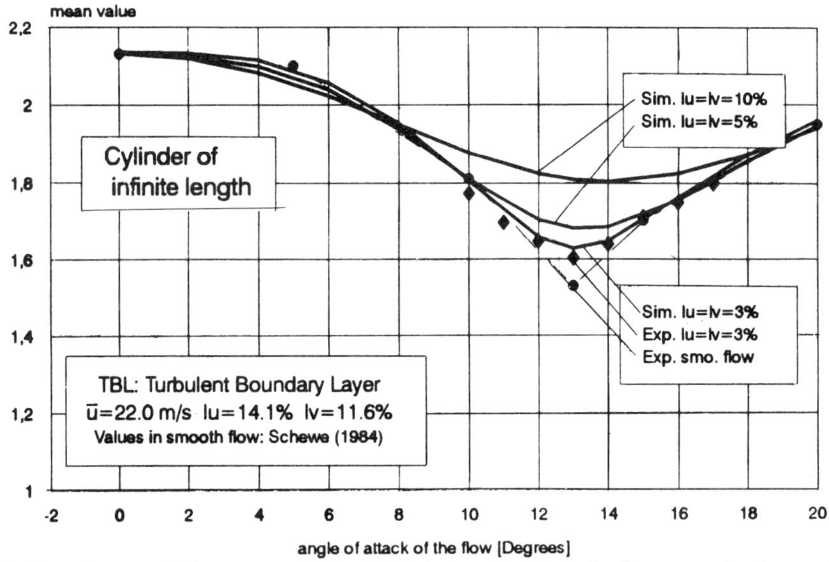

Fig. 7: Drag coefficients on a cylinder of infinite length with quadratic cross-section in smooth and turbulent flows

$$f_\Delta(\Delta) = \sum_n f_\varphi(\varphi'_n) \cdot \left|\frac{d\varphi'}{d\Delta}\right|_{\varphi'_n} \tag{15}$$

here n is the number of ambiguities at Δ and $\left|\frac{d\varphi'}{d\Delta}\right|$ is the Jacobian, here for the case of a scalar transformation. $\bar{\Delta}$ may be computed as the mean of $f_\Delta(\Delta)$ Then, the mean forces are given by

$$\frac{\bar{F}_{D.L}}{b \cdot \bar{q}} = c(\bar{\varphi}) + \bar{\Delta}_{D,L} \tag{16}$$

Experimental verifications show, that even small scales of turbulence shift a mean value (fig. 7). That can be modeled correctly, even for low- and high-turbulent flows.

FLOW - STRUCTURE INTERACTIONS

Stationary aerodynamic coefficients

The general aeroelastic response of buildings vibrating in turbulent wind is described by

$$\mathbf{g}(\ddot{\mathbf{q}}, \dot{\mathbf{q}}, \mathbf{q}) = \mathbf{f}(\ddot{\mathbf{q}}, \dot{\mathbf{q}}, \mathbf{q}, t) \tag{17}$$

The presence of response quantities on the right side of eq. 17 represents aeroelastic feedback, rendering even a linear structure to be non-linear, **f** contains both buffeting and aeroelastic components.

Forces induced by motion play an important role for the dynamic excitation of flexible, light structures, such as masts and chimneys (e.g. Waldeck (1990), long span horizontal structures, especially stayed cable bridges in the erection state (Scanlan (1993a)), and large span roofs (e.g. Daw and Davenport (1987)).

One class of force models that has been applied widely to structures both in smooth and turbulent flow is that based on quasi-stationary theory. The prime feature of that model is the assumption that the external forces acting on a body are not influenced by motion but that motion induces a further force component due to the actual state of deformation of the structure (Blevins (1990)). In the case of a line-like structures, interaction effects are decribed on a cross section with three degrees of freedom. These are the translational displacements in mean wind, x(t), and lateral directions, y(t), and the torsional rotation, θ(t). The overall wind forces arise from a relative flow around the vibrating structure determined from relative velocities

$$u_{rel}(t) = \bar{u} + u'(t) - \dot{x}(t), \quad v_{rel}(t) = v'(t) - y(t) \ . \tag{18a, b}$$

$\dot{x}$, $\dot{y}$ are the time derivatives of the translational displacements x(t) and y(t). The lateral response velocity affects the angle of attack of the wind. The current angle of attack of the flow is

$$\gamma(t) = \arctan\left\{\frac{v_{rel}(t)}{u_{rel}(t)}\right\} - \theta(t) \tag{19}$$

The calculation of an overall wind force acting on the relative flow direction follows

$$f(t)=q_{rel}(t)\ A\ c(\gamma(t)) \ , \tag{20}$$

where the relative dynamic pressure q_{rel} depends on the relative velocity between the incident flow and the structure, A is the reference area, which is b, or b^2 respectively for the aerodynamic torsional moment, and c(γ(t)) stands for an aerodynamic coefficient, which is solely a function of the angle of flow attack in frame of quasi-stationarity.

In connexion with the treatment of structures in the time domain, there is no need to separate between external and motion-correlated components, as long as the right side of eq. 17 contains the buffeting forces and the aeroelastic forces after eq. 20. Such formulation can be characterized as a fully aeroelastic approach. It allows an assessment of the stability of the structural response, yet not the structure itself. The latter problem can be dealt with when the aeroelastic force contributions in eq 17 are linearized and expressed as aeroelastic matrices multiplied by the vector of structural displacements (vector of aerodynamic stiffness forces) and its time derivatives (vectors of viscous aerodynamic damping forces or aerodynamic mass forces). Numerical problems can arise due to the fact that the tangent aeroelastic matrices are non-symmetric, unlike the tangent mass, damping and stiffness matrices of the linearized vector of restoring force **g** (Borri et al. (1995)).

Aerodynamic damping is the significant aeroelastic component which should be accounted for in the buffeting problem of line-like structures. Approximative estimates are usually written as damping ratios

$$D_x = \frac{\rho b^2}{\mu_m} \frac{\bar{u}}{f_x b} \frac{c_D}{4\pi} \text{ and } D_y = \frac{\rho b^2}{\mu_m} \frac{\bar{u}}{f_y b} \frac{c_D + dc_L / d\varphi}{8\pi} \qquad (21a, b)$$

and are derived by linearizing eq. 20 and separating the total force into buffeting and motion-correlated components. Eq.s 21a, b contain the time-average values of the aerodynamic damping ratios. The lateral aerodynamic damping contributions $dc_l/d\varphi$ can add to and substract from structural damping. That may lead to negative total damping, which then cause large amplitude oscillations and in the extreme case lead to aerodynamic instability.

Non-stationary buffeting coefficients

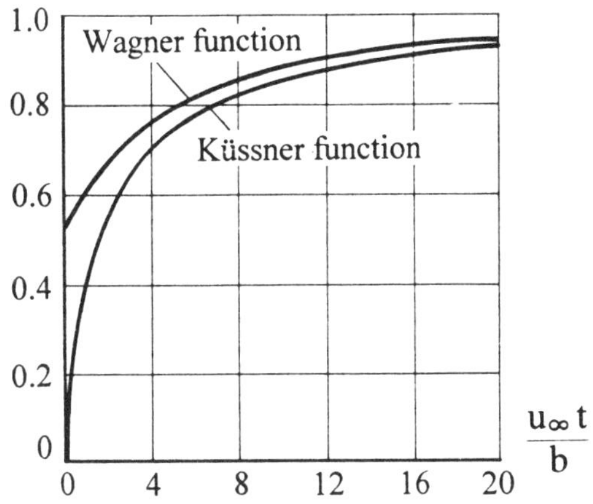

Fig. 8: Küssner and Wagner function (after Försching (1984))

Beyond some frequency limit quasi-stationarity is an inadequate representation of the variation of forces in time. The instationarities in the aerodynamic coefficients can be accounted for by indicial response functions. Indicial response functions incorporate the effects of time lag in development of aerodynamic forces behind the time histories u'(t), w'(t). In Scanlan (1993b) the indicial response functions are presented for drag, lift and aerodynamic moment on bridge decks and linked to the Wagner and the Küssner functions. The Wagner function (cf. fig. 8) accounts for the lag of lift on a flat plate initiated by a step function change of the angle of flow attack, but it is supposed (Scanlan (1993b)) that it also can treat a step function change u'(s). A vertical gust w'(s) and, analogeously, a vertical displacement z(s) of the bridge deck requires a Küssner-type function. The approach works in both the frequency and the time domain.

CONCLUSIONS

The presented review gives an overview about the buffeting analysis of non-linear structures in the ultimate limit state, where structural non-linearities are dominant. Contributions coming from structure-flow interactions are also considered. The following conclusion can be drawn:

- In general, the linear approach ensures a conservative design since the realistic, i.e. non-linear behaviour gives rise to additional load bearing capacities. However, this is not always true since softening of the structure may shift the dominant frequencies in the low frequency, high energy range of the load spectral density.
- Equivalent linearization is an attempt to maintain the benefits of the linear theory and at the same time to include the non-linear structural properties in terms of tangent restoring forces.
- Time history calculations using incremental-iterative algorithms can account for arbitrary non-linearities of restoring forces, interactive external forces. The buffeting loads may in this case be represented by AR or other simulation procedures in the time domain with good accuracy. Time domain approaches are therefore becoming increasingly important.

ACKNOWLEDGEMENT

Part of the present work was performed within of the research programme "Dynamics of Structures" at the Ruhr-Universität Bochum, supported by the "Deutsche Forschungsgemeinschaft". This support is gratefully acknowledged.

REFERENCES

Atalik, T.S. and Utku, T. (1976), "Stochastic linearization of multi-degree-of-freedom non-linear systems", Earthquake Engng. Struct. Dynamics, Vol. 4, pp. 411-420.

Beem, H. et al. (1994), " FEMAS90 - Finite element modules of arbitrary structures, User's manual", Institute for Statics and Dynamics, Dept. Civ. Eng., Ruhr-Universität Bochum, Germany.

Blevins, R.D. (1990), "Flow-induced vibrations", Van Nostrand Reinhold, New York.

Borri, C., Höffer, R., Zahlten, W. (1995) "A non-linear approach for evaluating simultanous buffeting and aerolastic effects on bridge decks", 9th ICWE '95, New Dehli, to be published.

Borri, C., Zahlten, W. (1991), "Fully Simulated Nonlinear Analysis of Large Structures Subjected to Turbulent Artificial Wind", Mech. Struct. & Mach., 19(2), pp. 213-250.

Bouc, R. (1967) "Forced vibrations of mechanical systems with hysteresis", Abstract, Proc. 4th Inter. Conf. Nonl. Osc., Prague, Czechoslovakia.

Cai, Y. and Chen, S.S. (1994), "Dynamics of elastic cables under parametric and external resonance", J. Eng. Mech., Vol. 120, No. 8.

Chen, Y., Ahmadi, G. (1992), "Wind effects on base isolated structures", J. Eng. Mech., Vol. 118, No. 8.

Comité International des Chimnées Industielle CICIND (1987), "Commentaries for the Model Code for Concrete Chimneys - Part A: The Shell", CICIND, 136 North Street, Brighton BN1R6, England.

Daw, D.J. and Davenport, A.G. (1987), "Aerodynamic damping and stiffness of a semi-circular roof in turbulent wind", Advances in Wind Engineering, Proc. of the ICWE, Aachen, July 6-10, 1987.

Försching, H.W. (1974), "Grundlagen der Aeroelastik", Springer-Verlag, Berlin.

Gerstoft, P. (1986), "A simplified method for dynamic analysis of a guyed mast", JWEIA, 23 (1986), pp.487.

Höffer, R., Niemann, H.-J.; Lou, J.X. (1993), "Modeling of artificial stochastic wind loads and interactions forces on oscillating line-like structures", Proc. of ICOSSAR '93, Innsbruck, Austria, 9.-13. August 1993, Balkema, Rotterdam.

Iwatani, Y. (1982), "Simulation of multidimensional wind fluctuations having any arbitrary power spectra and cross spectra" (auf japanisch), Journal of Wind Engineering, Japan Association for Wind Engineering, Vol. 11, pp. 5-15.

Kahla, N.B. (1994),"Dynamic analysis of guyed towers", Engng. Struct., Vol. 16, No. 4, pp. 293-301.

Kareem, A. (1987), "Wind effects of structures; a probabilistic viewpoint", Prob. Eng. Mech., Vol. 2, No. 4, pp. 166.

Kovacs, I., Andrä, H.-P. (1993), "Traglastnachweis von Turmbauwerken unterdynamischer Windbelastung", Bautechnik 70, Heft 11, S. 650-661, Verlag Ernst & Sohn.

Kovacs, I., Svensson, H.S.; Jordet, E. (1992), "Analytical aerodynamic investigation of cable-stayed Helgeland Bridge", J. Struct. Eng., Vol. 118, No. 1, January.

Kristensen, L., Lenschow, D.H.; Kirkegaard, P.; Courtney, M. (1989), "The spectral velocity tensor for homogeneous boundary-layer-turbulence", Boundary-Layer Meteorology 47, pp. 149-301.

Lutes, L.D. et al. (1984), "Stochastic fatigue damage accumulation", J. Struct. Eng., Vol. 110, No. 11, November.

Lu, W., Qian, J.; Liao, W. (1994), "Hyperbolic Cooling Tower Non-Linear Dynamic Response to Wind", East European Conference on Wind Engineering, 4-8 July 1994, Warsaw, Poland.

Maeda, J. and Makino, M. (1992), "Characteristics of gusty winds simulated by an A.R.M.A. model", JWEIA, 41-44, pp. 427.

Niemann, H.-J. (1990), "Dynamic response of cantilevered structures to wind turbulence", Structural Dynamics, Krätzig et. al. (eds.), Balkema, Rotterdam.

Peil, U. and Nölle, H. (1993), "On fatigue of guyed masts to wind load", Proc. of ICOSSAR '93, Innsbruck, Austria, 9.-13. August 1993, Balkema, Rotterdam.

Peinelt, R.H. (1989), "Tragwerksschwingungen unter nicht-Gauß'scher Windlast", in: Berz, G. (Hersg.), Konzepte und Anwendungen von Windlastnormen, pp. 437, WTG-Publikation Nr. 1, Aachen.

Petersen, C. (1971), "Aerodynamische und seismische Einflüsse auf die Schwingungen insbesondere schlanker Bauwerke", Fortschritt-Berichte der VDI-Zeitschriften Reihe II, Nr. II.

Petersen, C. (1992), "Chaotische Taumelschwingungen abgespannter Maste", Stahlbau 61, H.6, pp. 179, Verlag Ernst & Sohn, Berlin.

Pradlwarter, H.J. (1990), "On the Existence of 'true' equivalent linear systems for the evaluation of the nonlinear stochastic response due to nonstationary Gaussian excitation", Report 28-90, Institute of Engineering Mechanics, University Innsbruck.

Pradlwarter, H.J. et. al. (1991), "Design of industrial R/C chimneys", in: Schuëller, G.J. (ed.), Structural Dynamics, pp. 438, Springer-Verlag, Berlin.

Pradlwarter, H.J., Schuëller, G.I., Chen, X.-W. (1988), "Consideration of non-Gaussian response properties by use of stochastic equivalent linearization", Proc. Third International Conference on Recent Advances in Structural Dynamics, 18.-22. July 1988, Southampton, Wright-Patterson Air Force, pp. 737-752.

Roberts, J.B. and Spanos, P.D. (1991), "Random Vibrations and Statistical Linearization", John Wiley, New York.

Scanlan, R.H. (1993a), " Bridge buffeting by skew winds in erection states", J. Eng. Mech., Vol. 119, No. 2, February.

Scanlan, R.H. (1993b), "Problematics in Formulation of Wind-Force Models for Bridge Decks", J. Mech., Vol. 119, No.7, July.

Schewe, G. (1984), "Untersuchung der aerodynamischen Kräfte, die auf stumpfe Profile bei großen Reynolds-Zahlen wirken", Institut für Aeroelastik, DFVLR-Mitt.84-19, Göttingen.

Schrader, P. (1993), "Die statistische Stabilität gemessener integraler Längenmaße und anderer Windparameter", Dissertation, Fakultät für Bauingenieurwesen, Ruhr-Universität Bochum, Bochum, Germany.

Schuëller, G.I. (ed.) (1991),"Structural dynamics-Recent advances", Springer Verlag, Berlin.

Shinozuka, M. and Jan, C.M. (1972), "Digital simulation of random processes and application", J. Sound Vibr. 25(1), pp. 111-128.

Tonis, D. (1989), "Dynamisches Verhalten von Abspannseilen", in: Berz, G. (Hersg.), Konzepte und Anwendungen von Windlastnormen, WTG-Publikation Nr. 1, Aachen.

Waldeck, J.L. (1990), "Measurement and analysis of the dynamic effects of atmospheric wind on a 300 m high concrete chimney - Final Report", Internal Report 90/02, CSIR Building Technology, Pretoria, June 1990.

Wen, Y.K. (1976), "Method for random vibration of hysteric systems", J. Eng. Mech., Vol. 47, pp.150.

Yang, J.N., Lin, Y.K.; Samali, B. (1981), "Coupled Motion of Wind-Loaded Multi-Story Building", J. Eng. Mech., Vol. 107, No. EM6, December.

STATE OF THE ART VOLUME

WIND DAMAGE STUDIES

JAMES R. McDONALD
Civil Engineering Department,
Texas Tech. University,
Lubbock, TX 70409-1023, USA

SUMMARY

In the past 25 years, much as been learned about severe windstorms and their interaction with the built environment. Careful documentation of wind damage after a storm and follow-up analysis of the collected data reveals much information about the nature of the windfield and the performance of various structures and components.

INTRODUCTION

In the past 25 years, much has been learned about severe windstorms and their interaction with the built environment. The objective of this paper is to trace the history of wind damage studies, to identify significant lessons learned from the damage, and to identify needs for future research. Tornadoes and hurricanes cause more injuries, deaths and property damage than any other types of storms in the United States. Discussions herein are primarily restricted to these two weather events.

Tornadoes

Tornadoes are one of nature's most violent windstorms. Although they occur in other parts of the world (Australia, Japan, Eastern Europe), weather conditions are most favorable in the United States for tornado genesis. Cool, dry air moving across the Rocky Mountains overruns warm moist air from the Gulf of Mexico to create unstable conditions that produce severe thunderstorms. Under conditions that are not entirely understood, tornadoes extend from the cloud base to the ground and create a path of damage and destruction . Tornadoes vary greatly in intensity and size.

Maximum wind speeds in tornadoes were once believed to be 200-225 m/s (Golden, 1976). Because tornado wind speeds are almost impossible to measure, indirect methods have been relied upon for tornado wind speed estimates. Methods that have been used include: (1) photogrammetric analysis of tornado films, (2) analysis of cycloidal ground marks, (3) the appearance of damage (Fujita-Scale), and (4) analysis of structures located in the tornado path.

Lacking a means of direct measurement of tornado wind speeds, Dr. Ted Fujita at the University of Chicago introduced a scheme for rating tornado intensity based on appearance of damage (Fujita, 1971). The Fujita-Scale contains six intensity categories ranging from F0 to F5. F0 indicates a very weak tornado, while F5 suggests a very intense one. Fujita-Scale ratings are assigned by observing the most severe damage in a tornado path and comparing the damage to word descriptions or typical photographs of the Fujita-Scale damage categories. Table 1 contains the wind speed ranges and word descriptions of the six categories. The wind speeds are interpreted as peak gust (2 sec) at 10 m height in flat, open terrain. An advantage of the Fujita-Scale was that tornadoes in the historical tornado records prior to 1971 could be classified from word descriptions of the damage.

The Fujita-Scale has never been thoroughly calibrated. The F0 to F3 wind speeds are thought to be about right, but F4 and F5 categories generally overestimate the wind speeds to cause the damage associated with each category (Minor et al., 1977). Even with these limitations, the Fujita-Scale is the accepted way to estimate tornado intensity.

Structural analysis of buildings or other structures which were located in the tornado damage path is an effective way of estimating wind speeds, if carefully applied (Mehta et al., 1976). The wind pressures or forces required to cause the observed damage are first determined. The wind speeds to produce the wind pressures or forces are then calculated. Because the results can sometimes be misleading, Mehta (1976) developed a *credence level*, whereby accuracy of the wind speed estimate could be judged. Analysis of damaged structures produced the

TABLE 1. F-SCALE CLASSIFICATION OF TORNADOES BASED ON DAMAGE

Scale	Damage	Wind speed
(F0)	LIGHT DAMAGE	40-72 mph
	This speed range corresponds to Beaufort 9 through 11. Some damage to chimneys or TV antennae; branches broken off trees; shallow-rooted trees pushed over; old, hollow trees break or fall; sign boards damaged.	
(F1)	MODERATE DAMAGE	73-112 mph
	73 mph is beginning of hurricane wind speed (Beaufort 12). Surface peeled off roots; windows broken; trailer houses pushed or overturned; trees on soft ground uprooted; some trees snapped; moving autos pushed off road.	
(F2)	CONSIDERABLE DAMAGE	113-157 mph
	Roofs torn off frame houses (strong upright walls left standing); weak structure or outbuildings demolished; trailer houses demolished; railroad boxcars pushed over; large trees snapped or uprooted; light-object missiles generated; cars blown off highway; block structures and walls badly damaged.	
(F3)	SEVERE DAMAGE	158-206 mph
	Roofs and some walls torn off well-constructed frame houses; some rural buildings completely demolished or flattened; trains overturned; steel-framed hangar/warehouse type structures torn; cars lifted off the ground (may roll some distance); most trees in a forest uprooted, snapped or leveled; block structures often leveled.	
(F4)	DEVASTATING DAMAGE	207-260 mph
	Well-constructed frame houses leveled, leaving piles of debris; structures with weak foundations lifted, torn and blown off some distance; trees debarked by small flying debris; sand soil eroded and gravel flies in high winds; cars thrown some distance or rolled considerable distance, finally to disintegrate; large missiles generated.	
(F5)	INCREDIBLE DAMAGE	261-318 mph
	Strong frame houses lifted clear off foundation and carried considerable distance to disintegrate; steel-reinforced concrete structures badly damaged; automobile-sized missiles carried a distance of 100 yds or more; trees debarked completely; incredible phenomena can occur.	

first evidence that maximum wind speeds in tornadoes were not as large as once believed.

The general consensus today is that the highest tornado wind speeds are not likely to exceed 110 to 135 m/s. With the installation of more than 150 NEXRAD WSR-

88D Doppler radars in the United States by the end of 1996, direct measurements of tornado wind speeds will soon be possible.

The distribution of tornadoes by Fujita-Scale classification is shown in Figure 1. Clearly, the vast majority of tornadoes falls in the three weakest categories. This fact is very significant, because structures can be designed with conventional materials to resist the effects of tornadoes classified less than or equal to F2 without resorting to concrete bunkers or underground buildings.

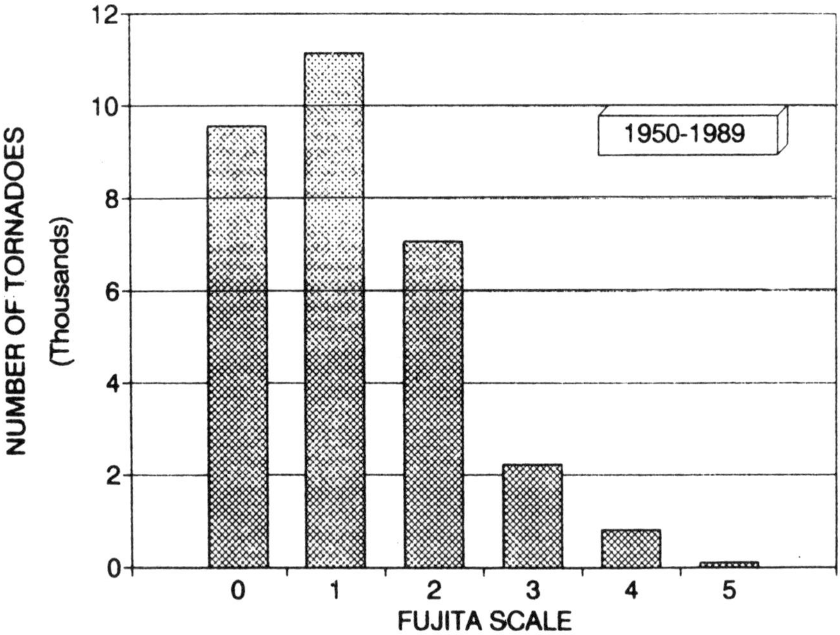

FIGURE 1. DISTRIBUTION OF TORNADOES BY FUJITA-SCALE IN THE UNITED STATES

Hurricanes

In the North American region, a hurricane is a severe tropical cyclone which develops over the warm North Atlantic Ocean, Caribbean Sea, Gulf of Mexico or Eastern North Pacific.

The intensities of hurricanes are rated according to the Saffir-Simpson Potential Scale (see Table 2). The Saffir-Simpson Scale places hurricanes in one of five intensity categories ranging from Category 1 (minimal) to Category 5 (catastrophic) based on wind speeds and central atmospheric pressure. Wind speeds and pressure measurements are obtained by aircraft penetrating the storm at an altitude of 3,000 m or more. The Saffir-Simpson Potential Scale is used primarily for issuing hurricane watches and warnings. Wind speeds are interpreted as "sustained" (one-minute winds) at 10 m height in flat, open terrain. Thus, wind speeds in the Fujita-Scale and Saffir-Simpson scale do not have the same averaging time.

TABLE 2. SAFFIR/SIMPSON DAMAGE POTENTIAL SCALE RANGES

Category	Central Pressure mb (in. Hg)	Winds (mph)	Surge (ft)	Damage
1	≥980 (28.94)	74-95	4-5	Minimal
2	965-979 (28.50-28.91)	96-110	6-8	Moderate
3	945-964 (27.91-28.47)	111-130	9-12	Extensive
4	920-944 (27.17-27.88)	131-155	13-18	Extreme
5	<920 (27.17)	>155	>18	Catastrophic

The size of a hurricane may be expressed in terms of the diameter of hurricane winds (33 m/s or larger) or the diameter of gale force winds (18 m/s). A typical hurricane will have hurricane force winds with a 160 km in diameter, and gale force winds with a 570-640 km diameter, although there is great variability from these values.

Wind Damage Studies Prior to 1970

Prior to 1970, very few examples of wind damage studies from an engineering perspective can be found. Damage was photographed and described in words, but little or no analysis was performed. Segner (1960) made one of the first estimates of tornado wind forces to cause structural damage from his study of the Dallas, Texas Tornado of 1957. Hurricane Camille was studied by personnel at the National Bureau of Standards (Dikkers et al., 1970). Dr. Ted Fujita (1992) used aerial photos of tornado damage to deduce wind field characteristics of the Fargo, North Dakota tornado of 1957 and the Palm Sunday tornadoes of 1965.

Lubbock Tornado of May 11, 1970

A devastating tornado rated F5 on the Fujita-Scale destroyed about one-eighth of the City of Lubbock, Texas (see Fig. 2). The $135 million damage was the largest on record to that point in time. The storm damage attracted the attention of meteorologists and engineers from a number of organizations around the country. Dr. Fujita (1970) was interested in verifying his theory of satellite vortices and found evidence of the phenomena in the damage patterns left by the storm. The National Academy of Engineering sent a team of engineers, headed by Dr. J. Neiles Thompson from the University of Texas at Austin, to document and evaluate the damage (Thompson et al., 1970). Faculty members from the College of Engineering at Texas Tech University organized quickly to document the damage through photographs and recorded observations.

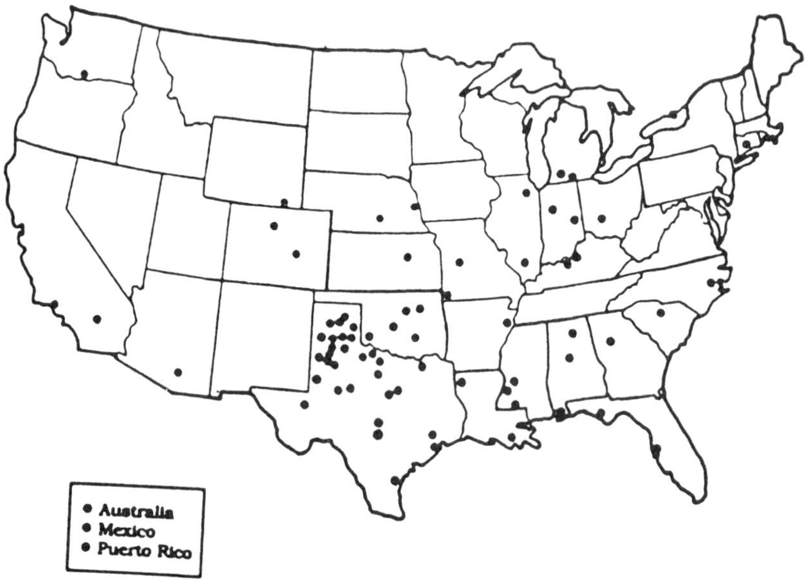

FIGURE 2. LOCATIONS OF IDR DAMAGE DOCUMENTATIONS (1970-1994)

The National Science Foundation (NSF) recognized an opportunity to learn from the Lubbock Tornado. Supported by NSF funding, the Texas Tech faculty group subsequently performed detailed engineering analyses on 98 structures affected by the storm. The document, published by Mehta et al. (1971), is the first comprehensive study of tornado damage from an engineering point of view.

Hurricane Celia, Corpus Christi, Texas

On August 30, 1970, Hurricane Celia struck the city of Corpus Christi and Padre Island, causing significant damage. The Texas Tech University faculty group dispatched a team to investigate the damage. Minor et al. (1972) observed surprising similarities in the damage caused by Hurricane Celia and the Lubbock tornado. Air flowing over a building (bluff body) created outward-acting wind pressures on all surfaces except the windward wall, which is inward-acting. Broken windows or other openings in the building envelope allowed air to flow into or out of the building, creating positive or negative internal pressures, respectively.

Nuclear Power Plant Design Criteria

In the late 1960s and 1970s, the nuclear power industry was expanding rapidly in the United States. The United States Nuclear Regulatory Commission (USNRC) established tornado design criteria for nuclear power plants based on federal regulations. The criteria provided for a 160 m/s design wind speed, a 21 kPa atmospheric pressure change and a set of monster tornado missiles (USNRC, 1974; USNRC, 1975). This criteria affected the design of all facilities critical to safe shutdown of the nuclear reactor.

Much of the research involving tornado damage documentation was directed toward justification of a less conservative tornado design criteria for nuclear power plants. Probability assessments of core melt-down caused by tornadoes consistently estimated exceedence probabilities less than 1×10^{-7} per year. Although the Nuclear Regulatory Commission refused to reduce the very conservative criteria, tornado technology, from an engineering perspective, is much better understood today because of research conducted for the USNRC.

Recent Hurricane Experiences

Hurricanes Hugo in 1989 and Andrew in 1992 significantly affected the attitudes of government, the insurance industry and the general public. Losses were estimated at $7 billion in Hugo and $30 billion in Andrew. These losses are intolerable to the insurance industry and the general economy of coastal states. The damage from these two storms called attention to future potential hurricane damage because of inadequate existing construction along the Atlantic and Gulf Coasts.

Because of potential losses to public and private property, attention has been directed to performance of various classes of structures, building code provisions and construction practices. Each of these issues relies on damage documentation to obtain a clear understanding of building performance in high winds.

TWENTY-FIVE YEARS OF WIND DAMAGE STUDIES

The Lubbock tornado of 1970 sparked a new interest in extreme windstorm damage investigations. Organizations involved in windstorm damage investigations and several significant storms of the last 25 years are reviewed in this section.

Research Organizations

The University of Chicago

The work of Dr. Ted Fujita and his group at the University of Chicago immediately comes to mind. Although Dr. Fujita is more interested in understanding the tornado phenomenon itself, rather than the performance of buildings, the extensive data collected by him forms a valuable resource for both meteorologists and engineers.

In order to understand wind-structure interaction, the nature of the tornado wind flow fields must be understood. Dr. Fujita has done a superb job of sorting through damage and debris patterns to deduce the tornado wind flow patterns. Using the Fujita-Scale and aerial photos of damage, he deduces tornado intensity and wind flow directions across and along the damage path. These wind field analyses are an important in evaluating damage to buildings. An example of Dr. Fujita's work is shown in Figure 3.

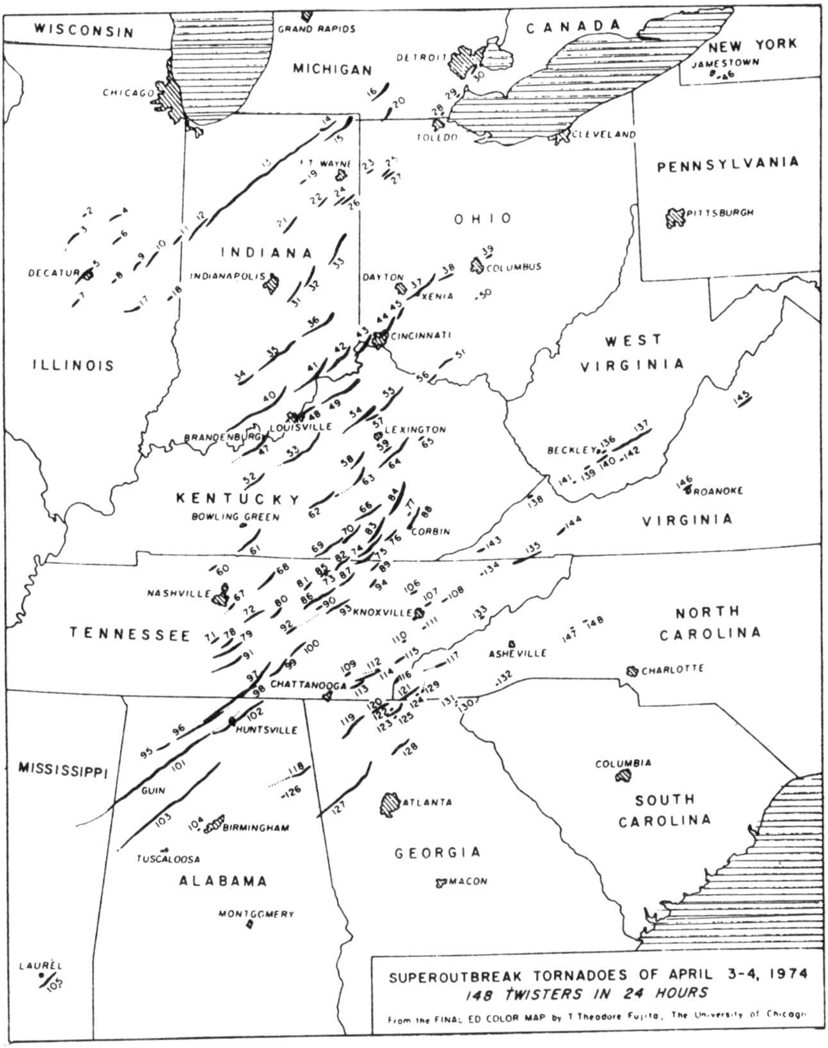

FIGURE 3. SUPEROUTBREAK TORNADOES OF APRIL 3-4, 1974

Tornado wind field models are another important contribution of Dr. Fujita (1978). He proposed models in 1977 and 1978 for engineering applications. DBT-77 is a single-vortex model, whereas DBT-78 is a multiple-vortex model. The models are useful in assessing wind field flow patterns over an observed damage path and can also be used for tornado missile trajectory simulations.

Texas Tech University

In 1972 the faculty group at Texas Tech University organized the Institute for Disaster Research (IDR). Although originally intended to deal with multiple hazards, the major emphasis over the years has been on wind. Following studies of the Lubbock Tornado and Hurricane Celia, IDR personnel continued to document windstorm damage at every opportunity from tornadoes, hurricanes and other extreme windstorms. To date, IDR has documented more than 70 windstorm events in the United States, Canada, Mexico, Puerto Rico, and Australia. Figure 2 shows locations of wind damage investigations by IDR personnel.

The wind damage archives at Texas Tech contain more than 10,000 slides and photographs, as well as other data. Over the years, the data have been analyzed to draw many conclusions relating to building performance in windstorms.

Clemson University

The wind engineering group at Clemson University also conducts windstorm damage investigations. They have studied both tornadoes and hurricanes over the last 10-15 years. The group compares the observed building damage to code requirements for wind loads in order to gage the effectiveness of the building code provisions. Correlation of property losses to measured or estimated wind speeds is another interest of the Clemson group. In recent years, the Clemson group has made good use of wind tunnel simulation has helped to better understand damage observed in the field.

Other groups

Other groups, agencies, organizations and associations conduct periodic damage investigations when the effort is consistent with their mission. Table 3 lists some of these organizations. Results of their studies are usually published in the technical literature or trade publications.

TABLE 3. OTHER AGENCIES CONDUCTING DAMAGE DOCUMENTATION

Agency or Group	Activities
National Severe Storms Forecast Center	Doppler radar signatures from tornadoes
University of Oklahoma	Tornado chasers
University of Florida	Hurricane damage, code compliance
University of Miami	Hurricane damage, effect of debris impact
Iowa State University	Effects of tornadoes on buildings and structures
Texas A & M University	Tornado and hurricane damage, code compliance
Lawrence Institute	Identify safe places for occupant protection
Ball State University	Tornado damage
University of Missouri-Rolla	Hurricane missile impact research
Southern Building Code Congress International, Inc.	Effectiveness of building codes, code compliance
International Congress of Building Officials	Effectiveness of building codes
Florida Concrete Products Association	Response of residential buildings in high winds

Significant Wind Storm Events

In addition to the windstorms already mentioned (Lubbock t and Hurricane Celia), a number of other tornadoes and hurricanes produced damage from which significant lessons were learned. Tornadoes are discussed first; then the hurricanes.

Wichita Falls, Texas Tornado; April 9, 1979

The path of this massive tornado was more than 2 km wide. The storm passed primarily through a residential area of the city. Houses, schools and light commercial construction were damaged.

McDonald et al. (1981) observed that people seeking shelter in small rooms or closets, survived. Glass et al. (1980) observed that half of the 46 deaths occurred in automobiles. The concept of an in-residence shelter was being developed at Texas Tech University (Kiesling and Goolsby, 1974). The Wichita Falls storm demonstrated the need for a strengthened small room or closet inside the house to withstand wind pressures and missile impacts.

Superoutbreak Tornadoes of April 3-4, 1974

A total of 147 tornadoes occurred within a 24-hour period spanning April 3 and 4, 1974. Figure 3 is a map prepared by Fujita (1975) showing the location of the tornado paths. This largest of outbreaks on record produced storms in all Fujita Scale categories. From the aerial surveys conducted by Dr. Fujita and others, the variations of Fujita Scale intensity along the length of the damage paths were determined. Knowledge of these variations for the different Fujita-Scale categories is needed for tornado hazard assessment.

Bossier City, Louisiana Tornado; December 1978

This F4 tornado touched down on the east bank of the Red River and proceeded through Bossier City. The tornado produced some of the most dramatic tornado missiles ever observed. Six steel wide-flange beams that supported the roof of Meadowview Elementary School were picked up by the tornado and transported various distances. One traveled 150 m and penetrated 2.4 m into the soft ground (see Fig. 4). Another traveled more than 330 m and perforated the roof of a house.

A cooperative effort between Fujita and McDonald produced a very accurate wind field definition over the school building by comparing damage observed in aerial and ground level photos. The tornado windfield was modeled by DBT-77. Tornado trajectory simulations using the tornado wind field model accurately predicted the impact points of the six missiles as shown in Figure 5 (McDonald, 1981).

FIGURE 4. STEEL WIND-FLANGE BEAM WEIGHING 326 kg PENETRATED 2.4 m INTO GROUND

Grand Gulf, Mississippi Tornado

The Grand Gulf, Mississippi tornado is the only one known to pass directly over a nuclear power plant (Fujita et al., 1978). The tornado intensity was rated F3, which is nowhere near the 160 m/s design wind speed. Since the plant was under construction, collapsing construction cranes caused much of the damage. Light gage metal warehouses, not designed to tornado criteria, also were heavily damaged. The most interesting damage occurred when the tornado passed directly over a hyperbolic cooling tower (see Fig. 6). The tower was complete to the 137 m level with another 23 m to go. The tower was approximately the same diameter as the tornado funnel. The tornado passed directly over the cooling tower, collapsed a crane inside the tower, and then reformed its vortex on the other side. A large segment of the concrete cooling tower shell was knocked out by the crane, but the tower did not collapse.

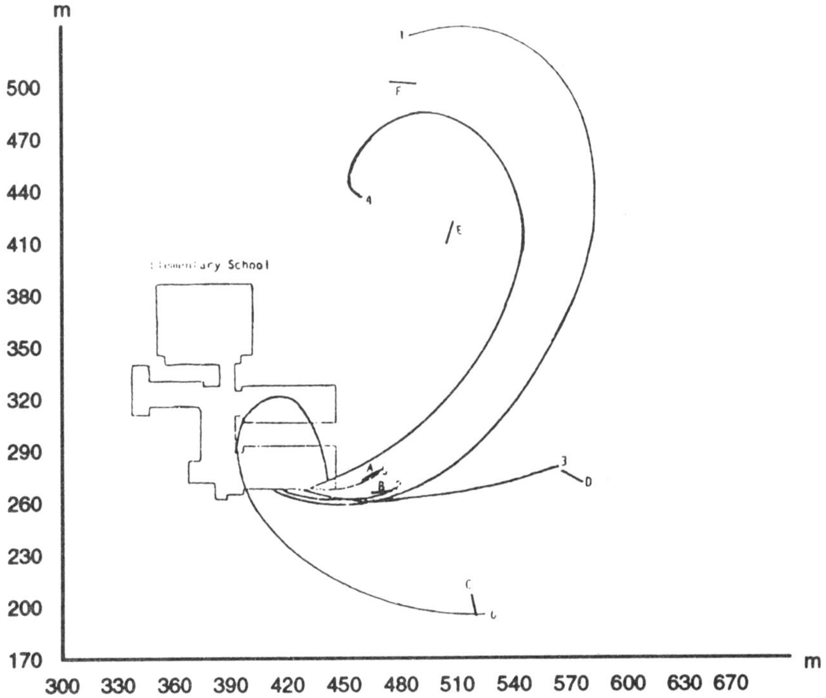

FIGURE 5. COMPUTER SIMULATION OF STEEL WIDE-FLANGE BEAMS IN BOSSIER CITY, LOUISIANA TORNADO

FIGURE 6. TORNADO DAMAGE TO COOLING TOWER AT GRAND GULF NUCLEAR POWER PLANT

Cyclone Tracy; Darwin, Australia

Cyclone Tracy virtually destroyed the city of Darwin. Houses were severely damaged, because the hurricane-force winds blew for more than four hours. A number of lessons were learned in the Darwin experience (Walker, 1975): (1) roof fastenings of sheet metal roof cladding failed in fatigue, (2) roofs were not adequately tied to top of walls and the load path from wall to foundation was not continuous, and (3) "high set" houses were unstable because of inadequate bracing.

These findings have been observed in numerous other damage experiences. The significance of the Australian experience is that the wind engineering community recognized the problems and immedicately did something about them. Research and testing was performed, codes were improved and enforced. As a result, significant increases in the wind resistance of buildings and structures have been realized in the last 20 years in Australia. Although the knowledge is there, similar actions have not taken place in the United States.

Hurricane Fredric

Hurricane Fredric presented an opportunity to compare observed damage to buildings and structures with wind speeds measured by anemometers in the area. The study by Mehta et al. (1983) provided a valuable calibration of wind speed and damage.

Hurricane Alicia

Hurricane Alicia crossed Galveston Island, Texas, causing extensive damage to residential and light commercial construction. The storm continued northward toward downtown Houston. Minor (1985) reported that gravel blown from the roofs of adjacent buildings caused extensive breakage of window glass in high-rise buildings (see Fig. 7). This experience and others have led code bodies in the United States to consider missile impact criteria for the design of buildings in hurricane-prone areas.

FIGURE 7. WINDOW GLASS BREAKAGE FROM FLYING DEBRIS IN HURRICANE ALICIA

Hurricane Hugo

Hurricane Hugo was a large powerful storm (Category 4) that hit the South Carolina coast near Charleston. The storm caused extensive wind and water damage to residential developments on barrier islands and the mainland.

The hurricane winds primarily affected roofs and roofing systems. Asphalt shingles, built-up roofs and single-ply membrane roofs all experienced considerable damage. McDonald and Smith (1990) observed that roof damage allowed rain and water to get inside the buildings and cause additional damage to property and interior finishes. Hurricane Hugo emphasized the need for improvement in roofing systems and the test methods for certifying their performance in high winds.

Post-storm investigations by Sparks (1990) revealed that many communities had no building codes in force and those communities that had codes, had poor records of enforcement. Follow-up studies showed that buildings were repaired or rebuilt using the same inadequate construction practices as before. Hurricane Hugo attracted a lot of attention because of the large dollar losses, but little was done to develop mitigation programs.

Hurricane Andrew

Hurricane Andrew was not as intense nor as large as Hurricane Hugo, but caused more than three times the dollar damage. Total loss is estimated at $25 billion. Insured losses reached $16 billion. Because the storm came ashore south of Miami, the damage was primarily to residences and light commercial buildings. Nine thousand manufactured homes were destroyed in South Florida.

Hurricane Andrew produced very little storm surge. Virtually all damage was caused by wind. Rain in the days and weeks after the storm increased the loss of personal property in houses and buildings that could not quickly be made waterproof.

The South Florida Building Code (SFBC) was thought to be the most stringent code in the country. Hurricane Andrew revealed a number of weaknesses in the SFBC, including inadequate design wind pressures at eaves, ridges and roof corners, lower effective design wind speeds than specified, poor practices of attaching asphalt shingles and tiles and allowing use of inappropriate building materials such as fiber board on exterior walls (Zollo, 1993).

Poor enforcement of the code led to additional weaknesses in construction. A study by Cook (1993) suggested that poor workmanship, short cuts and use of inferior quality materials all contributed to poor wind resistance of buildings and structures affected by the storm.

The South Florida Building Code has stringent requirements for strapping roof trusses to the top of the wall. These provisions were followed in the vast majority of cases. The practice kept roof structures in place even though roofing and roof deck might have been stripped off. By maintaining the roof integrity, most walls remained standing. The incredibly small number of deaths is attributed to the fact that buildings did not collapse (FEMA, 1993). Concrete masonry construction required in single-story residences also performed very well.

Minor and Behr (1993) determined that missiles and flying debris broke windows and perforated roof and wall cladding. The openings let in wind and water, which caused additional damage to property and interior finishes. Increase in internal pressure, owing to the openings, produced additional wind pressures on walls and roofs, sometimes creating additional structural damage.

The tremendous losses in Hurricane Andrew have gained the attention of building officials, insurance companies, government agencies and the general public. Finally, there seems to be a movement to seriously improve construction practices and make buildings more wind resistant. Code provisions are being reviewed and strengthened as needed. A requirement for impact resistant glass or shutters have been adopted in Dade and Broward Counties (1994). Enforcement of codes and education of people in the building trades is contributing to better construction in Florida.

LESSONS LEARNED FROM WIND DAMAGE STUDIES

Many lessons have been learned from 25 years of wind damage studies. A few of the more significant ones are reviewed in this section. Lessons learned from tornado damage, hurricane damage and from the performance of various categories of buildings are discussed.

Lessons from Tornado Damage

Structural analysis of damaged buildings, studies of cycloidal ground marks and photogrammetric analysis of tornado movie loops established upper bounds on tornado wind speeds. Instead of 270 or 360 m/s, as once was believed, the upper bound is thought to be in the range of 110 to 135 m/s.

Atmospheric pressure change in tornadoes is not a major damage mechanism (Minor et al., 1977). Natural porosity, openings or broken windows allow air to flow out of a building and quickly equalize the pressure. Outward acting aerodynamic wind pressures are the major damage mechanisms in tornadoes. As a result of this finding, the National Weather Service no longer advises the public to open windows when a tornado threatens. Instead of opening windows they are told to seek shelter in a small interior room or closet.

Objects and debris are readily picked up and transported by tornado winds. These so-called "tornado missiles" pose a threat of injury or death to persons caught out in the open during passage of a tornado. Damage investigations reveal the shape, size, and weight of a missile, its original location and impact point. Unless video or movies are obtained during passage of the tornado, nothing can be deduced about the missile trajectory. Computer simulation of missile trajectories have been widely published, but results vary because of the different assumptions made by the authors. For design, the maximum velocity (horizontal and vertical components) and maximum height achieved by the missile are needed. Computer simulations can be used to estimate reasonable values of speed and height for different classes of missiles. Damage surveys identify the kinds of missiles to expect. If a computer simulation can start the missile at its origin and predict the impact point, then the trajectory is a reasonable representation of the missile's behavior. Computer simulation of the steel beams transported by the Bossier City tornado reproduced the impact points shown in Figure 5.

Aerodynamic pressures are the principle damage mechanisms in tornadoes. Observations of damage caused by tornadoes show evidence of windward wall, side wall and leeward wall pressures, if the direction of the wind relative to the building is taken into account. Uplift pressures on hip and gable roofs and high local pressures at eaves, ridges and roof corners are observed in tornado damage as well as damage from straight-line winds. These observations led to the conclusion by DOE (1994) that the procedures of ASCE 7 (ASCE 1990) are valid for engineering design when dealing with tornado winds or straight-line winds.

In most structures damaged by tornadoes, areas could be defined within the building that would have provided reasonable protection of occupants during passage of the storm. McDonald (1990) called such areas "protective areas," rather than shelter areas, because buildings were not specifically designed to resist tornadoes. Through the process of observing performance of many different buildings, general guidelines for identifying protective areas have been developed.

The guidelines are used to identify protective areas in residences, schools, nursing homes and other occupied buildings. The general rule for protective areas is: select small interior rooms or closets with as many walls between occupants and the exterior as possible. Buildings susceptible to total disintegration or collapse such as mobile homes or light-gage framed structures and spaces with spans over 10 m (in auditoriums or gymnasiums) should be avoided.

Lessons from Hurricane Damage

Wind speeds in hurricanes are measured by aircraft flying through the storm or by anemometers located in the path. In the past, these measurements have been used to correlate damage and wind speed (e.g. Mehta et al., 1983). Controversies about wind speeds have arisen when there is an absence of anemometers in the path of the storm. In the case of Hurricane Andrew, there were virtually no reliable wind speed measurements in the storm path. Anemometers were either destroyed by the winds, or were not operational during the storm. Some operators literally turned off the instruments and went home!

Recent hurricane experiences revealed several weaknesses in local building codes. The weaknesses are primarily manifested in damage to the building envelope, including both roof and wall cladding. Windborne debris in hurricanes is not as formidable as in tornadoes, but can do a great deal of damage to window glass and tile roofs. Weak links in the wind load path are revealed as the storm passes over a building.

Failure of cladding takes the form of lost roofing material, and uplift of roof substrate (deck). Roofing system compliance tests used in practice today in the United States do not simulate the true wind loading. As a result, most roofing systems do not perform as they should.

Roof deck failures are attributed to inadequate nailing, inappropriate building materials, and poor quality control (see Fig. 8). Similar statements can be made about wall cladding. An example of weak links in the load path in Hurricane Andrew was inadequate bracing of gable end walls in residences. Prefabricated trusses had no ridge blocking, and no braces against inward or outward wind pressures acting on the gable end wall.

Debris impact was judged to be a major damage mechanism in Hurricane Andrew (see Fig. 9). Debris impact has two adverse consequences: (1) openings created in glazing allow wind and water inside the building and (2) internal pressures build up inside the building and increase the net outward loads of wall and roof structures.

FIGURE 8. TYPICAL ROOF DAMAGE IN HURRICANE ANDREW

As a hurricane passes over a location, the direction of the wind changes continuously. The building experiences variable wind angles of attack that eventually produce the worst loading on a particular component. The effect also reverses wind pressures during the course of the storm.

The slow translation of a hurricane means an area can be affected by hurricane force winds for three to six hours. The long time span is sufficient for fatigue effects to set in on some components, such as sheet metal attached by fasteners. The Australians observed this problem in Cyclone Tracy in 1974 and subsequently adopted code provisions to counteract the fatigue effects. At the present time, none of the U.S. model codes have fatigue mitigation provisions for wind-induced loads.

FIGURE 9. HOUSE DAMAGED BY DEBRIS IMPACTS IN HURRICANE ANDREW

Building Classifications

Twenty-five years of wind damage investigations have produced a data base from which conclusions relating to the performance of certain building classifications may be drawn. The building classification described herein is based on studies at Texas Tech University by Mehta et al. (1992).

Factors that influence the performance of buildings in high winds are: (1) environment, (2) structural system, (3) roof envelope, (4) wall envelope and (5) other considerations.

The environment around a building strongly affects the types and intensity of wind-induced loads. Factors include topography, debris exposure, presence of roof gravel, and type of storm exposure. Topography influences the magnitude of wind speeds near the earth's surface. Terrain exposures are generally classified as urban, suburban, open country and open water. Terrain features such as hills, valleys, or escarpments can either increase the wind speed as air travels up and over the feature or provide shielding as in the case of a long, deep valley.

When sources of debris and roof gravel are close to a building, its chances of damage are increased. Also, a building located on a barrier island off the coast of Florida or in tornado alley in Central Oklahoma has a higher risk of damage than one in regions with lower hurricane or tornado risks.

The structural system of a building is a measure of the degree of engineering attention paid to the design. For example, high-rise buildings (six or more stories) receive a large amount of engineering attention in the design of the structural system. These buildings perform very well in high winds. The Great Plains Life Building in Lubbock, Texas experienced wind speeds twice as large as the design wind speed (McDonald, 1970). The structural system was damaged, but did not collapse.

Heavy steel, concrete and reinforced masonry structural systems generally receive a high degree of engineering attention. These structural systems perform well in high winds, even when the loads exceed those specified by the building code. Inherent strength and stability provided by the American Institute of Steel Construction (AISC), the American Concrete Institute (ACI), and the Brick Institute of America (BIA) specifications, which are typically adopted by model building codes, result in structural systems that do not collapse in high winds.

Light steel construction, unreinforced masonry and light timber construction generally have less reserve strength and redundancy in the structure. The likelihood of complete structural collapse is much larger than in heavy steel, concrete and reinforced masonry structural systems.

Light timber construction of the type used in residential construction receives little or no engineering attention. Designs are essentially taken from handbook tables by persons who do not necessarily understand wind loading concepts. These structures are the most susceptible to wind damage and, in fact, are the source of most dollar damage. Manufactured homes and metal building systems receive a degree of engineering attention, but the effort is directed to optimization of the design for minimum cost. Slight wind force overloads sometimes lead to total destruction

The roof envelope plays a major role in the performance of buildings in high winds. Failure includes loss of roofing materials, uplift of the roof substrate (deck), or total uplift of the roof structure. Failure of skylights or overhead glazing also constitutes unacceptable performance of the roof envelope.

Failures of wall envelopes are due to removal of the cladding or perforation of the cladding by windborne debris. Failure of doors and windows in the wall envelope allows wind and water to enter the building. Windows fail by either the glass breaking or the window frame blowing out of its rough opening because of inadequate fasteners to hold it in place. Large overhead doors deform under wind load and disengage from their tracks. They also are susceptible to missile perforation, if not protected.

Need for Further Research and Documentation

Much has been learned about the performance of buildings, structures and facilities in the last 25 years. The nature of the storms are better understood, upper bound wind speeds have been established, and the significant characteristics of wind, such as profile and turbulence, have been identified. The interactions of wind and structure are reasonably well defined.

Information has been collected at various institutions. The data are in various forms and degrees of organization. There is not a central clearinghouse where researchers can go to find the damage data needed for a project. Funds have not been available to systematically catalog and organize data. As a result, a lot of information is lying on shelves in forms that are not suitable for current research needs.

In the future, standard methods should be established and standard forms for reporting data should be developed that are suitable for electronic media storage. A central clearinghouse should be established to catalog, store, maintain and retrieve data. Long-term, continuous funding is needed for this effort.

Other more specific data collections and analyses in post-storm activities should include: (1) correlation between structural performance in high winds and building code provisions, (2) correlation between damage and wind speed for various types of buildings and structures, (3) establishment of performance data bases for different buildings and individual components, and (4) collection of information for calibrating and benchmarking procedures that predict the performance of buildings and other structures.

Structural Performance and Building Code Provisions

Future damage investigations should carefully document the performance of specific buildings or structure types and evaluate the behavior in light of the building code provisions in effect at the time of construction. Buildings that were damaged and those that survived are both of interest. Researchers should answer the questions: Were code-specified design wind pressures adequate? Were code-specified loadings similar to the ones that caused damage? For example, if failure of a roof initiated at the corner, did the building code specify a roof corner load?

Damage from windborne debris should be carefully documented. What types of debris, how large, how heavy? What was the origin of debris? How far did it travel? What kind of damage was produced? What were the consequences of the debris impacts? What percentage of the building was damaged by wind pressures? What percentage by debris? Did the building code in force have prescriptive requirements (i.e. identify specific member strength and size, fastening rules, etc.)? Were the prescriptive provisions adequate? If not, what were the weaknesses? Did the code provide for continuous load paths from exterior surface to foundation? What were the weaknesses in the load path? Cladding attachment? Roof anchorage? Shear wall anchorage? Were the building materials used appropriate for the loading and climate? Were fasteners (nails, staples, anchor bolts) adequate? How did they compare with building code requirements?

Studies described above require meticulous investigations of individual buildings. The data must be gathered quickly after a storm, because rescue and cleanup operations destroy or invalidate the data.

Correlation Between Wind Speed and Damage

Damage functions, which are correlations between wind speed and some measure of damage, are needed for predicting future damage from postulated storms at other locations. Ideally, the wind speeds are measured by instruments that survive the storm. Because anemometer locations at best are widely scattered, numerical modeling of the storm (hurricane or tornado) is needed to produce wind speed isovels. The actual wind speed measurements are used for calibration of the numerical model. Correlations can then be made between modeled wind speeds and damage.

Consistent schemes for measuring damage are needed. Measures used in the past include percent damage (with total damage (100%) equal to destruction beyond repair), dollar amount and damage ratio (ratio of dollar damage of building and contents to estimated value of building and improvements). The functions should be defined for various categories of buildings and structures. The damage functions should be formulated from a statistical population. Hence, large amounts of data from different kinds of storms of different intensities are needed.

Performance Data Bases

The performance history of specific classes of buildings and structures and of specific materials, construction techniques, building geometries and architectural features are needed by insurance underwriters, building officials, city planners and designers. The data bases should contain information on type and extent of damage, storm severity, dollar losses, prescriptive requirements or code provisions. Some of the data will come from field investigations of the damage; others will come from insurance loss records. Cooperation is needed between field investigators and insurance companies or associations. Again, a central clearing-house for this type of data is needed.

Several years will be needed to collect the data described above. The collection of the data requires pre-storm planning and organization and the funds readily available in order to react quickly after the storm. Teams should be organized and trained ahead of time. Procedures, standards, and quality control should be established. Individuals and organizations are needed who have no special interests or axes to grind. Accurate unbiased data must be obtained.

CONCLUSION

Twenty-five years of wind storm damage studies by various organizations and agencies have produced sufficient technical knowledge to construct wind resistant buildings, structures and facilities. Unfortunately, in the United States, this knowledge has not been put to use, except for high-rise buildings and certain critical facilities such as nuclear power plants and corporate computer centers.

Existing knowledge from damage studies should be used in the future to construct more wind-resistant buildings. In order to avoid monumental losses as experienced with Hurricane Andrew, existing buildings, especially residences, should be upgraded (retrofitted) to improve their wind resistance.

With the understanding of building performance in high winds from damage investigations, new materials, new products and new construction procedures should be developed immediately. Educational programs are needed to make everyone involved in design and construction more aware of the proper approaches to wind-resistant construction.

Existing data and data from future damage investigations should be assembled in performance data bases for use by persons involved in wind damage mitigation. These data bases do not exist today, but are badly needed.

Wind damage investigations in the past have produced a large body of knowledge. Such efforts should continue in the future, because there is still much to be learned.

REFERENCES

ASCE (1990), "Minimum Design Loads for Buildings and Other Structures," ASCE 7-88, American Society of Civil Engineers, New York, NY, 134 pp.

Cook, R.A. (1993), "Hurricane Andrew: Recommendations for Building Codes and Building Code Enforcement," Florida Department of Community Affairs, 33 pp.

Dikkers, R.D., Marshall, R.D. and Thom, H.C.S. (1970), "Hurricanes Camille - August 1969," National Bureau of Standards, 10393, 66 pp.

DOE (1994), "Natural Phenomena Hazards Design and Evaluation Criteria for Department of Energy Facilities," DOE-STD-1020k, United States Department of Energy, Washington, DC.

FEMA (1993), "Building Performance: Hurricane Andrew in Florida," Federal Emergency Management Agency, FIA-22, Washington, DC, 93 pp.

Fujita, T.T. (1971), "Proposed Characterization of Tornadoes and Hurricanes by Area and Intensity," SMRP Research Paper No. 91, Department of Geophysical Sciences, The University of Chicago, Chicago, IL.

Fujita, T.T. (1975), "Superoutbreak Tornadoes of April 3-4, 1974," Department of Geophysical Sciences, The University of Chicago, Chicago, IL.

Fujita, T.T. (1978), "Workbook of Tornadoes and High Winds for Engineering Applications," SMRP Research Paper No. 165, Department of Geophysical Sciences, The University of Chicago, Chicago, IL.

Fujita, T.T. and McDonald, J.R. (1978), "Tornado Damage at the Grand Gulf, Mississippi Nuclear Power Plant Site: Aerial and Ground Surveys," NUREG/CR-0383, USNRC, Washington, DC.

Fujita, T.T. (1979), "The Bossier City Tornado of December 3, 1978," SMRP No. 169, Department of Geophysical Science, The University of Chicago, Chicago, IL, 33 pp.

Fujita, T.T.(1970), "The Lubbock Tornadoes: A Study of Suction Spots," *Weatherwise*, Vol. 23, No. 4, August.

Glass, R.I., et al. (1980), "Injuries from the Wichita Falls Tornado: Implications for Prevention," *Science*, Vol. 207, February, pp. 734-738.

Golden, J.H. (1976), "Assessment of Wind Speed in Tornadoes," *Proceedings*, Symposium on Tornadoes: Assessment of Knowledge and Implications for Man, Texas Tech University, Lubbock, TX, June 22-24, 1976, pp. 5-42.

Golden, J.H. and Davies-Jones, R.P. (1975), "Photogrammetric Wind Speed Analysis and Damage Interpretation of the Union City, Oklahoma Tornado, May 24, 1973," *Preprints*, Second U.S. National Conference on Wind Engineering Research, Fort Collins, CO.

"Impact Test Procedures" (1994), Dade County Building Code Compliance Office, Protocol, FL, pp. 201-94.

Kiesling, E.W. and Goolsby, D.E. (1974), "In-Home Shelters from Extreme Winds," *Civil Engineering Magazine*, Vol. 44, No. 9, September, pp. 105-107.

McDonald, J.R. (1970), "Structural Response of a Twenty-Story Building to the Lubbock Tornado," Institute for Disaster Research, Texas Tech University, Lubbock, TX, 43 pp.

McDonald, J.R. (1981), "Incredible Tornado-Generated Missiles," *Proceedings*, 4th U.S. National Conference on Wind Engineering Research, Seattle, WA, pp. 11-13.

McDonald, J.R. (1990), "Damage Mitigation and Occupant Safety," *The Tornado: Its Structure, Dynamics, Predictions and Hazards*, C. Church, Ed., Geophysical Monograph 79, American Geophysical Union, pp. 523-528.

McDonald, J.R. and Smith, T.L. (1990), "Performance of Roofing Systems in Hurricane Hugo," IDR Doc. #90D, Institute for Disaster Research, Texas Tech University, Lubbock, TX.

McDonald, J.R., Mehta, K.C. and Minor, J.E. (1981), "The Wichita Falls Tornado: A City's Response after the Storm," *Proceedings,* 4th U.S. National Conference on Wind Engineering Research, Seattle, WA, pp. 11-13.

Mehta, K.C. (1976), "Wind Speed Estimates: Engineering Analysis," *Proceedings,* Symposium on Tornadoes: Assessment of Knowledge and Implication for Man, Texas Tech University, Lubbock, TX, June 22-24, pp. 89-103.

Mehta, K.C., Cheshire, R.H. and McDonald, J.R. (1992), "Wind Resistance Categorization of Buildings for Insurance," *Journal of Wind Engineering and Industrial Aerodynamics*, Vols. 41-44, Elsevier Science Publishers, B.V. Amsterdam, pp. 2617-2628.

Mehta, K.C., McDonald, J.R., Minor, J.E. and Sanger, A.J., (1970), "Response of Structural Systems to the Lubbock Storm," Texas Tech University Storm Research Report, SRR 03, October, NTIS Acc. No. PB-204-983.

Mehta, K.C., Minor, J.E. and McDonald, J.R. (1976), "Wind Speed Analysis of April 3-4, 1974 Tornadoes," *Journal of the Structural Division*, ASCE, Vol. 102, pp. 1709-1724.

Mehta, K.C., Minor, J.E., and Reinhold, T.A. (1983), "Wind Speed-Damage Correlation in Hurricane Frederic, *Journal of Structural Engineering*, ASCE, Vol. 109, No. 1, January, pp. 37-49.

Minor, J.E. (1985), "Window Glass Performance and Hurricane Effects, *Proceedings* of a Specialty Conference on Hurricane Alicia: One Year Later," A. Kareem, Ed., ASCE, pp. 151-167.

Minor, J.E. and Behr, R.A. (1993), "Architectural Glazing Systems in Hurricanes: Performance, Design Criteria and Designs," *Proceedings*, 7th U.S. Conference on Wind Engineering, University of California at Los Angeles, June 27-30, 1993.

Minor, J.E., McDonald, J.R. and Mehta, K.C. (1977), *The Tornado: An Engineering Oriented Perspective*, NOAA Technical Memorandum, ERC NSSL-82.

Minor, J.E., Mehta, K.C. and McDonald, J.R. (1972)," Failure of Structures Due to Extreme Winds," *Journal of the Structural Division*, ASCE, Vol. 98, No. ST11, pp. 2455-2472.

Powell, M.D. and Houston, S. (1993), "Surface Wind Field Analysis in Hurricane Andrew," 20th Conference on Hurricane and Tropical Meteorology, May 10-14, San Antonio, TX, 4 pp.

Reinhold, T.A., Vickery, P.J. and Powell, M.D. (1992), "Wind Speeds in Hurricane Andrew: Myths and Reality," American Concrete Institute Annual Convention, San Juan, PR, October 26, 12 pp.

Segner, E.P., (1960), "Estimating of Minimum Wind Forces Causing Structural Damage in the Tornadoes at Dallas, Texas, April 2, 1957," U.S. Weather Bureau Research Paper No. 41, pp. 169-175.

Smith, T.L. (1993), "How Did PUF Roofs Perform During Hurricane Andrew?", *Professional Roofing*, January.

Sparks, P.R. (1990), "Development of the South Carolina Coast 1959-1989 Prelude to a Disaster," *Proceedings* of a Public Forum in Charleston, SC, September 13-15, pp. 1-7.

Thompson, J.N., Kiesling, E.W., Goldman, J.L., Mehta, K.C., Wittman, J. Jr., and Johnson, F.B., (1970), "The Lubbock Storm of May 11, 1970," National Academy of Sciences for the National Academy of Engineering, Washington, DC, 75 pp.

USNRC (1974), "Design Basis for Nuclear Power Plants," Reg. Guide 1.76, U.S. Nuclear Regulatory Commission, Washington, DC.

USNRC (1975), "Missiles Generated by Natural Phenomena," Section 3.5.1.4 NUREG-75/087 (Rev. 1), Standard Review Plan, Office of Nuclear Reactor Regulation, U.S. Nuclear Regulatory Commission, Washington, DC.

Walker, G.R. (1975), "Report on Cyclone 'Tracy' December 1974 - Effect on Buildings," Australian Department of Housing and Construction, Melbourne, Victoria, Australia (3 Vols.), 99 pp, 167 pp and 164 pp.

Zollo, R.F., Ed. (1993), "Hurricane Andrew: August 24, 1992, Structural Performance of Buildings in Dade County, FL," Department of Civil and Architectural Engineering, University of Miami, Coral Gables, FL.

STATE OF THE ART VOLUME

WIND FLOW AND CITY PLANNING

H.J. GERHARDT and C. KRAMER

Fluid Mechanics Laboratory,
Fachhoch-schule Aachen,
Goethestrasse 1, D-5100 Aachen, Germany

SUMMARY

The paper classifies recent trends in urban development and discusses their possible impact on the city climate. The background of and a typical example for an environmental impact assessment (EIA) will be presented. Finally, simple guidelines for city planners and architects will be given to prevent aerodynamic accidents.

INTRODUCTION

The macro-scale climate is locally changed by the presence of cities resulting in a city climate. Important parameters for the climate are temperature, humidity, and wind. The influence of a city on the city climate may be passive, e. g. changes in the surface roughness, or active, e. g. additional heat release. As long a cities grew slowly those changes in city climate were small and often imperceptible. At present, urban developments - at least in the populated western countries - take place on a larger scale. Therefore, the anticipated changes in the urban climate are larger. To avoid, or at least minimise, negative effects the interdependence between urban developments and city climate should be known to the city planners, city councils and developers. It will be discussed in this paper.

TRENDS IN URBAN DEVELOPMENTS

The growth of the urban population is continuing and even increasing in the industrial countries. This urbanisation has several reasons, amongst others

- structural changes in the economy and the growth of the economy lead to an increased need of office and industrial spaces.

- decreasing working hours leading to an increase in leisure time spent in holiday centres and adventure parks, etc.

- new life styles and work requirements exemplified by the architectural ideas of "living in a park" and "working in park".

- drastically increased individual traffic and goods transport requiring large areas for the traffic infrastructure.

- the disposal of waste, sewage, contaminated soil, etc. requiring large area intensive waste disposal units.

Figure 1 shows as a typical example the urban development in and around Frankfurt. The urbanised area between river Main and Taunus mountain range are given in 10 year steps starting from 1950. It is obvious that the area between the communities at the southern foot of the Taunus mountain range (Königstein, Kronberg, Oberursel) and the City of Frankfurt is increasingly urbanised. It may be expected that this whole area will become an interconnected urbanised area within the next 20 years.

For Germany alone the divers land uses result in an annual loss of natural, unurbanised area of approximately 0.25 % of the total surface area. It is obvious that a land conversion cannot continue at that rate. Therefore, the current trends in urban development and their consequences for the environment has to be analysed.

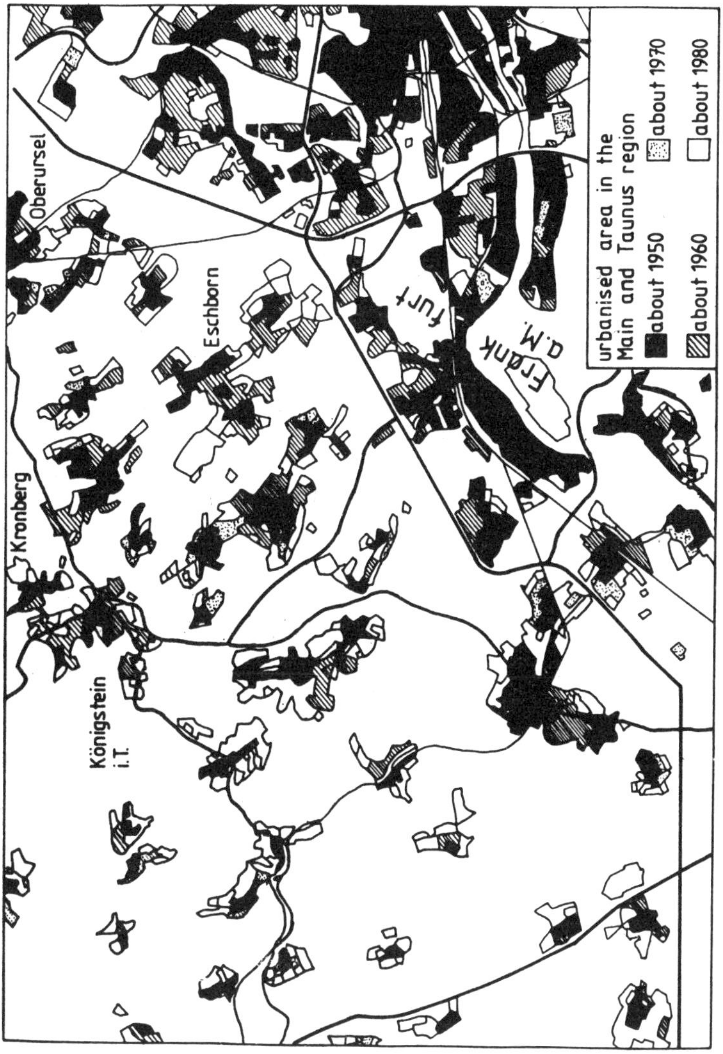
Oberursel
Eschborn
Kronberg
Königstein
i.T.
Frankfurt
a.M.
urbanised area in the
Main and Taunus region
about 1950
about 1960
about 1970
about 1980

Figure 1

Urban Extension (Outer Development)

Since the required space for new apartment blocks, office buildings, small and medium sized industry is not available in the historically grown cities of, in particular, the industrial countries, the necessary space can only be provided outside or at the fringes of the cities. Typical examples for the resulting outer developments are industrial estates, satellite cities for living purposes and, to a lesser degree, satellite cities for offices. Figure 2 shows an aerial view of a typical industrial estate development near the city of Aachen. It consists of various sized, low rise buildings set up seemingly without a master plan in a region previously used agriculturally. It may be expected that the area between the adjacent suburbs will be urbanised in the near future. Due to the large amount of sealed surface such developments will alter the local climate appreciably.

Figure 2

An example for a new large scale office area is the development at present going on in Eschborn near the city of Frankfurt, Figure 3. The development, started a few years ago, will replace an agriculturally used area by a satellite city consisting of medium to high rise office buildings. Eschborn is situated between the Taunus mountain range and Frankfurt, see Figure 1. A densely build area may block off any drainage flow or country breeze extending possibly from the Taunus mountain to the outskirts of Frankfurt.

Figure 3

In addition to the described outer developments inner urban extensions may occur. If the relatively few remaining vegetated areas are urbanised this may result in a twofold negative effect. The heat island effect will be increased and at the same time the ventilation decreased. In addition the beneficial local effects of such vegetated areas will be abolished.

Urban Redevelopment and Subsequent Intensification

In addition to the inner city extensions an increase in the density of buildings may be observed at present in major cities of the industrial countries. In particular in the city centres older buildings are replaced by new high rises. A typical example is given in Figure 4. It shows a wind tunnel model of the city centre of Frankfurt. The lighter buildings exist, the darker buildings are in the planning or construction stage. The new buildings range in height from 120 m to 215 m replacing 5 to 7 storey buildings.

Figure 4.

Another example of urban intensification is Berlin. Here, inner urban extension and urban intensification take place at a fairly large scale at the same time. The consequences of the massive urban extension and intensification will be discussed below, using the planned development at the Potsdamer Platz as an example

Urban Development and City Climate

The urban extension and urban intensification will change the urban climate. The radiated, thermal, moisture and aerodynamic properties of the surrounding environment will be altered. In particular, the radiative effects will lead to an increase in soil and air temperature, Oke (1987). The areas newly settled due to urban extension will be taken from rural areas. Those climatically relevant surfaces will be lost for climatic exchange processes like country breezes and ventilation in general. In addition the area of city climates will be enlarged. Urban intensification will increase the surface roughness. Thus, decreasing the wind flow, the relative humidity and increasing the heat island effect. Both, urban extension and intensification is accompanied by increasing traffic, mostly individual car traffic. The increase in traffic emission - heat and pollution - will intensify the negative effects of increased urbanisation. Finally, large solitary buildings and large building complexes may produce local wind problems, e. g. small pedestrian comfort or even wind velocities endangering pedestrians.

Clearly, urban expansions will modify the urban climate, see e. g. Oke (1987). Many city planners are, however, still not aware of this interdependence. On the other hand, the increasing cost of the city properties and of the construction and the demand for high quality usage necessitate a priori investigations into the influence of urban changes on the urban climate. Only proper planning may avoid future problems. Whereas investors are usually sensible to possible local problems, e. g. pedestrian comfort on their property and optimal placements of air conditioning inlets and outlets, they are far less interested in more global aspects, e. g. changes in city ventilation, humidity, etc. Here, regulations introduced by the city or the state seem to be necessary.

ENVIRONMENTAL IMPACT ASSESSMENT (EIA)

The objective of environmental impact assessments is to establish, specify and evaluate the effects of a project on:

- human, animal and plant life
- soil, water, air, climate and landscape

including the corresponding interaction and effects on cultural and material goods. EIAs shall estimate the possible environmental consequences of planning actions, and thus help to avoid, reduce or compensate negative environmental consequences. EIAs were introduced in the U.S.A. at the end of the nineteensixties, by the end of the nineteenseventies several 10.000 assessment had been conducted. In 1985 the European Community passed a guideline on environmental impact assessments of public and private projects. According to the EC-guideline large industrial installations (chemical, metal processing, food and paper industries), power plants (conventional and nuclear), mining installations, major infrastructural projects (airports, seaports, pipelines, waterways, motorways) and major urban building projects (hotel complexes, holiday villages, etc.) are subject to environmental impact assessment studies.

The guideline is inacted differently in the different countries of the EC. E. g. major urban development and building projects are usually covered by an EIA in the European Community, however, not in the German "Act on Environmental Impact Assessment". Furthermore this act does not regulate who has to provide the inspection documents, whether it is the project initiator, the governmental planning agency, or the investor, and which authority is to carry out the assessment procedure. The act finally rules that an EIA does not present a procedure in its own right, but that the assessment is to be made within the legal framework of the professional approval procedure.

An environmental impact assessment usually consists of three elements, namely

1) environmental relevance assessment (ERA)
2) scoping
3) in-depth assessment.

Generally the ERA is a feasibility study. It checks if the project under consideration is of environmental relevance. It is a simple checklist-like evaluation. The assessment deals with the site (flood area, ventilation, proximity to nature reserves, etc.) and the size and type of the project (height, area, volume, emission, traffic, etc.). If the proposed project is relevant for the environment in one or more items an in-depth assessment is necessary. The scoping determines the frame for this in-depth assessment. It specifies the items and the range to be investigated and the methods to be used.

The in-depth assessment contains the environmental assessment of the present situation, prognoses of the environmental development without and with realisation of the project, separate evaluations of the various items and a global evaluation, recommendations to avoid, to decrease or to compensate negative effects on the environment, and an easy-to-understand summary.

It is important to note that the results of an environmental impact assessment are legally not binding. They give **recommendations** which may be considered, but may be overridden by other considerations.

CASE STUDY "POTSDAMER PLATZ, BERLIN"

In 1934, the Potsdamer Platz used to be the busiest place in Berlin. Since the border between East- and West-Germany cut through the Potsdamer Platz the surrounding area became fallow land after 1945. The two parts of Berlin developed separately with two city centres, one in the Western and one in the Eastern part. After the German unification a master plan for Berlin has been developed in which the Potsdamer Platz area has a prominent place. As a link between the Eastern and Western city centres the plan calls for a major development with divers usage (living quarters, offices, shops, department stores, theatres, cinemas, etc.) and with high urban intensity.

Eight investors will realise on a total area of 510.000 m^2 and a built over area of 240.000 m^2 a floor space of 1.100.000 m^2 above ground level. The roof height is between 22 m and 37 m with a few higher buildings of height up to 104 m. Besides the usual traffic areas on street level there will be tunnels for regional and long distance trains, subways, and for road vehicles. The master plan of the development around the Potsdamer Platz was designed by the architects Hilmer and Sattler and is shown in Figure 5. The main investors are - counter clockwise from N to S - Herti (department stores), Sony, debis of the Daimler Benz AG and ABB.

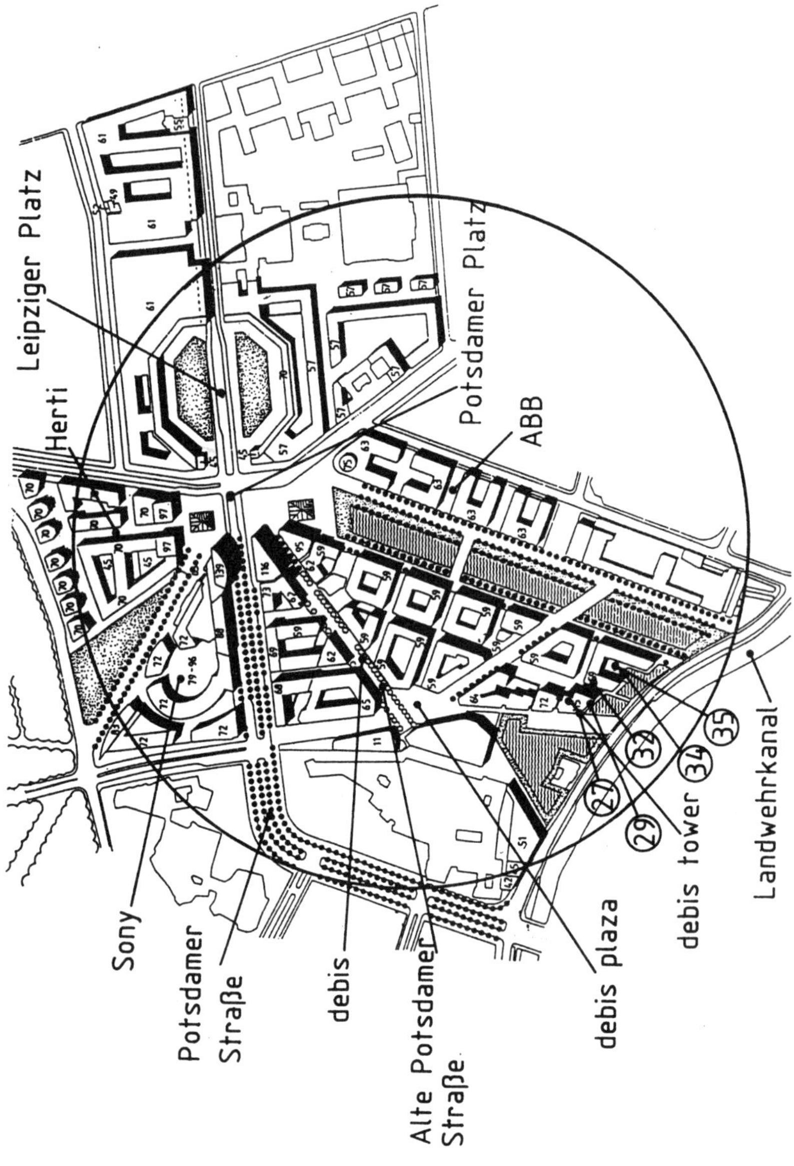

Figure 5

The Potsdamer Platz area is situated between the large central park area "Tiergarten" and the densely built city areas. Due to the particular location and the size of the planned development a major impact on the environment has to be expected. In particular the 10 year construction phase will effect the neighbouring areas. It was obvious that an EIA was initiated by the Berlin government and conducted by the consultancy office "Büro für Kommunal- und Regionalplanung" (BKR) in conjunction with experts for various relevant aspects - amongst others I.F.I.. The scoping phase revealed the following major areas for investigation:

- water and soil
- flora, fauna, biotope
- urban climate and immission field
- noise
- usage, in particular with respect to cultural and architectural properties

A very critical part of the EIA was concerned with the consequences of the construction on the ground water. During the 10 year construction phase the ground water level will be lowered appreciably. Conceivably this will have a strong impact on the old trees of the Tiergarten area. However, in this paper only the items city climate and immission field are dealt with. The following questions had to be answered:

- What are the effects of the planned development on the meso and micro scale wind field?
- Will a new heat island be created?
- Which positive and negative effects are expected from the planned park areas and lakes?
- How will the humidity be influenced by the development?

Description of the Present Situation

The area under consideration is at present fallow land and part of a ventilation aisle along the old railroad area running approximately N/S. It reaches from the Tiergarten in the North to the Tempelhof airport in the South, see Figure 6. The ventilation around the Potsdamer Platz is, except for wind direction E and NW, good and reaches almost rural quality. The reduction of wind velocities compared to rural areas is only 20 % at day time and 40 % at night time.

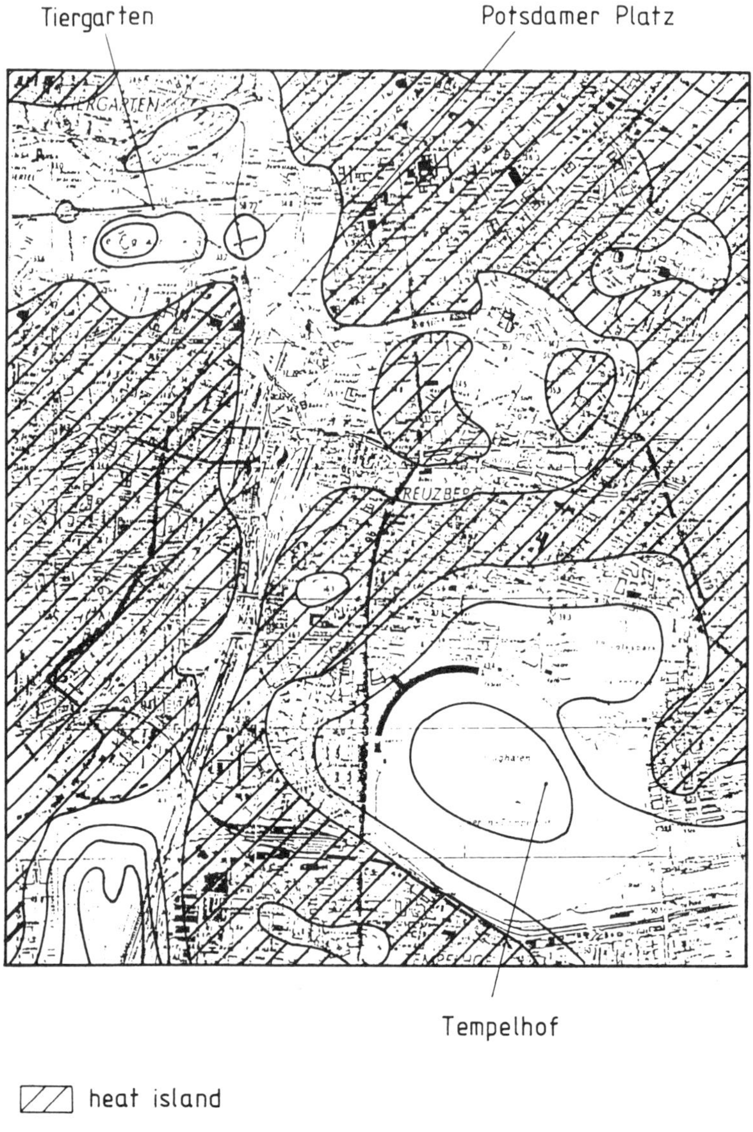

Figure 6

The thermal situation, too, in the fallow land around the Potsdamer Platz is comparable to the one in rural areas with relatively high temperature maxima and low temperature minima. Due to this temperature difference flows of cold air may develop. The temperature field in an area of approximately 33 m^2 around the Potsdamer Platz is shown in Figure 6 (after Ketzler et al. (1993)). The Tiergarten area is a cold island surrounded by a ring of heat islands. This ring is only interrupted by the ventilation aisle at the Potsdamer Platz. The air temperature in the Tiergarten is up to 9 K lower than in the heated surrounding. Typically, the relative humidity in the Tiergarten is 65 %, in the neighbouring heat island areas less than 60 %. The urbanised area in the vicinity of the Potsdamer Platz is characterised by a poor city climate. Surface inversions occur at the Potsdamer Platz statistically at 15 days per year, elevated inversions (elevation height between 200 m and 300 m) occur at approximately 240 days per year.

Evaluation of the Effect of the Planned Development

To quantify the effect of the planned development on the urban climate an experimental study was conducted in the I.F.I. boundary layer wind tunnel (cross section 1.75 m x 0.9 m). During phase 1 of the investigation the area shown in Figure 7 was modelled at a scale of 1:800. The 1.5 m turntable diameter is equivalent to a full scale diameter of 1.2 km. During phase 2 and 3 detailed investigations were conducted for various parts of the proposed development at scales of 1:500 and 1:200. In particular, the surface flow field, the immission situation and the wind forces on the major structures of the Sony, debis and ABB-developments were investigated. Surface wind velocities were determined using the sand erosion technique and Irwin probes, Irwin (1981). An immission prognosis for the anticipated traffic emissions has been elaborated. The traffic emission was simulated using propane as tracer gas. The tracer gas immission in the various streets was adjusted according to the traffic prognosis (Heusch/Boesefeld (1993)). Results of phase 1 will be published shortly (Gerhardt (1994)). The previously presented data and yet unpublished data concerning the global effect of the planned development may be summarised as follows.

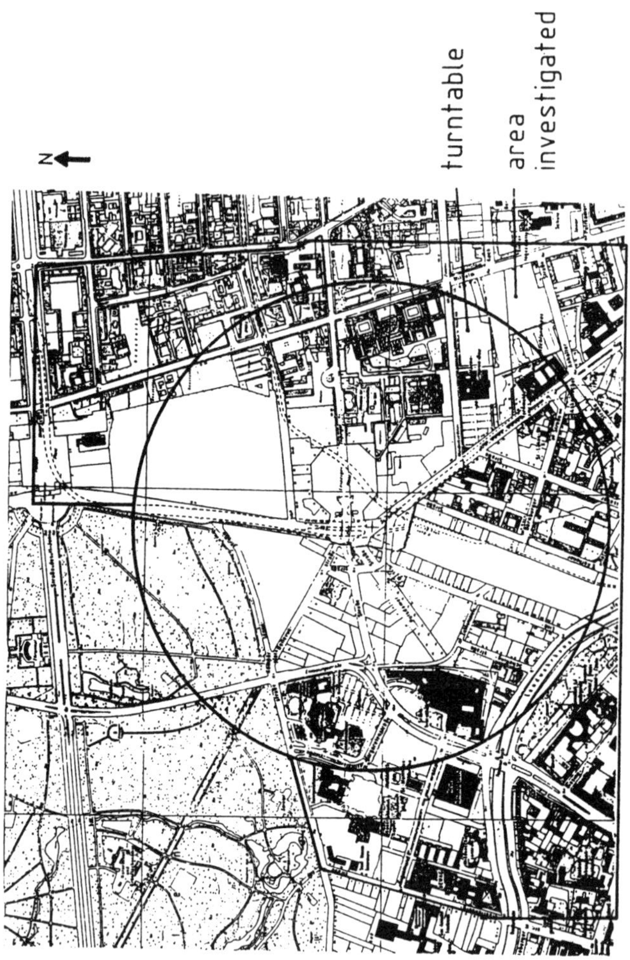

Figure 7

Wind flow

The proposed development will decrease the wind velocity in the urban canopy layer (UCL) at and near the Potsdamer Platz by 25 % to 50 %. Thus, the ventilation will be reduced and the air properties will be negatively effected. The intensity of the NS-ventilation aisle will be diminished or even interrupted. This will have negative consequences in particular for the urban area south of the Potsdamer Platz. Depending on wind direction greatly increased surface wind velocities will occur locally. Here, the pedestrian comfort is reduced leading possibly to a diminished quality of usage. Critical areas in this respect are the western and southern border of the proposed development, the Potsdamer Platz and parts of the Potsdamer Straße.

Temperature

Due to the urban intensification of the proposed development large heat emissions are anticipated. The emitted heat flow per unit area is expected to be larger than in the neighbouring heat islands. Due to the proposed development the ring of heat islands around the Tiergarten area will be closed. This will lead to an increase in temperature in the Potsdamer Platz area. Furthermore, the city quarters of Berlin which neighbour the Potsdamer Platz area to the E, SE and S will be cut off from the thermal equilibration effect of the Tiergarten.

The heat radiation around the Potsdamer Platz will be greatly changed. Due to the multiple reflections of the incident radiation on the building walls in the street canyons more heat will be stored in the building walls. In addition, less heat losses by the outgoing radiation have to be expected for the same reasons. The temperature increase will be enhanced by the decrease in ventilation capacity due to the decrease of the wind velocity in the urban canopy layer. Critical with respect to the temperature increase will be the courtyard areas. Since the cold air flow from the Tiergarten through the - presently fallow land - will be blocked by the new development a further increase in the temperature level has to be anticipated.

Immission field

Since the ventilation capacity will be decreased, and, at the same time, the traffic intensity will be increased an increase in the concentration of immission gas components, notably NO_2, has to be expected. This increase will not be compensated by the positive effect of the increasing number of cars with emission converters. Therefore, the presently already critical immission situation will be further worsened.

Conclusions

For the following reasons the planned development will have a large impact on the environment:

- Due to the large drainage during the construction phase (up to 25.000.000 m^3/h) the ground water level will be lowered by up to 1.7 m. This may lead to grave disturbances of the water supply of that vegetation of the Tiergarten which depends on the ground water.
- It has to be expected that a large part of the valuable vegetation of the fallow land and of the existing old trees may be destroyed. Consequently a number of relatively rare or endangered animals and plants will be lost.
- The development will form a new heat island closing the ring of already existing heat islands around the Tiergarten. The thermal equilibration effect of the Tiergarten for the neighbouring city quarters will be greatly reduced.
- The, at present, existing N/S-ventilation aisle will be blocked.
- Due to lack of pedestrian comfort some areas of the development may not be suitable for the anticipated usage (side walk cafés, theatres, shopping, etc.). Locally, the wind velocities in extreme gusts may even be dangerous for pedestrians.

Recommendations of the EIA

Here, only the recommendations concerning the urban climate will be discussed. It will be distinguished in recommendations

- to avoid and to decrease and

to compensate negative effects on the environment.

To retain a N/S-ventilation aisle is highly desirable - even if the width is reduced appreciably . Whereas the ventilation aisle will be retained in the southern part of the development area, see Figure 5, the realisation of the present plan with the very dense urbanisation north and east of the Potsdamer Platz will block the ventilation. It is recommended to open the area north of the Potsdamer Platz by the following measures:

- shortening the north-westerly and the south-westerly "arms" of the construction around the Leipziger Platz
- shifting the Herti complex in direction West

To avoid the closing of the heat island ring around the Tiergarten the following measures are necessary:

- decrease of the heat emissions by reducing the traffic and by very good thermal insulation properties of the buildings;
- decreasing the radiation intake by plants covering the facades and roofs.

The necessary reduction in local wind velocities may be achieved by reducing the roof height of some buildings and/or by local changes of the building envelop, e. g. using canopy roofs.

Compensation measures for the reduced or even completely blocked ventilation along the N/S-aisle seems to be essential. Thus, the preservation of the presently undeveloped area to the South and to the East of the Potsdamer Platz is necessary. Finally, any further intensification of the planned development should be ruled out.

Detailed Investigations

The results of the EIA lead to a number of detailed investigations commissioned by the major investors to improve the urban climate and in particular the local wind field. As an example for those investigations the study conducted at the southern part of the debis compound shall be discussed. The wind flow is dominated by the debis tower, a 21 storey high rise building. The roof height of the surrounding buildings is at most 28 m. The debis compound borders on a canal the so called "Landwehrkanal". The streets to either side of the Landwehrkanal are main traffic arteries with a total of approximately 10.000 vehicles per hour during peak traffic times. Due to the high traffic emissions and the building flow field, in particular around the debis tower, a large emission entrainment is anticipated at the southern part of the debis compound. Based on the wind tunnel tracer gas study NO_2-concentration in the southern part of the debis compound are predicted which will surpass the European limiting guidelines.

To prevent the entrainment of traffic emissions into the new development and, at the same time, to reduce pedestrian discomfort various wind shelters have been investigated. Figure 8 shows the southern part of the wind tunnel model (scale 1:200). It includes all wind shelters investigated. The wind fence Z simulates the effect of a hedge (height 2 m) in combination with a row of trees (height 15 m). Due to the wind fence Z the local wind velocities at the foot of the debis tower and in particular the entrainment of traffic emissions will be decreased appreciably. Figure 9 shows the effect of the wind fence on the local NO_2-concentration during peak traffic hours. The Figure gives the 98 %-value of the peak emissions normalised by the limiting value for five critical locations. Variant I denotes the going in position, for variant II the wind fence Z has been added, for variant III the wind fence Z, a canopy roof at the debis tower and vegetation inside building C3 (to the right of the debis tower in Figure 8) have been added. The wind fence Z will decrease the local NO_2-concentration appreciably.

In conclusion, the local wind and emission field may be improved by changes in the building geometry, by add ons and/or by wind shelters.

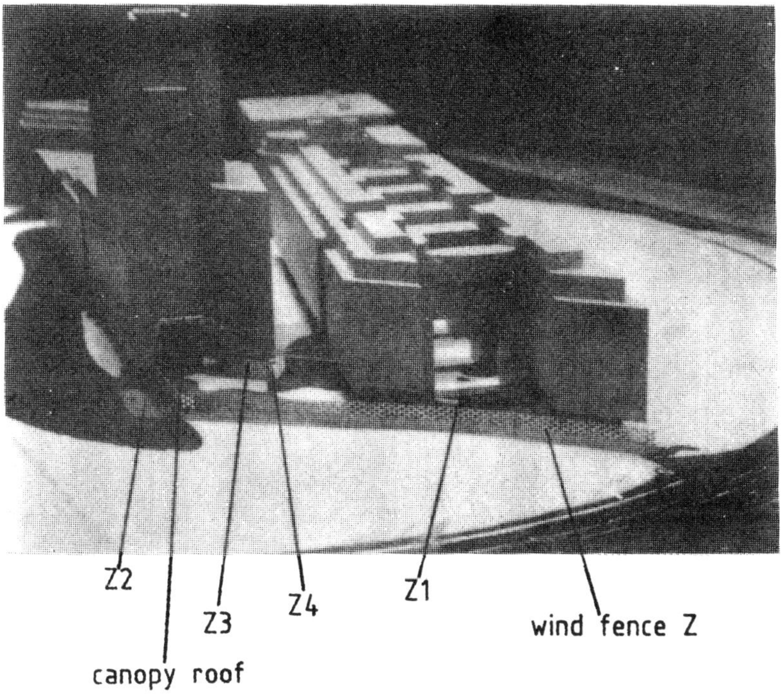

Figure 8

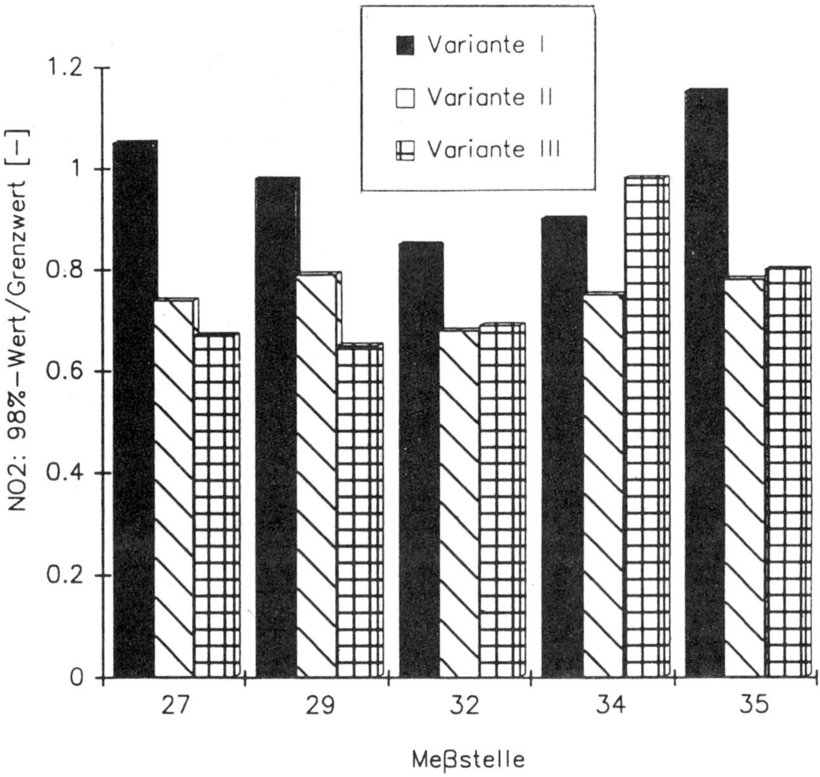

Figure 9

RECOMMENDATIONS FOR PLANNERS

Global effects of new developments are normally due to "aerodynamics mistakes" in the master plan. The following general recommendations should be observed for urban developments in both forms discussed, namely urban extension and urban redevelopment.

- Preserve **all** open (climatic relevant) areas extending from the rural areas or from larger inner city parks into the built-up areas.
- Orientate and structure developments according to the influence of ventilation flows.
- Avoid to disturb or cut off drainage flows in valleys. Typical disturbances are dams, barriers due to buildings or lines of trees with their axes perpendicular to the valley axis.

- Avoid building barriers with their axes perpendicular to the hill slopes.
- Reserve open space at the urban boundaries when extending the urban areas into rural areas.
- Decrease the urban density from the centre to the outskirts.

Local effects are normally due to building interferences. Helpful guidelines to avoid negative inferences have been published, e. g. Gandemer et Guyot (1976). Two quite common planning mistakes, which were pointed out already by Gandemer et Guyot may be seen in the debis compound at the Potsdamer Platz, namely

- a solitary high rise surrounded only by much lower buildings and
- a building barrier of height h ≥ 30 m with a fairly large lateral extent being exposed to the prevailing wind.

An example of a solitary high rise is the debis tower at the southern end of the debis compound, see Figures 5 and 6. With its height of approximately 85 m it overtops the surrounding buildings (h ~ 28 m) by some 50 m. The area of strong pedestrian discomfort is approximately 10 m around the building foot, it contains all pedestrian areas around and all entrance areas to the debis tower. To decrease the pedestrian discomfort a fairly large canopy roof above the entrances is necessary, see Figure 8.

The buildings along the western border of the new development (Sony and debis compound), see Figure 5, are very much exposed to the prevailing wind from west. Consequently large pedestrian level winds will occur in the neighbouring street. This negative effect is enhanced by the Venturi effect at the junctions of the streets running W/E (Potsdamer Straße) and S/W - N/E (Alte Potsdamer Straße) with the debis plaza. At these street junctions extremely large pedestrian lèvel winds are predicted based on the model scale investigation. Most critical was the junction of the Alte Potsdamer Straße and the debis plaza. By reshaping this junction and slightly increasing the height of the two buildings at the eastern side of the debis plaza the situation could be improved appreciably.

Whereas city planners and architects will obtain useful information for common building situations from books like the one by Gandemer et Guyot (1976), the wind effects around complex building situations should be evaluated in the planning phase by wind engineers. Usually such evaluations will prevent aerodynamic accidents and will ensure that the usage anticipated by the architectural design will be possible after realisation of the building complex.

LITERATURE

Gander, J. et Guyot, A. (1976), "Intégration Du Phénomène Vent Dans La Conception Du Milieu Bati", La Documentation Française

Gerhardt, H.J. (to be published), "Windklima und Stadtplanung - ein Fallbeispiel", Proc. D-A-CH '93, Windtechnologische Gesellschaft

Heusch/Boesefeld GmbH (1993), "Verkehrliche Grundlagen und Emissionsprognose, UVU Potsdamer/Leipziger Platz"

Irwin, P.A. (1981), "A Simple Omnidirectional Sensor for Wind Tunnel Studies of Pedestrian Level Winds", JWEIA 7, pp 219 - 239, Elsevier

Ketzler, G., Schleicher, H., Stützel, V. (1993), "Umweltverträglichkeitsuntersuchung Potsdamer/Leipziger Platz in Berlin - Fachbeitrag Stadtklima

Oke, T.R. (1987), "Boundary Layer Climates", Methum

SMALL SCALE MODELLING OF COMBINED WIND AND BUOYANCY EFFECTS

MICHAEL POREH

Department of Civil Engineering, Technion, Israel Institute of Technology, Haifa, 32000, Israel

SUMMARY

The requirements for physical simulation of flows with both wind and buoyancy effects are discussed. It is shown that only *approximate simulations* of such flows can be achieved in small-scale wind tunnel models, and that the required scaling criteria for such simulations have not been uniquely determined. The presented review of small-scale studies of plume rise in neutrally stratified and convective boundary layers, plume lift-off, large area fires, urban heat islands and smoke motion in buildings shows that different criteria are being proposed by different investigators for simulating identical problems. Thus, the accuracy of approximate modelling, which varies from one problem to the other, also depends on the choice of the scaling criteria. In spite of this unsatisfactory state of the art, approximate small-scale models appears to be an outstanding method for validating numerical codes and for studying applied wind engineering and environmental problems.

INTRODUCTION

The continuous and rapid growth of computer power creates many times a feeling and a hope that a comprehensive and rigorous Computational Fluid Dynamics model capable of solving all wind engineering problems will soon be available. A similar feeling existed a few decades ago among researchers that were busy solving the turbulent transport equation using the K-Theory. Such dreams ignore the inherent semi-empirical nature of all turbulent closures, which makes it necessary to validate predictions of new problems by comparison with experimental data.

The need to validate numerical models has been one of the major driving forces for the development of special wind tunnels for small-scale physical modelling, often called *fluid modelling*, of environmental and wind engineering phenomena. Indeed, the success of such modelling has been of great help in the development of better turbulent closures. But, it has also turned out that small-scale fluid models can be successfully used to get a better understanding of complex flows and to solve practical environmental and wind engineering problems. This success, and the thrilling excitement of scaling-down and observing natural phenomena in small-scale models, should not hide the fact that fluid modelling, too, has its own inherent limitations and uncertainties. Some of them are related to the approximate nature of the similarity obtained in small-scale models and to the problem of determining the scaling criteria for such simulations, which is discussed below.

THE APPROXIMATE FLUID MODELLING PROBLEM

To preserve similarity between a model (m) and a prototype (p), all the dimensionless numbers (Π_i), that can be formed from all the independent variables and the boundary conditions associated with the studied phenomena, must be equal in the model and the prototype:

$$\Pi_i(m)=\Pi_i\ (p). \tag{1}$$

This condition secures the desired similarity and determines the scaling criteria for the simulation, namely, the scaling laws for the different variables in the model.

In most fluid models, the values of some dimensionless parameters, whose effect is assumed to be relatively small, are not matched with their values in the prototype flow. Consequently, such models provide only *approximate simulations* of the studied phenomena. The accuracy of approximate simulations is not the same for different phenomena in the same flow field. It can also change considerably across the flow field. Because of the interaction between the different regions of the flow, the overall accuracy of such simulations is not always clear, as shown below.

The Rossby number, $U/L\Omega$, where Ω is the vertical component of the earth's angular velocity, for example, cannot be matched, and thus, the relative effect of the Coriolis acceleration is not reproduced correctly in such models.

Similarly, it is impossible to match in small-scale models the Reynolds of the prototype flow. The Reynolds number is a measure of the importance of turbulent fluxes relative to molecular fluxes. Thus, it is usually assumed that if the flow in the model is turbulent, namely

$$Re > Re_{crit}\ , \tag{2}$$

the effect of molecular fluxes can be neglected. This assumption is valid, for example, when one focuses on transport processes within the turbulent region of the constant stress layer, say $30 < zu^*/\nu < 0.15\delta$, where δ is the boundary layer thickness and u^* is the shear velocity. Thus, in this region,

$$du/dz=(1/k)u^*/z, \tag{3}$$

where k is a universal constant, independent of the viscosity and the small scale roughness of the ground surface. Integration of (3) yields the logarithmic law

$$u/u^*=(1/k)\ln z + C\,. \tag{4}$$

Near smooth walls, however, a viscous sublayer is formed. Its relative thickness increases as the Reynolds number of the model is decreased and, thus, one cannot simulate in small scale models transport phenomena across this sublayer. Moreover, although viscous effects above the sublayer are negligible, the value of C in Eq. 4 does depend on the viscosity of the fluid, and it is not reproduced correctly in such models. Fortunately, one can eliminate viscous effects above the sublayer by increasing the roughness in the model (Cermak, 1971; Snyder, 1972).

Besides possible Reynolds number effects on the mean velocities in certain regions of the model, a reduction of the Reynolds number attenuates the fluctuations of the kinetic turbulent energy of the small-scale eddies in the entire flow field. The resulting change in the turbulent energy spectrum is very significant whenever the studied phenomenon is affected by small-scale turbulence; relative diffusion near a small source, for example.

When a buoyancy emission is included in the boundary conditions, it is usually characterized by its exit velocity from the source, W, the diameter of the source, d, and the difference between its density and that of the ambient fluid, $\Delta\rho=\rho_a-\rho_s$. As explained earlier, the dimensionless numbers that describe the boundary conditions should also be equal in the model and prototype. A typical example is the densimetric *Froude number*

$$Fr=U/[(\Delta\rho/\rho_a)gL]^{1/2}. \tag{5}$$

According to Eq. 5, when $\Delta\rho/\rho_a$ is equal in the model and the prototype; the velocity scale reduction, $\lambda(U)$, where $\lambda(x)=x_m/x_p$ is equal to the scale reduction factor of the variable x, is proportional to $\lambda(L)^{1/2}$, and the Reynolds number is proportional to $\lambda(L)^{3/2}$. Now, it is generally agreed that the effect of $\Delta\rho/\rho_a$ on the flow field is small, as long as the densimetric Froude number and other dimensionless numbers describing the source remain constant. As seen from Eq. 5, by increasing the value of $\Delta\rho/\rho_a$ at the source, the value of the Reynolds number of the model can be increased. Similarly, it appears that the exit velocity and/or the diameter of the source in the model need not always be modelled exactly, as long as the equality of more significant dimensionless numbers is maintained, which makes it possible to further increase the Reynolds number of the model.

It appears, however, that when such relaxations are made solely on the basis of intuitive arguments and analogies, the determination of the required scaling criteria for simulation becomes an incomplete problem (Poreh and Kacherginsky, 1981). It is, thus, not surprising that researchers have proposed different criteria for simulating identical problems. As a result, the accuracy of approximate fluid modelling, which varies from one problem to the other and is not necessarily uniform across each flow field, also depends on the choice of the scaling criteria for each problem.

SCALING CRITERIA FOR FREQUENTLY SIMULATED FLOWS

Plume Rise in a Neutrally Stratified Atmosphere.

The trajectory, $\underline{z}=f(\underline{x})$, of a buoyant plume rising from a vertical stack in a neutrally stratified atmosphere is determined by the characteristics of the ambient flow and the initial properties of the plume. The ambient flow field can be described by a characteristic length, L, such as the height of the stack, a reference velocity U, the mass density of the ambient air ρ_a, and M_i dimensionless parameters describing the relative flow characteristics, boundary conditions and fluid properties. These parameters include the Reynolds number, Rossby number, and Prandtl number of the flow. The parameters describing the stack and the initial properties of the plume are: the exit velocity, W, the initial plume density, ρ_s, the initial specific weight deficiency, $\Delta\rho g$, the initial plume diameter, d, as well as N_i dimensionless parameters that describe the relative properties of the plume gases and their initial distributions at the stack, if these are not uniform. It follows, from dimensional considerations, that the plume trajectory can be described by the following relation:

$$\underline{z}/\underline{x} = F\{\rho_a U^2/(\Delta\rho gL),\ W/U,\ d/L,\ \Delta\rho/\rho_a,\ Wd/\nu,\ M_i,\ N_i\} \qquad (6)$$

An exact simulation of the flow field and the plume rise requires that the values of all the independent dimensionless parameters appearing on the right-hand side of

TABLE 1: Scaling laws used by different research groups for similarity of plume rise

	Exact simulation	Approximate simulations						
		Fan [1967], Hoult et al. [1977], Cermak [1971]	Melbourne [1968], Isyumov et al. [1976, 1979], Snyder [1972,1981] Near Field (2nd choice)	Snyder [1981] Far Field (1st choice)	Barrett [1973]	Skinner [1978], Ludwig [1976], Snyder [1981] Far Field	Poreh and Kacherginsky [1981] Necessary (2nd choice)	Poreh and Kacherginsky [1981] Recommended conditions
Matched dimensionless parameters	$\frac{\rho_a U^2}{\Delta\rho g L}\,\frac{U}{W}\,\frac{L^2}{d^2}$	$\frac{\rho_a U^2}{\Delta\rho g L}\,\frac{U}{W}\,\frac{L^2}{d^2}$	$\frac{\rho_a U^2}{\Delta\rho g L}\,\frac{L^2}{d^2}$	$\frac{\rho_a U^2}{\Delta\rho g L}\,\frac{L^2}{d^2}$	$\frac{\rho_s U^2}{\Delta\rho g L}$	$\frac{\rho_a U^2}{\Delta\rho g L}\,\frac{U}{W}\,\frac{L^2}{d^2}$	$\frac{\rho_a U^2}{\Delta\rho g L}\,\frac{U}{W}\,\frac{L^2}{d^2}$	$\frac{\rho_a U^2}{\Delta\rho g L}\,\frac{U}{W}\,\frac{L^2}{d^2}$
	$\frac{\rho_s W^2 d^2}{\rho_a U^2 L^2}$	--	$\frac{\rho_s W^2 d^2}{\rho_a U^2 L^2}$	$\frac{\rho_s W^2 d^2}{\rho_a U^2 L^2}$	$\frac{\rho_s W^2}{\rho_a U^2}$	$\frac{\rho_s W^2 d^2}{\rho_a U^2 L^2}$	$\frac{\rho_s W^2 d^2}{\rho_a U^2 L^2}$	$\frac{\rho_s W^2 d^2}{\rho_a U^2 L^2}$
	W/U	W/U	--	--	--	--	--	W/U
	d/L	d/L	d/L	d/L	Wd^2/UL^2	d/L	--	--
	ρ_s/ρ_a	--	--	--	--	--	--	--
$\lambda(U/(gL)^{1/2}$	1	$\lambda(\Delta\rho/\rho_a)^{1/2}$	$\lambda(\Delta\rho/\rho_a)^{1/2}$	$\frac{\lambda(\Delta\rho/\rho_a)^{1/2}}{\lambda(\rho_s/\rho_a)^{-1/2}}$	$\lambda(\Delta\rho/\rho_s)^{1/2}$	$\frac{\lambda(\Delta\rho/\rho_a)^{1/2}}{\lambda(\rho_s/\rho_a)^{1/4}}$	--	$\lambda(\Delta\rho/\rho_s)^{1/2}$
and its value for $\Delta\rho/\rho_a\vert_p = 0.2$, $\Delta\rho/\rho_a\vert_m = 0.8$	1	2	2	1	4	$2\sqrt{2}$		4
$\lambda(W/U)$	1	1	$\lambda(\rho_s/\rho_a)^{-1/2}$	$\lambda(\rho_s/\rho_a)^{-1/2}$	$\lambda(\rho_s/\rho_a)^{-1/2}$	$\lambda(\rho_s/\rho_a)^{-1/2}$		1
and its value for $\Delta\rho/\rho_a\vert_p = 0.2$, $\Delta\rho/\rho_a\vert_m = 0.8$	1	1	2	2	2	2		1

Eq. (6) be equal in the model and prototype. This requirement cannot be fulfilled even when one attempts to simulate only the ambient flow. However, the role of the Rossby number may, in many cases, be ignored. When the Reynolds number of the flow in the model is sufficiently large, its effect, as well as those of the Prandtl number and the Schmidt number, may also be neglected. In such cases,

$$\underline{z}/\underline{x} = F[\rho_a U^2/(\Delta\rho g L), \quad W/U, \quad d/L, \quad \Delta\rho/\rho_a] \,. \tag{7}$$

Equation 7 implies that the required criteria for simulating plume rise is the equality of the four dimensionless numbers on its right hand side, namely:

$$\lambda[\rho_a U^2/(\Delta\rho g L]=1, \quad \lambda(W/U)=1, \quad \lambda(d/L)=1 \quad \text{and} \quad \lambda(\Delta\rho/\rho_a)=1. \tag{8}$$

Simulations that satisfy these criteria are called *exact simulations*, although they do not fully reproduce all the molecular effects.

Now, although Eq. 7 may be formulated in an infinite number of forms, for example:

$$\underline{z}/\underline{x} = F\{U^3 L/[(\Delta\rho/\rho_a) g W d^2], \quad W/U, \quad d/L, \quad \Delta\rho/\rho_a\} \,, \tag{9}$$

the equality of the dimensionless numbers in Eq. (7) secures the equality of those in Eq. 9 and in any other form of this general relationship. When the requirement $\lambda(\Delta\rho/\rho_a)=1$ is relaxed the number of the dimensionless numbers that have to be matched is reduced to three. It is observed, however, that the set of three numbers one obtains by neglecting $\Delta\rho/\rho_a$ in Eq. 7 is not the same as the set which one obtains if it is neglected in. Eq. 9. This example demonstrates that the required criteria for approximate simulations cannot be determined solely by dimensional analysis. Indeed, different criteria have been used by different investigators for the simulation of plume rise, as shown in Table 1 (Poreh and Kacherginsky, 1981).

It may rightly be argued that the problem could be eliminated if the above assumption was confirmed in a careful tests, in which the other remaining dimensional parameters remained constant. Unfortunately, this is not the case. Supporters of the different criteria usually base their choice on partially intuitive arguments and a limited amount of experimental data. Thus, until a comprehensive experimental verification program is carried out in relatively large-models, it would be difficult to rigorously prove which scaling criteria give more accurate simulations of a the plume rise, or any other similar phenomenon. It seems, however, that most researchers support the conclusion that approximate simulations of plume rise should ensure the equality of the *dimensionless buoyancy flux* and the *dimensionless momentum flux* of the source, which are the first numbers on the right hand side of Eq. (9). Poreh and Kacherginsky (1981) have, thus, suggested, that only these two scaling criteria are required for approximate simulations, and that one should use a third condition $\lambda(W/U)=1$, mainly to increase the Reynolds number of the flow in the model. Isyumov and Tanaka (1979), on the other hand, concluded that it is advantageous to use exact simulations even at the expense of somewhat reduced similarity of the overall flow region.

Plume Rise in Convective Boundary Layers

The simulation of plume rise in a Convective Boundary Layer (CBL) is more difficult, because of the difficulties in simulating correctly the CBL itself and because the flow in the CBL is characterized by several velocity scales.

The breakthrough in the understanding of the nature of the CBL is primarily due to the water-tank experiments and numerical modelling in the Seventies by the joint works of Deardorff and Willis (See References).

The averaged CBL is usually idealized as composed of several layers (Wyngaard, 1985): The *surface layer*, where wind-shear plays a dominant role. The *free convection layer*, where wind-shear is no longer important but height z above the ground is a significant parameter. It usually extends to $0.1z_i$, where z_i is the thickness of the CBL. The *mixed layer*, where the turbulence structure is independent of both height and shear and is primarily controlled by the convectively-generated turbulence. The *interfacial layer*, which separates the mixed layer and the usually stably-stratified air above.

The structure of the lower layers is best explained using the Monin-Obukhov (MO) similarity law. The buoyancy-induced local turbulent velocity scale w(z) is given by $w(z)=(z\sigma)^{1/3}$, where σ is the buoyancy flux per unit area. The mechanically generated turbulence in this layer is of the order of shear velocity, u_*, and thus

$$(w(z)/u_*)^3 = \sigma z/u_*^3 = -z/(kL_{mo}) \,, \tag{10}$$

where $L_{mo}= -u_*^3/(k\sigma)$ is the MO Length and k=0.4 is von Karman's constant. The MO similarity has been successfully used to formulate scaling criteria for wind-tunnel simulations of surface layers, as it secures the same ratio of w/u_* for equal values of $-z/L_{mo}$ in the model and the atmosphere.

One notes from Eq. 10 that the MO Length is also a measure of the thickness of the surface layer, as the mechanical turbulence in the CBL is dominant only in the layer $z<|L_{mo}|$. Thus, when $-z_i/L_{mo}$ becomes large, the relative thickness of the surface layer decreases and the entire effect of the shear-generated turbulence on the structure of the CBL can be ignored. This is the basis for the *convective scaling* proposed by Deardorff (1970), which assumes that the turbulence characteristics of the CBL depend, primarily, on z_i and σ or, equivalently, on z_i and the convective velocity scale, w*, which is defined as

$$w^*=(z_i\sigma)^{1/3} \tag{11}$$

Water-tank simulations are then valid only when the role of the shear-produced turbulence is relatively small, which is usually the case for $-z_i/L_{mo}>10$ or $w^*/u^*>3$. These conditions usually correspond to values of $U/w^*< 7$, as it is usually assumed that $u^*\cong 0.05$ U. For rough surfaces, larger values of u^*/U can be obtained, of the order of 0.1, and the effects of the mean wind and shear stresses would diminish only at values of $U/w^* \lesssim 3.5$.

It has also been suggested that surface roughness can affect the surface layer even when U/w* is small, as the large convective eddies induce local mean winds of random orientation and speed near the surface, which in turn produce shear and turbulent kinetic energy. These effects are not allowed for in either the convective similarity hypothesis or the M-O similarity hypothesis. The convectively-mixed layer similarity could also be affected by processes in the interfacial layer (Dreidonks and Tennekes, 1984). For lack of space, such effects will be ignored in the following discussion.

Other important questions concerning possible effects of several parameters on the turbulent structure of the CBL and on the diffusion of buoyant plumes within this layer are still unanswered (Deardorff, 1985; Briggs, 1985; Poreh and Cermak, 1984). Improvement of the state of the art depends on the availability of experimental data. Field data appears to be prohibitively expensive. Water-tank experiments, where both the mean motion and shear are missing, cannot answer some of these questions either. On the other hand, it appears simulations in wind tunnels, where control of the convective and shear velocities, temperatures, gradients and the heat flux from the floor can be achieved, would be very helpful in depicting some of these effects that cannot be simulated in water tanks (Poreh and Cermak, 1985; Plate and Rau, 1986; Caughey, 1984). However, their realization critically depends on the dimensions and the characteristics of these wind tunnels, and the choice of the scaling criteria for such simulations (Poreh et al., 1991; Meroney and Melbourne, 1992), that will be discussed below.

Obviously, these criteria should determine the scaling of the major characteristics of the CBL (U, u_*, w_*, and z_i) and the characteristics of the buoyant emissions (W, $\Delta\rho$, d, the momentum flux M, and the buoyancy flux B at the source). As the effect of $\Delta\rho/\rho$ in such simulations is neglected, one faces the previously discussed problem of selecting the appropriate scaling criteria. However, this time, it is more complicated, as the CBL is characterized by at least two velocity scales, U and w_*, rather than one. It will be assumed, for the moment, that u_* is determined by U and that the effect of the interfacial layer can be neglected.

Meroney and Melbourne (1992), suggest that such simulations require the equality of

$$MR = M/(\rho U^2 z_i^2)\ , \tag{12}$$

where M is the momentum flux from the source, and of

$$Fr_m = F/(U^3 d), \tag{13}$$

where F is the buoyancy flux from the source (proportionality constants in the definition of M and F will be ignored in this paper). They noted, however, that the *modified Froude number*

$$F^* = F/(z_i U w_*^2) \tag{14}$$

had been successfully used to characterize the motion of plumes in CBL (Briggs, 1975; Kerman, 1979; Venkatram, 1980). Clearly, unless a similarity of d/z_i and w_*/U is maintained, the equality of both Fr_m and F^* is not secured. It is questioned by the author, whether an exact matching of d/z_i is always essential for such simulations, as its value is usually a small number. The equality of w_*/U might be desired, but it is clear from the success of the water-tank simulations, where $U=0$, that it is not really essential, certainly not for $w_*/U<0.1$, unless an undistorted vertical/horizontal scaling of the motion in CBL is desired.

Melbourne et al. (1993), on the other hand, suggested to distinguish between two phases of diffusion. They proposed that similarity in the *initial phase* is secured by the matching of MR and

$$F/(U^3 z_i). \tag{15}$$

For the *second phase* they proposed to match the value of MR and F*. These two requirements are satisfied simultaneously when a similarity of w_*/U is maintained.

Poreh et al. (1991) focused on wind-tunnel simulations of cases in which the effect of shear at ground level is significant, as other cases can be simulated in water-tanks. For such cases, it is essential to maintain the similarity of w_*/u_*. They proposed, however, to increase the ratio u_*/U in the model by increasing the roughness of the wind tunnel. The distorted similarity of w_*/U and of the vertical/horizontal scaling of the CBL, they suggested, should not effect the turbulent processes and would make it possible to examine the far field of diffusion in relatively short wind tunnels.

Lift-off of Buoyant Plumes

The behavior of buoyant plumes from ground level sources is often similar to that of plumes from stacks. However, in many cases, the vertical momentum of the source is small and, in the presence of winds, the plume is carried downwind along the surface. It has been observed by Briggs (1973) and Meroney (1979), however, that at a certain distance from the source the buoyant plume lifts-off, as shown schematically in Fig. 1.

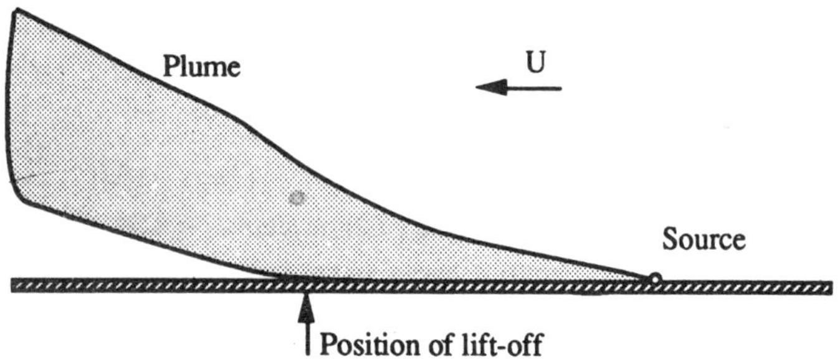

Fig. 1 Schematic description of lift-off

Since the ground level concentrations of pollutants from such sources decreases drastically downwind of the lift-off point, it is of great interest to determine whether and where lift-off occurs.

The results of only a few studies of lift-off have so far been reported (Meroney, 1979; Hall et al., 1980; Poreh and Cermak, 1988). Perhaps, because the problem is also related to the behavior of accidental emissions from nuclear power plants. Both

Meroney (1979) and Hall et al. (1980) have assumed that lift-off is controlled by the buoyancy flux from the source B, the wind speed U, and a suitable length scale L. Thus, the dimensionless concentrations downwind the source were assumed to depend on $B^*=B/U^3L$:

$$C^*=CUL^2/Q=F[B/U^3L] \tag{16}$$

where Q is the strength of the source. An implicit requirement for simulation is that the Reynolds number of the model is sufficiently large. To satisfy this requirement, the buoyancy flux in the model should be increased. Since the maximum possible value of $\Delta\rho/\rho_a$ in wind-tunnel models, is of the order of 0.85, this is accomplished by increasing the exit velocity W. In doing so, the momentum flux from the source, which is proportional to $\rho_s QW$, is increased. It is obvious that an increase of the vertical momentum flux from the source would augment the lift-off in the model. Poreh and Cermak (1988) showed, however, that even a horizontal momentum flux at the source, which discharges the buoyant fluid in the direction of the wind, critically affects lift-off.

Figure 2 shows the effect of the *dimensionless horizontal momentum flux* $M^*=\rho_s U_s^2 d^2/(\rho_a U^2 L^2)$ on the *dimensionless concentration* profiles, C^*, at a given distance downwind of the source (x_m=1.0 m; x/δ=1.67, where δ is the thickness of the BL) for two values of the dimensionless buoyancy flux $B^*= B/U^3L$. The data shows that the plume stays closer to the ground for larger values of M*.
Figure 3 shows that the ratio of the ground level concentration, C(0), to the maximum concentration, C_{max}, at a given distance downwind from the source (x_m=1.5 m; x/δ=1.67) depends on the ratio $B^*/(M^{*1/2})$.

Large Area Fires and Heat Islands

The mechanics of Large Area Fires (LAFs), whether forest fires or urban fires, the motion of fire plumes, the strength of the fire generated winds, and the interaction of fire plumes with atmospheric winds and circulation, have been studied intensively during the Eighties, when a threat of a nuclear war and the possibility of a *Nuclear Winter* were suggested (Carrier et al., 1985; Cotton, 1985). During that period, the possibility of simulating the flow field induced by LAFs in small-scale fluid models was investigated by Poreh et al.(1985, 1986,1988). Now, it had been shown by Williams (1969) that an exact small scale simulation of all the processes that take place in fires is impossible.

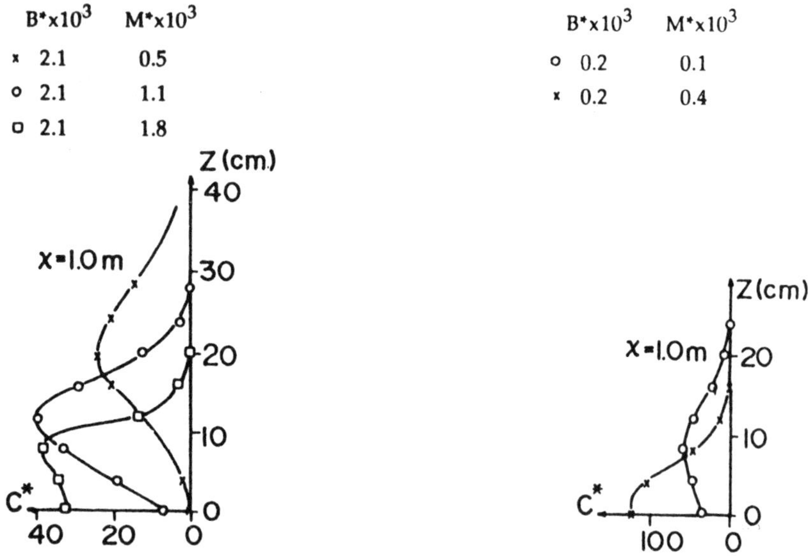

Fig. 2 The effect of the horizontal momentum flux on the concentration profiles, for two values of the buoyancy flux.

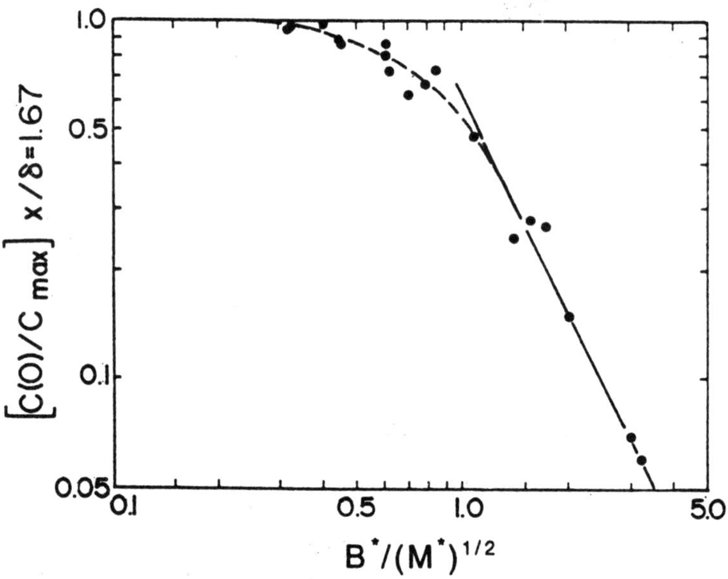

Fig. 3 The dependence of C(0)/Cmax at a given distance downwind from the source (x_m=1.5 m, x/δ=1.67) on $B^*/(M^{*1/2})$.

Poreh et al. (1985) observed, however, that the thickness of the *combustion layer* in LAFs is very small, relative to the horizontal length scale of such fires. In such cases, they claimed, the flow induced by the fire is similar to the flow induced by an area source of buoyancy, and neither the thickness of the combustion layer nor the method by which the buoyancy is produced is of importance.

The above assumption implies that the velocity and temperature field above the combustion layer is determined by only two parameters: The convective buoyancy flux from the fire, B, or equivalently the buoyancy flux per unit area s, and the size of the fire, say the radius, R, for a circular fire. Namely,

$$u_i(x_i)/(\sigma R)^{1/3}=F(x_i/R) \qquad (17)$$

and

$$g'(x_i)/(\sigma^{2/3}R^{1/3})=B(x_i/R) \qquad (18)$$

The above universal functions are valid, of course, for large Reynolds numbers only. Poreh et al. (1991) suggested that the flow above buoyancy sources is turbulent when the Reynolds number

$$\sigma^{1/3}R^{4/3}/\nu > 6{,}000\ . \qquad (19)$$

In the presence of ambient winds (U_a), atmospheric vorticity (ω), and atmospheric stratification $G=(d\theta/dz)g/\theta$, where θ is the potential temperature), the scaling of these boundary conditions should be:

$$\lambda(U_a)=\lambda(\sigma)^{1/3}\lambda(L)^{1/3},\ \lambda(\omega)=\lambda(\sigma)^{1/3}\lambda(L)^{-2/3},\ \text{and}\ \lambda(G)=\lambda(\sigma)^{2/3}\lambda(L)^{-4/3} \qquad (20)$$

To model the effect of swirling on the plume, it is required that centripetal accelerations acting on the lighter-than-air plume, V_ϕ^2/r, where V_ϕ is the circumferential velocity, be scaled in the model as g'. This is satisfied when

$$\lambda(\Delta\rho/\rho_a)=1,\quad \text{or} \quad \lambda(\sigma)^2=\lambda(L)\ . \qquad (21)$$

The question when may the above requirement be relaxed is of great importance. Such a relaxation is accepted in horizontal flows when the densimetric Froude number, $U/(g'h)^{1/2}$, is smaller than unity. Replacing U by V_ϕ, g' by $\rho V_\phi^2/R$ and L by R one finds that for rotating flows, this requirement may be relaxed when

$$\Delta\rho/\rho_a<1. \qquad (22)$$

Three possible methods of creating small-scale area sources of buoyancy have been used in previous studies.

The use of hot plates for simulating area fires was proposed by Nielsen (1965) and Long (1967). Parker et al. (1968) have used a hot plate to simulate the 335 m by 320 m experimental Flamebeau Fire. The primary disadvantage of generating the buoyancy flux in small-scale models by hot plates is that an unknown portion of the heat supplied to the plate, as high as 42%, is lost by radiation. In addition, the size of the model must be very large and the surface temperature of the hot plate must be very high, to ensure a turbulent flow and sufficiently high velocities that can be readily measured. Close to the surface of a hot plate a *conduction layer* is formed, where heat transfer is controlled by the molecular diffusivity k. It can be shown that its relative thickness, h_c/R is of the order of $[(\alpha^3/\sigma)1/4]/R$ (Thomas and Townsend, 1957), where α is the thermal diffusivity. When Eq. (19) is satisfied, its value will be of the order of 0.001, and its effect on the rest of the flow can be neglected.

Small area sources of buoyancy can also be created by small fires. It is, however, necessary to ensure that the thickness of the combustion layer in the model is relatively small and that, at the same time, the induced velocity field is turbulent. Yokoi (1959) studied the temperature field above small-scale alcohol pool-fires and above a circular fire made out of many little wicks of alcohol lamps, which he termed discontinuous heat sources. Although the relative thicknesses of the combustion layers in these model-fires were not as small as desired, these fires produce a fair approximation of area sources of buoyancy.

Poreh et al. (1986) have modelled LAF by injection of helium through a porous plate at ground level. The thickness of the layer where the helium is mixed with less than six times air by mass, in such models, is relatively small, so that the effect of the value of $\Delta\rho$ of helium on the flow field can be assumed to be small. It has also been shown that the effect of the momentum flux in the models is limited to a very thin layer.

Figure 4 shows the distribution of the dimensionless parameter B along the vertical axis of various helium models and alcohol-fire models (The values of B_0 denoted at the abscissa are for pure helium). The results support the basic thesis that the overall structure of the flow field above area sources of buoyancy is determined by their size and intensity and is independent of the way this buoyancy is produced.

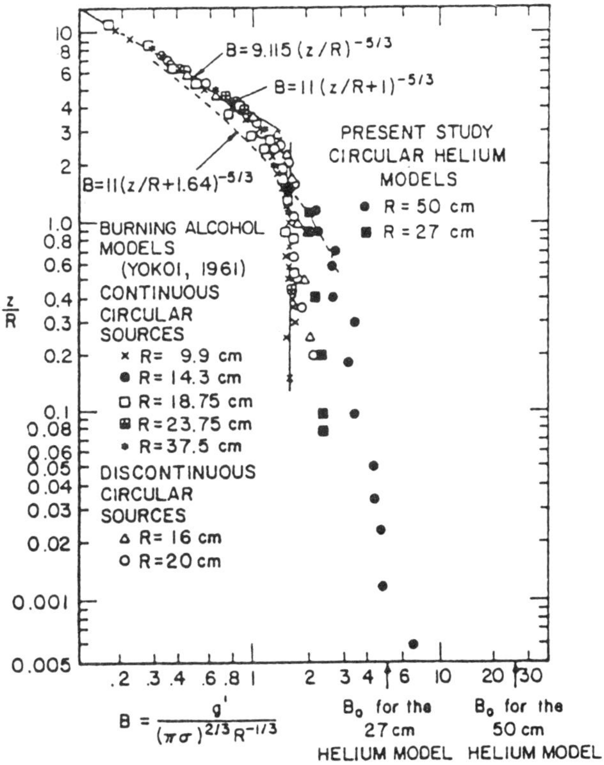

Fig. 4 Vertical distribution of the dimensionless number B along the axes of various circular buoyancy sources.

Now, it has been shown by Poreh (1992) that the flow and temperatures above urban *heat islands* are many times determined by the size and the buoyancy flux of the urban area. Thus, small scale models of LAF are also models of such heat islands. Using the results from LAF models, many new characteristics of heat islands were identified. It was also proposed that other types of heat islands could be simulated in small scale models, and the scaling criteria for such simulations were discussed.

Smoke Movement and Stack Induced Ventilation in Buildings

The movement of smoke generated by fires in buildings is of considerable practical importance, as smoke is recognized as the major killer in fire situations. Smoke often migrates to building locations remote from the fire space, threatening life and damaging properties. The movement of smoke also determines the response of alarm systems that use smoke or heat sensors (Klote and Milke, 1992). Now, the smoke movement is determined by a large number of parameters. For a given building configuration, the buoyancy flux generated by the fire is, usually, the most important parameter. The ambient wind speed, U, and its direction become important factors, however, for large wind speeds and relatively small buoyancy fluxes. The use of sprinkles, which cool the smoke plume, makes such cases quite frequent.

The prediction of smoke movement in buildings is usually done using simplified numerical models. Such models have basic shortcomings. Some of them may be overcome by small-scale physical models, that simulate the motion of the smoke in the building.

The principles of physical modelling of smoke movement in buildings have been discussed by Cannon and Zukoski (1976), Quintiere et al. (1988) and Poreh et al. (1992). It is clear from these studies that, only approximate simulations of this movement is possible. The scaling criteria and shortcomings of such simulations will be reviewed below.

Assuming that the effective gravitational acceleration $g' = g\Delta\rho/\rho$ and the velocity u_i at each point are functions of the buoyancy flux B, a length scale L, the thermal diffusivity α, the Prandtl number, $Pr=\alpha/\nu$, and the ambient wind speed U (and direction), it may be written that:

$$g^* = g'/(B^{2/3}L^{-5/3}) = F[\, U/(B^{1/3}L^{-1/3}),\ \alpha/(B^{1/3}L^{2/3}),\ Pr]. \quad (23)$$

The dimensionless velocity at each point, $u_i/(B^{2/3}L^{-5/3})$, will also be a function of the same dimensionless variables. The number $\alpha/(B^{1/3}L^{2/3})$ may be considered as a representative Peclet number of the heat flow. The Peclet and the Reynolds numbers in small-scale models cannot be matched with those in the prototype. If it is assumed that the effect of both Re and Pr can be neglected, one may write that

$$g^* = F[U/(B^{1/3}L^{-1/3})]\ . \quad (24)$$

As it is impossible to maintain the same Reynolds numbers in the model and the prototype, the conductive heat transfer to the walls is not properly simulated. Clearly, the film resistance in the models will be larger than in the full scale building. Since loss of heat to walls may, in many cases, reduce the stack effect, small-scale models may give an over optimistic simulation of smoke removal by this effect.

For zero or small wind speed one finds that when a steady state is reached,

$$g^*=g'/(B^{2/3}L^{-5/3})= \text{constant} \tag{25}$$

For time dependent conditions, g* will be a function of the dimensionless time

$$t^*=t/(B^{-1/3}L^{4/3})\ . \tag{26}$$

Figure 5a shows measurements of temperature rise versus time in two simulations with different values of U and B but with equal values of $B^{1/3}L^{-1/3}$. Figure 5b shows the same data in dimensionless variables.

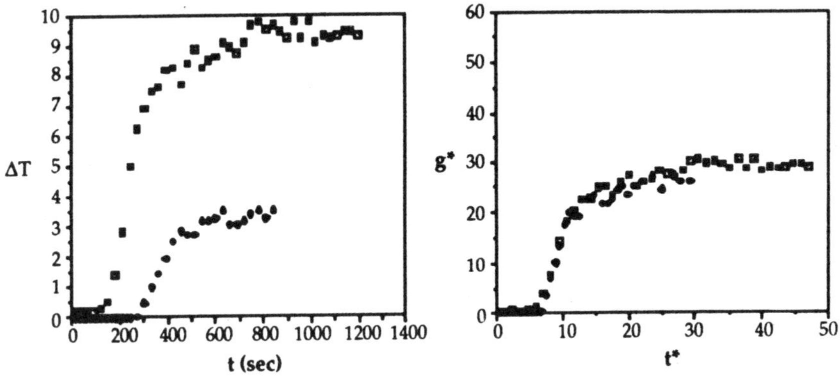

(a) (b)

Fig 5 The variation of the temperature rise ΔT and of g* in a water-tank model plotted versus t and t* (U= 3.8 and 6.3 cm/s, and $U/(B^{1/3}L^{-1/3})=19$).

When Eq. (25) is valid, the simulation of the flow and temperature fields for a given geometry can be made at a reduced geometrical scale and a desired value of B. The way B is produced is of no significance. In other words, as in the case of LAFs, the fire is assumed to be source of buoyancy, which drives the flow and, theoretically, the experimental methods described earlier for generating buoyancy sources, can be used in such simulations.

Now, some of the heat flux produced by a fire is converted into a convective heat flux. A large portion of it, 30-40%, may be transferred to the structure or be lost by radiation. The same is true in small-scale models, in which the prototype fire is simulated by a small scale fire or a heat source. However, as the ratio of the various heat fluxes in such models is not the same as in the prototype fire, such models do not offer any advantage in this respect.

One should also note that the size of the combustion region could be an important factor that might effect smoke motion both in the prototype building and the model. This is particularly true if the height of the hall or room, where the fire burns, is relatively small. When the size of the fire decreases and/or the burning rate increases, the height of the combustion region increases. The volumetric flux at a given height z, which is an important parameter in the design of smoke control systems, will also depend on these parameters.

An implicit assumption in the above dimensional analysis is that $\Delta\rho/\rho$ in the model and prototype need not be equal to each other for correct simulation. The limitations of such a relaxation have been discussed earlier, and it appears that they are not very significant in the modelling smoke movement in building.

A clear limitation of such small models is that the flow through small openings and cracks in the inner and outer envelopes of the different spaces in the building, which have small Reynolds numbers, cannot be correctly simulated. To overcome this limitation, one could use a single orifice to model all the cracks at each wall of a given space. The size of the orifice is determined so that the effective leakage area in the building is scaled to match the total *effective leakage area* of each wall in the model at the representative Reynolds number of the experiment. Clearly, the modelling of these flows is very rough, however, it is should be recalled that the estimate of the effective leakage areas in real buildings is very poor too.

Cannon and Zukoski (1976) have tried to augment the heat losses to the wall in the model by cooling the walls of the walls. Such a method is not be included in most physical models, both due to difficulties in its implementation and due to estimates that it is relatively small after the first phases of the fire and smoke spreading.

The ventilation and heat removal from industrial buildings are many times governed by the same mechanism as the motion of smoke. As suggested by Poreh et al. (1992) wind tunnel simulations of wind and buoyancy induced ventilation of such buildings is possible if the total buoyancy flux inside the buildings is known. The scaling criteria for such simulations are identical to those proposed above.

CONCLUSIONS

It has been shown that various complex flow phenomena, which are created or affected by buoyancy sources, have been simulated in small-scale models. It appears, however, that most of these simulations are not exact, and that the required scaling criteria for such approximate simulations cannot be rigorously determined. As a result, different criteria are being used by different investigators for simulating identical problems, and the accuracy of the simulations, which varies from one problem to the other and is not necessarily uniform across each flow field, also depends on the choice of the scaling criteria.

It has been concluded that only an extensive experimental study, aimed at evaluating the proposed criteria for each case, will be able to rigorously prove which of them minimizes the inaccuracies of approximate simulations.

Now, the question rises, whether one could really depend on results obtained in approximate simulations. Clearly, a universal approval cannot be given to all approximate simulations. However, it should be realized that such simulations have a basic quality, which is not always shared by other methods of investigation. They might not have the desired values of all the dimensionless numbers required for an exact simulation of the particular full-scale flow one would like to study. On the other hand, they always depict the characteristics of a real flow, and satisfy all the relevant laws of physics. Thus, they can always be used for validation of numerical models that could be executed at the exact conditions of these physical models. Experience indicates, that their use for studying real life problems is also justified, as the boundary conditions of such problems vary over a very wide range.

It remains, however, our duty to carefully asses the accuracy of each approximate simulation and try to improve it.

ACKNOWLEDGEMENTS

The work has been sponsored by a grant from the Fund for the Promotion of Research at the Technion. The author thanks Dr. J. E. Cermak for reviewing this work and for his important comments.

REFERENCES

Barrett, R.V. (1973), "Use of the Wind Tunnels to Investigate the Influence of Topographical Features on Pollution from a Tall Stack", Chimney Design Symp., University of Edinburgh, April 1973.

Briggs, L.A. (1973), "Lift-off of Buoyant Gas Initially on the Ground", ADTL Contribution File No. 87, Nov. 1973, Air Resources Atmospheric Turbulence and Diffusion Lab., NOAA, Oak Ridge, Tennessee.

Briggs, G.A. (1975), "Plume Rise Predictions in "Lectures on Air Pollution and Environmental Impact Analysis", Workshop Proceedings, Boston, Mass., Sept.29-Oct.3, 1975 American Meteorological Society, pp. 59-111.

Briggs, G.A. (1985), "Analytical Parameterizations of Diffusion: The Convective Boundary Layer", Journal of Climate and Applied Meteorology, Vol. 24, No. 11, pp. 1167-1186.

Cannon, J.B. and Zukoski, E.E. (1976), "Turbulent Mixing in Vertical Shafts under Conditions Applicable to Fires in High RiseBuildings". Technical Report, Guggenheim Jet Propulsion Center, Calif. Inst. of Tech., Pasadena.

Carrier, G.F., Fendell, F.E. andFeldman, P.S. (1985), "Firestorms", Journal of Heat Transfer, Vol. 107, pp. 19-28.

Caughey, F.J. (1984), "Observed Characteristics of the Atmospheric Boundary Layer", Atmospheric Turbulence and Air Pollution Modelling (ed.F.T.M. Nieuwstadt and van Dop), D. Reidel Publ. Co., Boston, pp. 107-158.

Cermak, J.E. (1971),"Laboratory Simulation of the Atmospheric Boundary Layer". AIAAJ., 6 pp. 1746-1754.

Cotton, W.R. (1985), "Atmospheric Convection and Nuclear Winter", American Scientist, Vol. 73, pp. 275-278.

Deardorff, J.W. (1970), "Convective Velocity and Temperature Scales for the Unstable Planetary Boundary Layer and for Rayleigh Convection", J. Atmos. Sci., 27, pp. 1211-1213.

Deardorff, J.W. and Willis, G.E. (1975), "A Parameterization of Diffusion into the Mixed Layer", Journal of Applied Meteorology, 14, pp. 1451-1458.

Deardorff, J.W. (1985), "Laboratory Experiments on Diffusion: The Use of Convective Layer Scaling", Journal of Climate and Applied Meteorology, Vol. 24, No. 11, pp. 1143-1151.

Dreidonks, A.G.M. and Tennekes, H.(1984), "Entrainment Effects in the well Mixed Atmospheric Boundary Layer", Boundary-Layer Met. 30, 75-105.

Fan, L.N. (1967), Turbulent Bouyant Jets in Ambient Stratified Fluids, Calif. Inst. of Technology Rep. KH-R-15, June 1967.

Hall, D.J., Barrett, C.F. and Simmonds, A.C. (1980), "Wind Tunnel Model Experiments on a Bouyant Emission from a Building", Warren Spring Laboratory, Department of Industry, Gunnels Wood Road, Stevenage, Hertfordshire SG1 2BX, 1980.

Hoult, D.P., O'Dea, S.R., Touchtan, G.L. and Ketterer, R.J. (1977), "Turbulent Plume in a Turbulent Cross-Flow; Comparison of Wind-Tunnel Test with Field Observations", J. Air Pollut. Control Assoc., 27, pp. 56-60.

Isyumov, N., Jandali, T. and Davenport, A.G. (1976), "Model Studies and the Prediction of Full-Scale Levels of Stack Gas Concentration", J. Am. Pollut. Contrl Assoc., 26, pp. 956-964.

Isyumov, N. and Tanaka, H. (1979), "Wind Tunnel Modelling of Stack Gas Dispersion - Difficulties and Approximations", Preprint Volume, Fifth Int. Conf. on .Wind Engineering, Colorado State University, Fort Collins, CO, July 1979, pp. VIII-3, 1-15.

Kerman, B.R. (1979), "A Similarity Model for Maximum Ground Level Concentration in a Freely Convective Atmospheric Boundary Layer", Boundary Layer Meteorology, Vol. 16, pp. 395-408.

Klote, J.H. and Milke, J.A. (1992), "Design of Smoke Management Systems", ASHRAE, 1992.

Long, R.R. (1967), "Fire Storms, Fire Research Abstracts and Reviews", Vol. 9, pp. 55-68.

Ludwig, G.R. and Skinner, G.T. (1976), "Wind Tunnel Modelling Study of the Dispersion of Sulfur Dioxide in Southern Allegheny County", Pennsylvania, Calspar Corporation, P.O. Box 400, Buffalo, NY 14255, December 1976.

Melbourne, W.H. (1968), "Wind Tunnel Modelling of Bouyant Chimney Plumes", Proc. 3rd Australian Conference on Hydraulic and Fluid Mechanics, Sydney, November 1968, pp. 81-85.

Melbourne, W.H., Taylor, T.J. and Grainger, C.F. (1993), "Dispersion Modelling in Convective Wind Flows", Department of Mechanical Engineering, Monash University, Clayton 3168, Victoria Australia, Submitted to Atmospheric Environment, 1993.

Meroney, R.N. (1979), "Lift-off of Buoyant Gas Initially on the Ground", J. Industrial Aerodynamics, October 1979, 5, Nos. 1 and 2.

Meroney, R.N. and Melbourne, W.H. (1992), "Operating Ranges of Meteorological Wind Tunnels for Simulation of Convective Boundary Layer Phenomena", Boundary Layer Meteorology, 61, pp. 145-179.

Nielsen, H.J. (1965), "Fire Storm Environmental Model", IITRI. Project Rep. Contract No. 0CD-PS-64-50, 1965.

Parker, W.J. (1967), "Urban Mass Fire Scaling Considerations", U.S. Naval Radiological Defence Laboratory, USNRDL-TR-67-150, 1967.

Parker, W.J., Corlett, R.C. and Lee, B.T. (1968), "An Experimental Test of Mass Fire Scaling Principles", U.S. Naval Radiological Defence Laboratory, NRDL-TR-68-117, 1968.

Plate, E.J. and Rau, M. (1986), "Concept of a Thermal Stratified Boundary Layer", Wind Tunnel Proceedings of the 3rd International Workshop on "Wind and Water-Tunnel Modelling of Atmospheric Flow and Dispersion", Lausanne, 1986.

Poreh, M. and Kacherginsky, A. (1981), "Simulation of Plume Rise using Small Wind-Tunnel Models", J. of Wind Engineering and Industrial Aerodynamics, 7, pp.1-14.

Poreh, M. and Hassid, S. (1982), "Simulation of Buoyancy and Wind-Induced Ventilation", Proceedings, International Workshop on Wind-Tunnel Modelling for Civil Engineering Applications, Gaithersburg, Maryland, Academic Press.

Poreh, M. and Cermak, J.E. (1984), "Wind Tunnel Simulation of Diffusion in a Convective Boundary Layer", Boundary Layer Meteorology, Vol. 30, pp.431-455.

Poreh, M. and Cermak, J.E. (1985), "Study of Neutrally Buoyant Plumes in a Convective Boundary Layer with Mean Velocity and Shear", Preprint Volume AMS Seventh Symposium on Turbulence and Diffusion, Boulder, CO, pp. 119-122.

Poreh, M., Stout, J.E., Peterka, J.A. and Cermak, J.E. (1986), "Physical Modelling of Large-Area Fire Plumes", Technical Report Defence Nuclear Agency DNA-TR-85-364, 1986.

Poreh, M., Stout, J.E., Peterka, J.A. and Cermak, J.E. (1986), "Criteria for Physical Simulation of Large-Area Fire Plumes", In Advancements in Aerodynamics, Fluid Mechanics and Hydraulics, Published by the American Society of Civil Engineers, New York, pp.959-966.

Poreh, M. and Cermak, J.E. (1988), "Lift-off of Buoyant Horizontal Plumes", Preprint Volume, Eighth Symposium on Turbulence and Diffusion, AMS, San Diego, Calif., April 1988.

Poreh, M. and Cermak, J.E. (1988), "Scale-Model Simulations of Large Area Fire Plumes", International Symposium on Scale Modelling, Tokyo, Japan, 1988.

Poreh, M., Rau, M. and Plate, E.J. (1991), "Design Considerations for Wind Tunnel Simulations of Diffusion within the Convective Boundary Layer", Atmospheric Environment Vol. 25A, No.7, pp. 1251-1256.

Poreh, M. (1992), "Analysis of Heat Islands Using Small Scale Models", Proceedings, The 2nd International Conference on Urban Thermal Environment, Fokuoka, Japan, Sept. 1992.

Poreh, M., Solomon, Y., Hassid, S. and Trebukov, S. (1992), "Simulation of Heat and Smoke Movement in Israeli Buildings", Technical Report, National Building Research Institute, Technion, Haifa, May 1992.

Quintiere, J.G. (1988), "Scaling Applications in Fire Research", International Symposium on Scale Modelling, Tokyo, Japan, 1988.

Skinner, G.T. and Ludwig, G.R. (1978), "Physical Modelling of Dispersion in the Atmospheric Boundary Layer", Rep. No. 201, 1978, Calspar Corporation, P.O. Box 400, Buffalo, N.Y. 14255.

Snyder, W.H. (1972), "Similarity Criteria for the Application of Fluid Models to the Study of Air Pollution Meteorology", Boundary Layer Meteorol., Vol. 3, pp. 113-134.

Snyder, W.H. (1981), "Guidelines for Fluid Modelling of Atmospheric Diffusion", EPA-600/8-81-009, April 1981.

Thomas, D.B. and Townsend, A.A. (1957), "Turbulent Convection Over a Heated Horizontal Surface", Journal of Fluid Mechanics, Vol. 2, pp. 473-492.

Venkatram, A. (1980), "Dispersion from an Elevated Source in a Convective Boundary Layer", Atmospheric Environment, Vol. 14, pp.1-10.

Williams. F.H. (1969), "Scaling Mass Fires, Fire Research Abstract and Reviews", Vol. 11, Pt. 1, pp. 1-23.

Willis, G.E. and Deardorff, J.W. (1974), "A Laboratory Model of the Unstable Planetary Boundary Layer", J. Atmos. Sci., pp. 1297-1307.

Willis, G.E. and Deardorff, J.W. (1976), "A Laboratory Model of Diffusion Into the Convective Planetary Boundary Layer", Quart. J. Roy. Meteor. Soc., 102, pp. 427-445.

Willis, G.E. and Deardorff, J.W. (1978), "A Laboratory Study of Dispersion from an Elevated Source within a Modeled Convective Planetary Boundary Layer", Atmos. Environ., 12, pp. 1305-1311.

Willis, G.E. and Deardorff, J.W. (1983), "On Plume Rise within a Convective Boundary Layer", Atmos Environ., 17, pp. 2435-2447.

Wyngaard, J.C. (1985), "Structure of the Planetary Boundary Layer and Implications for its Modelling", Journal of Climate and Applied Meteorology, Vol. 24, No. 11, pp. 1131-1142.

Yokoi, S. (1959), "Upward Convection Current from a Burning Wooden House", International Symposium on the Use of Models in Fire Research. Pub. 786 National Academy of Sciences, National Research Council, Washington, D.C. 1959.